AF689931

Handbook of Artificial Intelligence in Water Management

The *Handbook of Artificial Intelligence in Water Management* explores the intersection of artificial intelligence (AI) technologies and water management strategies aimed at achieving net zero emissions. It delves into the cutting-edge applications of AI in various facets of water resource management, ranging from optimizing water distribution systems to enhancing wastewater treatment processes.

Explaining how AI-enabled water quality monitoring systems enable faster response times to water quality issues, this book presents a wide range of AI-driven innovations from smart sensors and real-time monitoring systems to intelligent decision support tools. By leveraging advanced AI algorithms, machine learning techniques, and predictive analytics, water managers can optimize water usage, minimize energy consumption, and reduce greenhouse gas emissions across the entire water cycle for an efficient, resilient, and environmentally sustainable future. This book includes case studies and real-world examples for implementing innovative approaches.

This book is a valuable resource for advanced undergraduate students, researchers, and industry professionals interested in utilizing AI to mitigate environmental impacts, enhance resilience, and achieve net zero emissions in water management.

Handbook of Artificial Intelligence in Water Management

Towards Net Zero Emissions

Edited by

Soufiane Haddout and Tonni Agustiono Kurniawan

CRC Press
Taylor & Francis Group
Boca Raton London New York

CRC Press is an imprint of the
Taylor & Francis Group, an **informa** business

Designed cover image: Getty Images

MATLAB® and Simulink® are trademarks of The MathWorks, Inc. and are used with permission. The MathWorks does not warrant the accuracy of the text or exercises in this book. This book's use or discussion of MATLAB® or Simulink® software or related products does not constitute endorsement or sponsorship by The MathWorks of a particular pedagogical approach or particular use of the MATLAB® and Simulink® software.

First edition published 2026
by CRC Press
2385 NW Executive Center Drive, Suite 320, Boca Raton FL 33431

and by CRC Press
4 Park Square, Milton Park, Abingdon, Oxon, OX14 4RN

CRC Press is an imprint of Taylor & Francis Group, LLC

© 2026 selection and editorial matter, Soufiane Haddout and Tonni Agustiono Kurniawan; individual chapters, the contributors

Reasonable efforts have been made to publish reliable data and information, but the author and publisher cannot assume responsibility for the validity of all materials or the consequences of their use. The authors and publishers have attempted to trace the copyright holders of all material reproduced in this publication and apologize to copyright holders if permission to publish in this form has not been obtained. If any copyright material has not been acknowledged please write and let us know so we may rectify in any future reprint.

Except as permitted under U.S. Copyright Law, no part of this book may be reprinted, reproduced, transmitted, or utilized in any form by any electronic, mechanical, or other means, now known or hereafter invented, including photocopying, microfilming, and recording, or in any information storage or retrieval system, without written permission from the publishers.

For permission to photocopy or use material electronically from this work, access www.copyright.com or contact the Copyright Clearance Center, Inc. (CCC), 222 Rosewood Drive, Danvers, MA 01923, 978-750-8400. For works that are not available on CCC please contact mpkbookspermissions@tandf.co.uk

For Product Safety Concerns and Information please contact our EU representative GPSR@taylorandfrancis.com. Taylor & Francis Verlag GmbH, Kaufingerstraße 24, 80331 München, Germany.

Trademark notice: Product or corporate names may be trademarks or registered trademarks and are used only for identification and explanation without intent to infringe.

ISBN: 978-1-032-90229-6 (hbk)
ISBN: 978-1-032-90327-9 (pbk)
ISBN: 978-1-003-54709-9 (ebk)

DOI: 10.1201/9781003547099

Typeset in Times
by codeMantra

Contents

Preface

Water is essential for life, yet the availability of clean, safe water has become increasingly uncertain due to various global challenges such as climate change, population growth, and environmental degradation. As these pressures continue to mount, innovative solutions are required to safeguard this vital resource and ensure its availability for future generations. This book, *Handbook of Artificial Intelligence and Water Resource Management*, presents a collection of groundbreaking research and innovative applications of artificial intelligence (AI) in the management of water resources, specifically focusing on its role in conservation, sustainability, and resilience. This book draws on a wide array of expert contributions that examine the use of AI technologies in different areas of water management, including floodplain prediction, water quality monitoring, urban water management, precision agriculture, and aquatic ecosystem preservation. Each chapter offers insights into the potential of AI to address the unique challenges associated with water scarcity, water pollution, and climate-induced water insecurity. This book serves as an interdisciplinary resource, offering practical examples of how AI is being utilized to enhance decision-making, improve resource allocation, and support the global push toward achieving net zero emissions. From predictive modeling for flood management to AI-driven smart water infrastructure solutions, the chapters highlight the transformative impact of AI in the water sector.

Acknowledgments

We would like to express our deepest gratitude to the contributors and collaborators who have generously shared their expertise and insights in making this book a reality. Their passion for sustainable water management and commitment to advancing artificial intelligence (AI) technologies have been a constant source of inspiration. Each chapter is a testament to their dedication to improving the way we manage one of the most precious resources on Earth. Special thanks to the researchers, scientists, and industry professionals who have worked tirelessly to explore the intersections between AI and water conservation. Without their hard work and contributions, this book would not have been possible.

We would also like to acknowledge the support of our colleagues for their guidance and encouragement throughout the writing process. Their invaluable feedback and advice have helped shape the direction and content of this book. Finally, our heartfelt thanks go to our families and friends for their unwavering support and understanding.

About the Editors

Dr. Soufiane Haddout is a researcher in the Department of Physics, Faculty of Science, at Ibn Tofail University in Morocco, renowned for his expertise in estuarine research. He has made significant contributions to the field, particularly in understanding sediment dynamic processes, microplastics, heavy metals, water quality, energy, and modeling in estuarine environments. Dr. Haddout is a prolific author, with five published books and one currently accepted and awaiting publication. His scholarly output includes more than 70 refereed journal articles, with Dr. Haddout serving as the first author on 40 of these papers. Dr. Haddout's collaborative spirit is evidenced by his extensive partnerships with over 30 universities and institutions worldwide, including those in the USA, the UK, Japan, Singapore, the Philippines, Brazil, South Korea, Vietnam, Croatia, India, and Mozambique. He is an expert reviewer with the AEIC-Academic Exchange Information Centre in China and regularly reviews project proposals for international universities. His academic excellence has been recognized by over 200 certificates from international conferences and renowned journals. A sought-after speaker, Dr. Haddout has delivered numerous keynote and invited talks at international conferences. He also plays a significant role in the academic community as a Technical Committee member for various conferences. His contributions were acknowledged with the 2022 Best Research Award at the ICEAT conference, particularly for his work on bi-variate and CM plotting of the sediment dynamic process in estuaries. Dr. Haddout holds prestigious editorial positions in several leading journals. He is on the editorial board of *Marine Georesources & Geotechnology* (Taylor & Francis) and *Marine Development* (Springer), serves as an Associate Editor of the *Journal of River Basin Management* (JRBM) (Taylor & Francis), and is the founder and Editor-in-Chief of the newly launched journal *Estuarine Management and Technologies* with PENSOFT publisher. He is also Co-Chair of the scientific session at the International Conference of the Coastal and Estuarine Research Federation in the USA (2022) and the Lead-Chair of the 1st, 2nd, and 3rd International Conferences on Climate Change and Ocean Renewable Energy (CCORE 2022–2024). Additionally, he will serve as the Lead-Chair of the 1st International Conference on Technological and Research Advancements in Coastal and Estuarine Systems (TRACE 2025). Beyond his research and editorial responsibilities, Dr. Haddout coordinates significant international initiatives. He organized the 1st International Round Table Webinar on "A Multi-criteria Decision Making Tool for the Management of Water Bodies in Developing Countries Towards Climate Change Resilience" in 2022 and the 2nd International Roundtable Webinar on "Microplastic and Nanoplastic Pollution in Different Environments: Freshwater, Marine, Air, and Soil" in 2023. He also serves as a Guest Editor of the Special Issue "State-of-the-Art in Estuarine Modelling" in *Regional Studies in Marine Science*, Elsevier, "Microplastics and Estuaries" in the *Journal of Environmental Management.* As an active reviewer, he contributes to reputed journals. Dr. Haddout's work not only advances scientific understanding but also addresses practical environmental challenges, positioning him as a leading figure in estuarine research and water management. His efforts continue to impact academia, policy, and international collaboration in significant ways. His dedication to advancing knowledge and fostering collaboration has made him a key player in the global effort to manage and protect estuarine and coastal environments.

Dr. Tonni Agustiono Kurniawan is a recognized global leader in tackling complex environmental problems that have significant societal relevance and positive impact in the world. His research interests are in the areas of wastewater treatment and solid waste management. His degrees include a B.Sc. in Chemistry from the Bogor Agricultural University (Indonesia), an M.Sc. in Environmental Technology from the Thammasat University (Thailand), and a Ph.D. in Applied Chemical Technology from the Hong Kong Polytechnic University (China). He is the author of over 210 peer-reviewed publications in ISI-rated journals, including 26 conference proceedings, 9 book chapters, and 12 monographs in waste to energy and wastewater treatment. Dr. Kurniawan pioneered the fields by using digital technology in water treatment and waste recycling. His impactful and visionary work has contributed to the field of study. He has also shaped the research agenda by providing insights into how to transform unused waste into composites as low-cost adsorbents in tackling water pollution. His pivotal research in the 2000s included how bacteria transform wastewater into electricity while treating contaminated water. His work has not only provided rigor and relevance to build capacity through his authoritative insights, inspiration, and leadership in the fields of study but also bridged the gap between academia and industrial practices in water treatment. Dr. Kurniawan received the 2010 Green Talent Award from German Federal Ministry of Education and Research (BMBF), the 2013 Young Scientist Award from the World Economic Forum (WEF) (Switzerland), and the 2014 Scope-Zhongyu (France) for his outstanding research contribution to the fields of study. His influential breakthrough work on sustainable wastewater reuse using digitalization has augmented water recycling and provided eco-friendly options for wastewater treatment. Through the development of inexpensive functional materials, he has made critical contributions to photocatalytic technologies to treat contaminated water laden with recalcitrant pollutants. This research direction has inspired the global water industry to apply novel technologies for environmental remediation. With research experience spanning 15 years, over Euro 1 million has been awarded to him as the principal investigator by the European Commission (Brussels), Japan Society for Promotion of Science, Asian Development Bank, Academy of Sciences for Developing World (Italy), and Kurita Water and Environment Foundation (Japan). His focus on sustained scientific research is evident from more than 210 articles in refereed journals, 25 articles in conference proceedings, 6 monographs, and 9 book chapters. To date, Kurniawan is the first author of 25% of the works with an h-index of 53 and citations of over 13,500 (Scopus) while being the corresponding author of one-third of the same works. The scientific contributions are tangible manifestations of his competence and dedication to water profession. To improve dialogue between stakeholders and policy about sciences, he was selected to participate in "the 60th Nobel Laureate Meetings" and the 3rd BioVision Forum in Lindau (Germany) and in Lyon (France) in June 2010 and March

2011, respectively. The meetings dedicated to the Nobel Laureates in Chemistry and Life Sciences annually invite the world's best talents for the transfer of knowledge between generations of scientists through panel discussions, debates, and networking. This represents a platform for leading researchers in the fields to engage in a cross-generational dialogue on central issues of our future such as climate change, renewable energy, and food security.

List of Contributors

Luisito C. Abueg
Department of Economics
University of the Philippines Los Baños
Los Baños, Philippines

Ovide Decroly Wisnu Ardhi
Department of Information System
School of Postgraduate Studies
Universitas Diponegoro
Semarang, Indonesia
and
Department of Informatics Engineering
Vocational School
Universitas Sebelas Maret
Surakarta, Indonesia

Guda Bhargavi
Department of Agronomy
SR University
Warangal, India

Saikat Biswas
Division of Agriculture, Faculty Centre of Agriculture, Rural and Tribal Development
Ramakrishna Mission Vivekananda Educational and Research Institute
Howrah, India

Hossein Bonakdari
Department of Civil Engineering
University of Ottawa
Ottawa, Ontario, Canada

Eric Jay F. Bulalacao
Land and Water Resources Engineering Division
Institute of Agricultural and Biosystems Engineering (IABE),
College of Engineering and Agro-industrial Technology (CEAT),
University of the Philippines Los Baños
Los Baños, Philippines

Suresh Babu C.
School of Agriculture
SR University
Warangal, India

Eric B. Casila
Institute of Agricultural and Biosystems Engineering
College of Engineering and Agro-Industrial Technology
Los Baños, Philippines

Joan Cecilia C. Casila
Land and Water Resources Engineering Division
Institute of Agricultural and Biosystems Engineering (IABE),
College of Engineering and Agro-industrial Technology (CEAT),
University of the Philippines Los Baños
Los Baños, Philippines

Fong Chi Chan
Universiti Teknologi Malaysia
Johor Bahru, Malaysia

Yoon Tung Chan
Universiti Teknologi Malaysia
Johor Bahru, Malaysia

Rupa Das
Usha Martin University
Jharkhand, India

Isa Ebtehaj
Department of Soils and Agri-Food Engineering
Université Laval
Quebec City, Quebec, Canada

Doaa A. El-Emam
Environmental Sciences Department
Damietta University
New Damietta City, Egypt

Yi Ka Fong
Universiti Teknologi Malaysia
Johor Bahru, Malaysia

Soufiane Haddout
Department of Physics
Ibn Tofail University
Kenitra, Morocco

Sherwin P. De Jesus
Land and Water Resources Engineering Division
Institute of Agricultural and Biosystems Engineering (IABE)
College of Engineering and Agro-industrial Technology (CEAT)
University of the Philippines Los Baños
Los Baños, Philippines

Hong Yee Kek
School of the Built Environment
Universiti Teknologi Malaysia
Johor Bahru, Malaysia

Kanta Kobayashi
Department of Civil and Environmental Engineering
Tokyo Metropolitan University
Hachioji, Japan

Chunchu Suchith Kumar
Department of Agronomy
School of Agriculture
SR University
Warangal, India

Pritha Kundu
Division of Agriculture, Faculty Centre of Agriculture, Rural and Tribal Development Ramakrishna Mission Vivekananda Educational and Research Institute
Howrah, India

Tonni Agustiono Kurniawan
College of Ecology and the Environment
Xiamen University
Fujian, China

Joseph Conrad M. Li
Land and Water Resources Engineering Division
Institute of Agricultural and Biosystems Engineering (IABE)
College of Engineering and Agro-industrial Technology (CEAT)
University of the Philippines Los Baños
Los Baños, Philippines

Neroju Manasa
Department of Agronomy
School of Agriculture
SR University
Warangal, India

Mohamad Nur Hidayat Mat
Universiti Malaysia Pahang Al-Sultan Abdullah
Pahang, Malaysia

Claudette Joyce B. Mendina
Land and Water Resources Engineering Division
Institute of Agricultural and Biosystems Engineering (IABE)
College of Engineering and Agro-industrial Technology (CEAT)
University of the Philippines Los Baños
Los Baños, Philippines

Elmy Johana Mohamad
Universiti Tun Hussein Onn Malaysia
Parit Raja, Malaysia

José Luis Molina
IGA Research Group
Higher Polytechnic School of Engineering
University of Salamanca
Ávila, Spain

Saithala Mounika
Department of Agronomy
Professor Jayashankar Telangana Agricultural University
Hyderabad, India

Mude Ramesh Naik
Department of Agronomy
RSM Division
ICAR-NAARM
Hyderabad, India

Banoth Bixapathi Nayak
Department of Agronomy
School of Sciences
Woxsen University
Kamkole, India

Mohd Hafiz Dzarfan Othman
Advanced Membrane Technology Research Centre (AMTEC)
Universiti Teknologi Malaysia
Johor Bahru, Malaysia

Arunabha Pal
Department of Soil Science
Centurion University of Technology and Management
R. Sitapur, India

Thulisekari Prasanna
Department of Agronomy
SR University
Warangal, India

Mohammad Uzair Anwar Qureshi
Department of Civil Engineering
University of Ottawa
Ottawa, Ontario, Canada

Ruzairi Abdul Rahim
Universiti Tun Hussein Onn Malaysia
Parit Raja, Malaysia

Chidigae Sudha Rani
Agricultural Research Station Tandur
Telangana, India

Talakanti Rajeshwar Reddy
Department of Agronomy
Jawaharlal Nehru Krishi Vishwavidyalaya
Jabalpur, India

Jerome P. Romero
Land and Water Resources Engineering Division
Institute of Agricultural and Biosystems Engineering (IABE)
College of Engineering and Agro-industrial Technology (CEAT)
University of the Philippines Los Baños
Los Baños, Philippines

Gandi Rupa
Department of Agronomy
Jawaharlal Nehru Krishi Vishwavidyalaya
Madhya Pradesh, India

Gaini Sairam
Department of Agronomy, PJTAU
College of Agriculture
Telangana, India

Menna A. Shata
Botany and Microbiology Department
Damietta University
New Damietta City, Egypt

Burra Shyamsunder
Department of Agronomy
Guru Nanak University
Hyderabad, India

Tri Retnaningsih Soeprobowati
Department of Biology
Cluster for Paleolimnology (CPalim)
Universitas Diponegoro
Semarang, Indonesia

Liliane M. Suarez
Agrometeorology, Bio-structures and Environment Engineering Division, IABE, CEAT University of the Philippines Los Baños
Los Baños, Philippines

Chourat Sudhakar
Agricultural Research Station Tandur
Telangana, India

Huiyi Tan
Universiti Teknologi Malaysia
Johor Bahru, Malaysia

Durgam Venkatesh
Department of Agronomy, PJTAU
College of Agriculture
Telangana, India

Keng Yinn Wong
Universiti Teknologi Malaysia
Johor Bahru, Malaysia

Sien Jie Wong
Universiti Teknologi Malaysia
Johor Bahru, Malaysia

Kok Sin Woon
Carbon Neutrality and Climate Change Trust
Hong Kong University of Science and Technology
Hong Kong, China

Enrico Zacchei
Department of Mechanical Engineering
Polytechnic School of University of Salamanca (USAL)Avila, Spain

Introduction: Understanding Artificial Intelligence

Soufiane Haddout and Tonni Agustiono Kurniawan

DEFINING ARTIFICIAL INTELLIGENCE AND ITS VARIOUS APPLICATIONS

Artificial intelligence (AI) is a branch of computer science that focuses on creating intelligent machines capable of performing tasks that typically require human intelligence. These tasks include reasoning, learning, problem-solving, perception, language understanding, and decision-making. AI systems can be classified into various categories, including:

- **Narrow AI (Weak AI)**: Designed for specific tasks, such as voice assistants (e.g., Siri, Alexa), facial recognition, and recommendation systems.
- **General AI (Strong AI)**: Hypothetical AI capable of performing any intellectual task that a human can do, demonstrating human-like cognitive abilities.
- **Super AI**: A speculative concept where AI surpasses human intelligence, possessing superior problem-solving and decision-making capabilities.

AI applications span multiple industries, transforming traditional processes and increasing efficiency. Some key areas where AI is making a significant impact include:

- **Healthcare**: AI-powered diagnostics, robotic surgery, drug discovery, and personalized medicine (Topol, 2019).
- **Finance**: Algorithmic trading, fraud detection, credit scoring, and risk assessment (Brenner & Meyll, 2020).
- **Manufacturing**: Automation, predictive maintenance, and quality control (Lee, Kao, & Yang, 2014).
- **Education**: Intelligent tutoring systems, personalized learning experiences, and automated grading (Luckin et al., 2016).
- **Transportation**: Autonomous vehicles, traffic management, and route optimization (Litman, 2020).
- **Retail**: Chatbots, personalized recommendations, and inventory management (Agrawal, Gans, & Goldfarb, 2018).
- **Security**: Cybersecurity threat detection, facial recognition, and surveillance systems (Brundage et al., 2018).

EVOLUTION OF AI AND ITS CURRENT STATE OF DEVELOPMENT

The history of AI can be divided into several phases, each marked by significant advancements and breakthroughs.

1 The Birth of AI (1950s–1970s)

The concept of AI was first formally introduced by Alan Turing, who proposed the Turing Test as a criterion for machine intelligence (Turing, 1950). Early AI research focused on symbolic reasoning and rule-based systems. Key developments included:

- The Dartmouth Conference (1956), where AI was officially recognized as a field of study (McCarthy et al., 1956).
- The creation of early AI programs, such as the Logic Theorist and General Problem Solver (Newell & Simon, 1956).
- Limited computational power and unrealistic expectations led to the first "AI Winter" in the 1970s (Crevier, 1993).

2 Knowledge-Based Systems and Expert Systems (1980s–1990s)

AI research shifted toward knowledge-based systems, where expert systems like MYCIN (medical diagnosis) and XCON (computer configuration) were developed (Feigenbaum, 1984). However, limitations in scalability and adaptability led to another decline in AI enthusiasm.

3 Machine Learning and the Rise of Data-Driven AI (2000s–Present)

The emergence of machine learning (ML) and deep learning revolutionized AI research. Key developments include:

- The use of big data and improved computing power to train complex neural networks (Domingos, 2015).
- Breakthroughs in image and speech recognition, natural language processing (NLP), and reinforcement learning (LeCun, Bengio, & Hinton, 2015).
- The rise of AI frameworks such as TensorFlow and PyTorch (Abadi et al., 2016).

4 Current State and Future Prospects

AI continues to evolve, with advancements in areas such as:

- **Deep Learning**: Enhanced neural network architectures improving performance in vision and language tasks (Goodfellow, Bengio, & Courville, 2016).
- **AI Ethics and Explainability**: Growing focus on bias reduction, transparency, and fairness in AI systems (Binns, 2018).

- **Edge AI**: AI deployed on devices with minimal reliance on cloud computing (Shi et al., 2016).
- **Quantum AI**: The integration of quantum computing for faster AI problem-solving (Preskill, 2018).

AI's future will likely involve increased integration into everyday life, improving automation, decision-making, and human–AI collaboration. However, ethical considerations, privacy concerns, and regulatory challenges must be addressed to ensure responsible AI development (Russell & Norvig, 2020).

REFERENCES

Abadi, M., et al. (2016). *TensorFlow: A System for Large-Scale Machine Learning*. OSDI.

Agrawal, A., Gans, J., & Goldfarb, A. (2018). *Prediction Machines: The Simple Economics of Artificial Intelligence*. Harvard Business Review Press.

Binns, R. (2018). Fairness in Machine Learning: Lessons from Political Philosophy. *Proceedings of the Conference on Fairness, Accountability, and Transparency*.

Brundage, M., et al. (2018). *The Malicious Use of Artificial Intelligence: Forecasting, Prevention, and Mitigation*. arXiv preprint arXiv:1802.07228.

Crevier, D. (1993). *AI: The Tumultuous History of the Search for Artificial Intelligence*. Basic Books.

Domingos, P. (2015). *The Master Algorithm: How the Quest for the Ultimate Learning Machine Will Remake Our World*. Basic Books.

Feigenbaum, E. A. (1984). *Knowledge Engineering: The Applied Side of Artificial Intelligence*. Annals of the New York Academy of Sciences.

Goodfellow, I., Bengio, Y., & Courville, A. (2016). *Deep Learning*. MIT Press.

LeCun, Y., Bengio, Y., & Hinton, G. (2015). Deep Learning. *Nature*, 521(7553), 436-444.

Lee, J., Kao, H. A., & Yang, S. (2014). Service Innovation and Smart Analytics for Industry 4.0 and Big Data Environment. *Procedia CIRP*.

Litman, T. (2020). *Autonomous Vehicle Implementation Predictions*. Victoria Transport Policy Institute.

Luckin, R., & Holmes, W. (2016). *Intelligence Unleashed: An Argument for AI in Education*. Pearson Education.

McCarthy, J., Minsky, M., Rochester, N., & Shannon, C. (1956). *A Proposal for the Dartmouth Summer Research Project on Artificial Intelligence*.

Preskill, J. (2018). Quantum Computing in the NISQ Era and Beyond. *Quantum*, 2, 79.

Russell, S., & Norvig, P. (2020). *Artificial Intelligence: A Modern Approach*. Pearson.

Shi, W., et al. (2016). Edge Computing: Vision and Challenges. *IEEE Internet of Things Journal*, 3(5), 637–646.

Topol, E. (2019). *Deep Medicine: How Artificial Intelligence Can Make Healthcare Human Again*. Basic Books.

Turing, A. (1950). Computing Machinery and Intelligence. *Mind*, 59(236), 433–460.

Part I

Water Quality and Resource Management

1 SDG and Water
Artificial Intelligence Nexus

Saithala Mounika, Banoth Bixapathi Nayak, Mude Ramesh Naik, and Guda Bhargavi

1.1 INTRODUCTION

Water is an indispensable resource and a cornerstone of life on Earth, playing a critical role in sustaining ecosystems, supporting human activities, and driving economic development. From agriculture and industry to domestic use and ecosystem health, water is integral to nearly every aspect of life, making it the lifeblood of our planet. Despite covering 71% of the Earth's surface, only 2.5% of water is freshwater, and less than 1% is readily accessible for human use. Water resources are increasingly under pressure due to pollution, urbanization, climate change, and population growth. Over 25 countries, representing one-quarter of the global population, face extremely high water stress annually, and around four billion people experience severe water scarcity for at least one month each year (World Resources Institute, 2023). Alarmingly, global water demand is projected to outstrip supply by 40% by 2030 (World Economic Forum, 2023). The *2023 Sustainable Development Goals Report* highlights the growing severity of water scarcity, driven by rising global temperatures. In 2020, 2.4 billion people lived in water-stressed countries, and by 2022, 2.2 billion lacked access to safely managed drinking water, with 703 million lacking even basic water services. Urban water scarcity is projected to double, with the number of affected residents rising from 930 million in 2016 to between 1.7 and 2.4 billion by 2050. This escalating crisis underscores the urgent need for sustainable water management practices and innovative solutions to mitigate the impacts of water scarcity on ecosystems, human well-being, and global development.

1.2 GLOBAL DISTRIBUTION OF WATER

Table 1.1 shows the total renewable water resources and per-capita availability by region globally. The world's freshwater resources are unevenly distributed across continents, with the Americas holding the largest share at 45%, followed by Asia at 28%, Europe at 15.5%, and Africa at 9% (FAO, 2024). The availability of freshwater per person varies by continent, with America leading at 24,000 m^3/year, followed by Europe at 9,300 m^3/year, Africa at 5,000 m^3/year, and Asia at 3,400.1 m^3/year. Water demand is expected to increase significantly in the coming years, particularly in Sub-Saharan Africa, where demand is expected to increase by 163% by 2050 (FAO, 2024). In contrast, water demand in wealthier countries like North America

DOI: 10.1201/9781003547099-2

TABLE 1.1
Total Renewable Water Resources

Area	Regional Renewable Water Resources ($10^9 m^3$/year)	Total Renewable Water Resources per Capita (m^3/inhab/year)		
		2019	2020	2021
Australia and New Zealand	819	27,011.69	26,646.99	26,372.72
Central Asia	228.753	3,110.16	3,061.29	3,013.97
Central and Southern Asia	2,698.728	1,334.97	1,320.10	1,306.67
Eastern Asia	3,425.318	2,052.88	2,049.42	2,048.78
Eastern and South-Eastern Asia	8,430.813	3,614.59	3,600.57	3,591.13
Europe	6,760.1275	9,082.12	9,081.51	9,094.27
Europe and Northern America	12,428.6845	11,133.81	11,113.51	11,110.75
Landlocked Developing Countries	2,806.301	5,345.62	5,217.48	5,094.53
Latin America and the Caribbean	13,871.994	21,446.76	21,282.62	21,144.38
Least Developed Countries	6,083.169	5,806.04	5,667.82	5,534.63
Northern Africa	149.442	604.97	594.40	584.36
Northern Africa and Western Asia	522.022	984.67	970.20	957.01
Northern America	5,668.557	15,239.37	15,158.33	15,104.92
Oceania	1,648.55	38,088.40	37,520.36	37,049.59
Oceania (excluding Australia and New Zealand)	829.55	63,998.63	62,833.79	61,718.07
Small Island Developing States	1,335.306	19,042.23	18,872.48	18,711.81
South-eastern Asia	5,230.505	7,878.49	7,804.81	7,739.77
Southern Asia	2,480.225	1,273.21	1,259.24	1,246.69
Sub-Saharan Africa	3,858.561	3,570.50	3,478.50	3,390.83
Western Asia	372.58	1,315.95	1,299.83	1,285.94
World	42,758	5,502.09	5,448.78	5,401.73

Source: AQUASTAT, 2024.

TABLE 1.2

Total Water Stress—Sector Wise

Area	Agricultural Water Withdrawal as % of Total Renewable Water Resources			Agricultural Sector Contribution to Water Stress (%)			Industrial Sector Contribution to Water Stress (%)			Municipal Sector Contribution to Water Stress (%)			Water Stress (%)		
	2019	2020	2021	2019	2020	2021	2019	2020	2021	2019	2020	2021	2019	2020	2021
Australia and New Zealand	1.67	1.43	1.76	3.72	3.21	3.94	1.07	1.10	1.14	0.82	0.77	0.77	5.61	5.09	5.85
Central Asia	42.15	40.73	37.46	59.68	57.94	57.90	5.32	5.64	6.15	4.90	5.27	5.66	76.49	74.37	69.71
Central and Southern Asia	41.33	41.21	40.93	62.94	63.47	67.70	1.69	1.73	1.86	4.77	4.86	5.21	70	70.57	74.77
Eastern Asia	12.97	12.77	12.77	28.66	27.69	27.69	8.89	7.54	7.54	8.68	8.83	8.83	46.22	44.06	44.06
Eastern and South-Eastern Asia	10.36	10.28	10.28	22.45	22.04	22.04	4.25	3.75	3.78	4.52	4.62	4.62	31.22	30.4	30.44
Europe	1.27	1.26	1.26	2.54	2.53	2.53	3.71	3.62	3.63	2.22	2.23	2.24	8.49	8.39	8.41
Europe and Northern America	2.14	2.13	2.14	4.33	4.35	4.34	5.88	5.84	5.85	2.25	2.27	2.27	12.48	12.46	12.47
Landlocked Developing Countries	6.33	6.22	5.99	12.68	12.56	12.60	0.82	0.86	0.90	1.01	1.06	1.09	15.34	15.12	14.65
Latin America and the Caribbean	1.99	1.99	2.00	4.51	4.51	4.52	0.56	0.57	0.57	0.93	0.93	0.94	6	6.01	6.03
Least Developed Countries	2.96	2.96	2.96	6.40	6.40	6.40	0.12	0.12	0.12	0.54	0.54	0.54	7.06	7.06	7.06

(Continued)

TABLE 1.2 (*Continued*)
Total Water Stress—Sector Wise

Area	Agricultural Water Withdrawal as % of Total Renewable Water Resources			Agricultural Sector Contribution to Water Stress (%)			Industrial Sector Contribution to Water Stress (%)			Municipal Sector Contribution to Water Stress (%)			Water Stress (%)		
	2019	2020	2021	2019	2020	2021	2019	2020	2021	2019	2020	2021	2019	2020	2021
Northern Africa	74.19	74.04	74.04	98.95	98.99	98.99	5.56	5.56	5.56	15.98	16.00	16.00	120.49	120.55	120.55
Northern Africa and Western Asia	51.87	49.48	47.67	68.62	65.71	62.96	4.41	4.03	4.17	11.25	11.74	11.22	84.28	81.51	79.74
Northern America	3.18	3.18	3.18	7.59	7.60	7.60	9.93	9.91	9.91	2.67	2.67	2.67	20.19	20.19	20.19
Oceania	0.83	0.71	0.88	1.97	1.70	2.08	0.59	0.61	0.63	0.46	0.44	0.44	3.02	2.75	3.15
Oceania (excluding Australia and New Zealand)	0.01	0.01	0.01	0.02	0.02	0.02	0.06	0.06	0.06	0.08	0.08	0.08	0.16	0.16	0.16
Small Island Developing States	1.29	1.29	1.29	2.26	2.28	2.29	0.68	0.64	0.65	0.73	0.74	0.74	3.66	3.66	3.68
South-eastern Asia	8.21	8.21	8.21	17.60	17.60	17.60	1.13	1.16	1.21	1.70	1.73	1.74	20.43	20.49	20.55
Southern Asia	41.08	41.08	41.08	69.82	70.69	75.92	1.46	1.48	1.59	5.25	5.32	5.71	76.53	77.49	83.22
Sub-Saharan Africa	2.01	2.04	2.04	4.64	4.71	4.72	0.53	0.52	0.52	1.03	1.03	1.05	6.2	6.26	6.3
Western Asia	42.92	39.63	37.09	56.15	52.04	48.20	3.92	3.39	3.58	9.30	9.97	9.23	69.37	65.43	62.92
World	6.77	6.71	6.68	13.15	13.06	13.27	2.82	2.72	2.79	2.37	2.41	2.46	18.39	18.23	18.55

Source: AQUASTAT, 2024.

and Europe is expected to plateau. By 2050, 87 out of 180 countries are expected to have less than 1,700 m^3/year of annual renewable water resources per capita. The number of countries with less than 500 m^3/year is expected to increase from 25 in 2015 to 45 by 2050 (FAO, 2024).

Several sectors such as agriculture, energy production, manufacturing and industry, domestic use, health care, tourism, and recreation depend on water. These sectors not only utilize the available water but also reduce water quality and utility by adding pollutants and other non-acceptable waste materials into water bodies and underground water reservoirs in different ways, such as through surface runoff containing residues of pesticides, fertilizers, herbicides, and animal waste from agricultural production systems; industrial discharges such as heavy metals (e.g., mercury, lead), chemicals, solvents, and radioactive waste; sewage and wastewater containing pathogens, pharmaceuticals, and nutrients contributing to eutrophication; plastics and microplastics from packaging, consumer products, and waste mismanagement; oil and hydrocarbons from spills, runoff, and marine transport; and thermal pollution from heated water released during industrial processes affecting aquatic ecosystems. These contaminants reduce water quality and contribute to water stress as mentioned in Table 1.2. Water stress not only affects the availability but also exacerbates the impacts of climate change.

1.3 CLIMATE CHANGE

Climate change is disrupting weather patterns, leading to more frequent and intense extreme events that result in unpredictable water availability, exacerbating water scarcity, and contaminating water supplies. These disruptions in the water cycle make access to safe drinking water increasingly difficult. Between 2001 and 2018, around 74% of natural disasters were water-related, including droughts and floods (UNICEF, 2024). Climate change is expected to amplify both the frequency and severity of such events. In addition to extreme weather, climate change also impacts other crucial aspects of the water cycle, such as shrinking ice sheets, rising sea levels, ocean acidification, and altered rainfall patterns, which contribute to crop failures and natural disasters. In addition to climate change, factors such as urbanization, population growth, industrialization, intensive agricultural practices, and inadequate waste management also contribute to water pollution. This leads to a range of issues, including health risks, disruption of aquatic ecosystems, declining agricultural productivity, food security challenges, and negative economic impacts.

1.4 SUSTAINABLE DEVELOPMENT GOALS

Recognizing the severity of these challenges, the United Nations High-Level Political Forum on Sustainable Development adopted the 2030 Agenda for Sustainable Development in September 2015, which includes 17 Sustainable Development Goals (SDGs). These goals urge nations and regions to collaborate on scientific and technological innovation to address the SDGs and build a more sustainable and equitable future. Among these, SDG 6 (Clean Water and Sanitation) focuses on providing reliable, sustainable, and affordable solutions for universal access to clean water and sanitation.

Surface water bodies, including lakes, rivers, and reservoirs, are also experiencing significant global shifts, with one in five river basins showing high surface water level fluctuations over the last five years. While numerous methods exist to address water management challenges, artificial intelligence (AI) is emerging as a revolutionary force that addresses these challenges across different sectors.

According to the Global Sustainable Development Report (GSDR, 2019), the digital revolution, driven by continuous progress in areas such as AI, connectivity, information digitization, additive manufacturing, virtual reality, machine learning (ML), blockchain technology, robotics, quantum computing, and synthetic biology, will be a crucial factor in fostering sustainable development in the coming years. AI will play a pivotal role in reaching the SDGs and may assist in addressing major global challenges (ITU, 2017). AI thus plays a significant role in achieving multiple SDGs and its targets related to water management by improving efficiency, sustainability, and equity, which aligns with SDG 6 (Ensure access to water and sanitation for all), SDG 2 (End hunger, achieve food security and improved nutrition, and promote sustainable agriculture), SDG 13 (Take urgent action to combat climate change and its impacts), SDG 14 (Conserve and sustainably use the oceans, seas, and marine resources for sustainable development), and SDG 15 (Protect, restore, and promote sustainable use of terrestrial ecosystems, sustainably manage forests, combat desertification, and halt land degradation and halt biodiversity loss) (Figure 1.1).

These range of critical issues including water scarcity, pollution, water stress, and climate change across various sectors make effective and sustainable water management an urgent necessity.

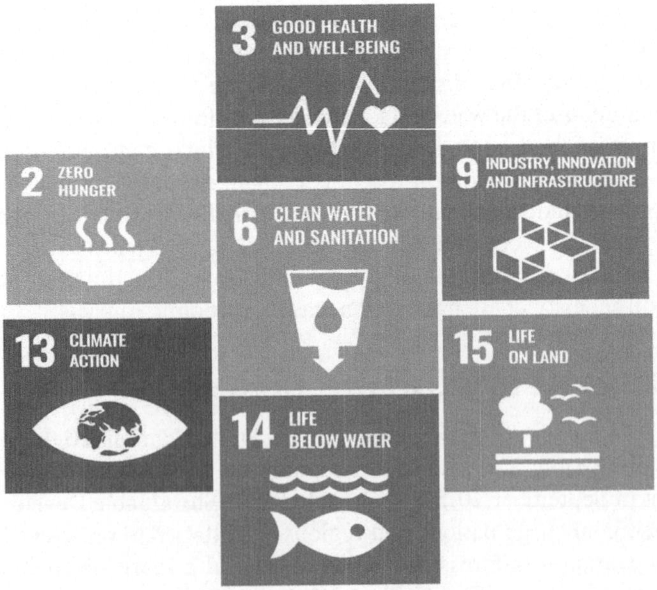

FIGURE 1.1 SDGs related to AI in water management.

1.5 CONVENTIONAL METHODS AND STRATEGIES FOR WATER MANAGEMENT

Water management is crucial for sustaining ecosystems, supporting human livelihoods, and driving economic growth. Key strategies for effective water management include water conservation, rainwater harvesting, wastewater reuse, groundwater management, desalination, watershed management, and climate-resilient practices (FAO, 2021; Wasko & Sharma, 2017). Improving water use efficiency is fundamental to sustainable management. In agriculture, techniques such as drip and precision irrigation deliver water directly to crops, reducing evaporation and runoff, which are among the major approaches (Zhang et al., 2019; Gupta & Gupta, 2021). Rainwater harvesting and wastewater reuse are effective methods to reduce dependence on traditional water sources. Rainwater collected from rooftops can supplement domestic and institutional needs, while large-scale reservoirs and check dams store rainwater for groundwater recharge and agricultural use (Mishra et al., 2020; Zhou et al., 2018). Wastewater treatment and reuse have also gained prominence, with treated wastewater being used for irrigation and industrial cooling (Sahoo et al., 2020; Wu et al., 2020). Natural systems like constructed wetlands ensure water quality with minimal chemical intervention (Mahlein et al., 2018).

Groundwater management and desalination technologies are essential for addressing water scarcity. Practices such as artificial aquifer recharge help stabilize groundwater levels by diverting excess surface water into aquifers, while regulated abstraction prevents over-extraction (Mukherjee et al., 2018). Desalination, using advanced methods like reverse osmosis, converts seawater and brackish water into potable water, providing a viable solution for agriculture, industry, and urban needs in regions with limited freshwater resources (Pahlevan et al., 2019; Huntington et al., 2016).

Integrated watershed management and advanced irrigation practices are critical for sustainable water use. Reforestation and erosion control in watersheds help maintain water quality by reducing sedimentation and nutrient runoff (Gebbers & Adamchuk, 2010; Zhou et al., 2021). Community-driven watershed initiatives foster sustainable practices by involving local stakeholders. Advanced irrigation methods, such as soil moisture monitoring and alternate row irrigation, optimize water usage in agriculture without compromising yields (Zhang et al., 2019; Zhou et al., 2021). Climate-resilient strategies and policy interventions are essential for adapting to changing environmental conditions. Investments in flood mitigation infrastructure and drought preparedness plans, such as groundwater banking and diversified supply systems, enhance resilience (Wasko & Sharma, 2017; Gupta & Gupta, 2021). Policies like water pricing mechanisms encourage conservation by incentivizing efficient use, while pollution control regulations safeguard water sources from contaminants (FAO, 2021; Zhang et al., 2019). Urban and industrial settings benefit from water-efficient appliances like low-flow faucets and dual-flush toilets, alongside water reuse systems that conserve freshwater resources (Wang et al., 2019). Public engagement, through education and incentives, plays a significant role in promoting sustainable water use, particularly in areas with high demand (Wu et al., 2020; Zhou et al., 2018).

Combining technological advancements and effective policy frameworks, these integrated approaches are essential for addressing global water challenges. By ensuring sustainable use, these strategies help meet growing demands while preserving water resources for future generations (FAO, 2021; Wasko & Sharma, 2017).

1.6 ROLE OF ARTIFICIAL INTELLIGENCE IN WATER MANAGEMENT

AI is transforming water management by providing innovative, data-driven solutions that address challenges across agriculture, urban water systems, industry, and disaster management (Table 1.3). The integration of AI with digital technologies such as remote sensing, Internet of Things (IoT) sensors, and smart algorithms has enabled precision, efficiency, and sustainability in managing water resources. Among the various sectors, agriculture remains a focal point, as it is both the largest consumer of freshwater and the most vulnerable to water scarcity (FAO, 2021).

1.7 ENHANCING WATER MANAGEMENT IN AGRICULTURE

AI-driven solutions have revolutionized water management in agriculture by introducing precision and real-time adaptability to irrigation practices. Precision agriculture, powered by AI, combines data from soil sensors, satellite imagery, and weather forecasts to determine the exact water requirements of crops at different growth stages (Gebbers & Adamchuk, 2010). By delivering the right amount of water at the right time, AI minimizes wastage, enhances yields, and improves the resilience of crops to drought or water stress. This approach is particularly beneficial in water-scarce regions, where every drop of water counts (Zhang et al., 2019).

Automated irrigation systems further amplify these benefits by integrating AI algorithms with remote sensing technologies. These systems monitor soil moisture levels, plant health, and environmental conditions to adjust irrigation schedules dynamically. Unlike traditional irrigation methods, which rely on fixed routines, AI-enabled systems are responsive to real-time changes, ensuring optimal water use while reducing over-irrigation or water deficits (Mishra et al., 2020). Moreover, AI facilitates pest and crop health monitoring by analyzing plant data to detect early signs of diseases or pests. By enabling timely interventions, this not only conserves water but also minimizes the overuse of chemicals, further protecting groundwater quality (Mahlein et al., 2018).

TABLE 1.3
AI-Water Management and SDGs

AI-Enabled Water Management Technologies	Contributed to SDGs
Monitoring of water reservoirs and dams	SDG 6, SDG15, SDG 9
Predictive maintenance of water infrastructure	SDG 6, SDG 2, SDG 9
Prediction of water demand and consumption	SDG 6, SDG 14, SDG 15, SDG 13
Water quality monitoring	SDG 3, SDG 6, SDG 2, SDG 13
Monitoring and prediction of water-related disasters	SDG 6, SDG 11, SDG 13

1.8 URBAN WATER SYSTEMS AND SMART INFRASTRUCTURE

In urban areas, AI plays a critical role in improving water supply, distribution, and infrastructure management. Smart water grids equipped with IoT sensors and AI algorithms enable real-time monitoring of water flow, pressure, and quality. These systems detect anomalies such as leaks or irregular usage patterns, allowing utilities to take immediate corrective actions (Zhou et al., 2018). By reducing water losses and improving distribution efficiency, AI ensures sustainable water delivery to urban populations.

AI-driven predictive models analyze historical usage data, weather forecasts, and demographic trends to forecast water demand accurately. Such insights help optimize supply allocation and resource planning, ensuring that urban centers are prepared to meet future water needs (Gupta & Gupta, 2021). Additionally, AI-based leak detection and predictive maintenance systems identify infrastructure vulnerabilities, flagging potential issues before they escalate. By preventing leaks and ensuring timely maintenance, these tools not only conserve water but also extend the lifespan of critical infrastructure (Wu et al., 2020).

1.9 AI IN INDUSTRIAL WATER USE AND WASTEWATER MANAGEMENT

Industries, especially those in water-intensive sectors like manufacturing, food processing, and energy generation, are increasingly adopting AI to optimize water usage. AI-powered systems streamline industrial processes by monitoring water consumption and identifying opportunities for reuse or reduction (Zhou et al., 2021). In wastewater treatment, AI algorithms analyze pollutant levels in real time and adjust chemical dosing, filtration, and aeration processes to maximize efficiency while minimizing energy use (Sahoo et al., 2020). Predictive maintenance further ensures that pumps, pipes, and treatment plants operate smoothly, preventing failures that could disrupt water quality or availability (Wang et al., 2019).

1.10 MONITORING WATER QUALITY AND POLLUTION CONTROL

AI, combined with remote sensing and advanced sensors, has emerged as a powerful tool for monitoring water quality and controlling pollution. By analyzing satellite imagery and sensor data, AI algorithms can detect pollutants in rivers, lakes, and coastal areas, including algal blooms, sediment loads, and chemical contaminants (Pahlevan et al., 2019). ML models also help trace pollution sources, enabling authorities to identify and mitigate issues promptly. Real-time water quality monitoring systems equipped with AI analyze parameters like pH, turbidity, and contamination levels, providing instant feedback to safeguard drinking water supplies and ecosystems (Gholizadeh et al., 2016).

1.11 AI FOR DISASTER MANAGEMENT AND CLIMATE RESILIENCE

AI plays a vital role in predicting and mitigating water-related disasters such as floods and droughts. By analyzing hydrological data, river flows, and meteorological forecasts, AI models provide early warnings with greater accuracy (Wasko & Sharma, 2017). These systems empower communities and governments to prepare for extreme events, reducing the impact on lives, agriculture, and infrastructure. Furthermore, AI helps assess the vulnerabilities of water infrastructure to climate change, suggesting adaptive measures that enhance resilience against prolonged droughts, rising sea levels, and unpredictable rainfall patterns (Mukherjee et al., 2018).

The use of AI in water management also strengthens climate resilience by providing actionable insights into long-term trends. For example, AI-powered models analyze satellite data to forecast climate-induced events, helping planners design sustainable strategies for water resource management in a changing environment (Huntington et al., 2016). This proactive approach ensures the efficient allocation of resources while mitigating the risks associated with water scarcity or excess.

1.11.1 Advantages of AI over Conventional Methods

AI revolutionizes water management by leveraging large datasets, identifying patterns, and enabling real-time, data-driven decisions, areas where conventional methods often fall short. Key advantages include the following.

1.11.1.1 Predictive Power and Real-Time Monitoring

AI enhances demand forecasting by analyzing historical and real-time data to predict water usage patterns, minimizing shortages, and optimizing distribution. It also improves weather and climate analysis, incorporating dynamic data to forecast rainfall, droughts, or floods with greater precision than traditional methods. In monitoring systems, AI integrated with IoT sensors detects leaks or pressure changes in real time, enabling rapid repairs and reducing water wastage. For water quality, AI algorithms continuously analyze parameters like pH, turbidity, and contamination, allowing immediate detection of pollutants compared to periodic sampling and slower lab-based responses.

1.11.1.2 Efficiency and Precision in Agriculture

AI-driven irrigation systems adapt to crop-specific needs by analyzing soil moisture, weather data, and plant growth stages. This dynamic approach ensures water is applied precisely when and where needed, unlike traditional irrigation schedules prone to inefficiencies. AI also facilitates data-driven resource allocation, conserving water while maintaining optimal crop yields.

1.11.1.3 Enhanced Decision-Making and Adaptability

AI optimizes water distribution by analyzing supply-demand dynamics and adjusting flow in real time, improving efficiency over manual, static systems. Predictive maintenance algorithms analyze equipment performance to preempt failures, reducing downtime and costs compared to reactive maintenance.

1.11.1.4 Scalability and Continuous Learning

AI systems continuously learn from new data, improving accuracy and adapting to evolving climate, demand, and water quality conditions. Their scalability allows

TABLE 1.4
AI-Driven Water Management Projects throughout the Globe

AI Platform	Key Function	AI Application	Sector	Country	Outcome	Reference	Aligned SDG
WINT—Water Intelligence	Real-time leak detection and water control	AI detects leaks, reduces water loss, and automates shutoff	Urban Infrastructure	Global	Conserves water and prevents property damage from leaks.	WINT Water Intelligence (2022)	SDG 6: Clean Water and Sanitation
BasinScout Platform	Groundwater and watershed management	Satellite data and machine learning	Water Resource Management	USA	Identifies areas for improving water quality and quantity.	The Nature Conservancy (2023)	SDG 6: Clean Water and Sanitation
Detecting Sewage Leaks	Sewage leak identification	Machine learning for flow pattern analysis	Environmental Monitoring	UK	Uncovered 926 incidents of untreated sewage discharge in rivers.	Environment Agency UK (2022)	SDG 6: Clean Water and Sanitation
ConserWater	Precision irrigation and crop water management	AI combines satellite data to estimate water needs	Agriculture	Global (NASA data)	Optimizes irrigation without hardware, reduces water waste, and improves yields.	ConserWater Technologies (2021)	SDG 2: Zero Hunger and SDG 6: Clean Water and Sanitation
RUBSEE European Project	Waste management monitoring	AI and computer vision	Waste Management	EU	Optimizes processes, ensuring regulatory compliance and operational efficiency.	RUBSEE Project Report (2023)	SDG 12: Responsible Consumption and Production
U-Net in Microscopic Analysis	Image segmentation for water treatment	CNN and U-Net algorithms for detection/tracking	Wastewater and Diagnostics	Global	Enhances detection and monitoring for disease, water quality, and contaminants.	Xu et al. (2020)	SDG 6: Clean Water and Sanitation
Rainfall Forecasting	Long-term rainfall prediction	Artificial Neural Networks (ANN)	Weather Forecasting	Australia	Supports flood, drought, and water resource management with improved accuracy.	Australian Bureau of Meteorology (2022)	SDG 13: Climate Action

(Continued)

TABLE 1.4 (*Continued*)
AI-Driven Water Management Projects throughout the Globe

AI Platform	Key Function	AI Application	Sector	Country	Outcome	Reference	Aligned SDG
RemixWater	Seawater desalination and water reuse	AI for system optimization and fouling detection	Urban Water Supply	Japan	Reduces energy costs and improves water system management.	RemixWater Project, Japan (2021)	SDG 6: Clean Water and Sanitation
WasteShark	Water body cleanup and monitoring	Autonomous vessel collects data and waste	Marine Ecosystem Management	Global	Removes hazardous waste, monitors water quality, and recycles biomass.	RanMarine Technology (2022)	SDG 14: Life Below Water
LAWUNA	Monitoring macro-pollutants	Real-time imagery via drones and smartphones	Freshwater Contamination	Global	Aids in shoreline cleanup and provides open-access environmental data.	LAWUNA Environmental Project (2022)	SDG 6: Clean Water and Sanitation
Google Flood Alerts	Flood prediction and warning	AI processes historical data and terrain analysis	Disaster Management	India	Issues accurate flood warnings, reducing damage and saving lives.	Google Flood Forecasting Initiative (2021)	SDG 11: Sustainable Cities and Communities
Water Demand Forecasting	Irrigation water demand prediction	ANN, Bayesian Framework, Genetic Algorithms	Agriculture	Spain	Enhanced water demand predictions by 11% compared to traditional models.	Martinez et al. (2020)	SDG 6: Clean Water and Sanitation
SVM for Water Quality	River water quality monitoring	Support Vector Machines (SVM) models	River Management	Iran	Accounts for 87% variability in Water Quality Index (WQI), improving water quality management.	Hashemi et al. (2021)	SDG 6: Clean Water and Sanitation
Smart Dam Monitoring (DMS)	Real-time dam monitoring	AI-based deformation and stress monitoring	Infrastructure Safety	Italy (Enel Green Power)	Prevents structural failures by early anomaly detection.	Enel Green Power Report (2023)	SDG 9: Industry, Innovation, and Infrastructure

(*Continued*)

TABLE 1.4 (*Continued*)
AI-Driven Water Management Projects throughout the Globe

AI Platform	Key Function	AI Application	Sector	Country	Outcome	Reference	Aligned SDG
Groundwater Quality (ANN)	Groundwater quality assessment	ANN and Multiple Linear Regression	Groundwater Resource	India (Shivganga Basin)	Monitors water quality using physicochemical parameters during pre/post-monsoon.	Mishra et al. (2022)	SDG 6: Clean Water and Sanitation
AI-based Smart Irrigation System	Real-time irrigation management	AI uses weather and soil data to automate irrigation	Agriculture	Tamil Nadu, India	Reduces water use and improves crop yield by delivering precise irrigation.	Tamil Nadu Agricultural University Report (2023)	SDG 2: Zero Hunger and SDG 6: Clean Water and Sanitation
AI for Cauvery Basin Management	River basin water resource monitoring	AI-based hydrological models and satellite imagery	Water Resource Management	Karnataka, Tamil Nadu, India	Optimizes water resource allocation across agricultural and urban sectors.	CWC (Central Water Commission) (2022)	SDG 6: Clean Water and Sanitation
Water Quality Prediction for Ganga River	Water quality assessment	AI models predict WQI based on water parameters	River Water Management	Uttar Pradesh, West Bengal, India	Helps monitor pollution levels and informs cleanup efforts for the Ganga.	CPCB Ganga Water Report (2022)	SDG 6: Clean Water and Sanitation
Crop Water Requirement Estimation	Predicting water needs for crops	AI integrates satellite imagery and weather data	Agriculture	Maharashtra, India	Reduces water wastage and improves crop productivity in water-scarce regions.	ICAR-AICRP on Water Management (2023)	SDG 2: Zero Hunger and SDG 6: Clean Water and Sanitation
Flood Forecasting for Brahmaputra	Flood prediction and early warning system	AI processes hydrological and meteorological data	Disaster Management	Assam, India	Improves flood preparedness, saving lives and livelihoods in flood-prone regions.	IMD, Assam Flood Report (2022)	SDG 13: Climate Action

deployment across agriculture, energy, industry, and urban water management, providing integrated solutions unlike conventional sector-specific approaches. AI's real-time adaptability, learning capabilities, and precision make it a transformative tool in water management, addressing challenges such as climate change and urbanization with unmatched efficiency and responsiveness.

1.11.2 Limitations of AI

Despite its advantages, AI has limitations. It requires high-quality data, which can be difficult to obtain in regions with limited infrastructure. Developing and maintaining AI models demands specialized expertise, and systems may struggle to adapt to unexpected events or rapid changes in water conditions. For sustainable and effective water management, AI must be integrated with human judgment and domain knowledge.

1.12 GLOBAL EFFECTS TOWARD SUSTAINABLE WATER MANAGEMENT

The application of AI for sustainable water management is rapidly advancing globally, with numerous AI-driven projects aimed at addressing key challenges in water conservation, agriculture, and environmental protection. Table 1.4 lists some of these projects, each aligned with specific SDGs to promote long-term sustainability in these sectors.

1.13 CONCLUSION

The integration of AI with water management is pivotal in advancing SDGs, particularly SDG 6 (clean water), SDG 2 (sustainable agriculture), SDG 9 (resilient infrastructure), and SDG 13 (climate action). AI-driven innovations like smart water grids, precision agriculture, remote sensing, and early warning systems optimize resource allocation, enhance monitoring, and support predictive modeling, addressing water scarcity, distribution efficiency, and quality amidst climate change. These technologies provide a sustainable pathway to meet global water demands and secure future water resources. However, effective policy and regulatory frameworks are essential to fully integrate AI into water infrastructure and development efforts. AI enhances water management by analyzing large datasets to optimize usage, predict demand, detect leaks, and improve disaster preparedness. By combining AI with traditional methods, water management becomes more adaptive, efficient, and sustainable. In agriculture, AI improves irrigation and crop health, while in urban and industrial systems, it optimizes infrastructure and reduces water losses. As climate change challenges global water security, AI-driven solutions offer resilience, efficiency, and sustainable management of water resources (FAO, 2021; Zhang et al., 2019; Mukherjee et al., 2018).

1.14 WAY FORWARD

To fully leverage AI for sustainable water management, it is essential to establish supportive policies, encourage cross-sector collaboration, and invest in research and infrastructure. Prioritizing high-quality data and AI literacy will improve predictive models, ensuring more efficient resource management, particularly in water-stressed regions. A holistic, multidisciplinary approach is necessary to address the complex challenges of water scarcity and climate change. By advancing AI-driven solutions, scaling innovations, and fostering investment, we can enhance water security, drive sustainability, and contribute to achieving the SDGs for a resilient future.

1.15 DECLARATION OF GENERATIVE AI IN SCIENTIFIC WRITING

I hereby declare that all the contents of this book chapter have been independently researched, authored, and compiled by me without using any generative AI tools.

REFERENCES

Australian Bureau of Meteorology. (2022). Rainfall forecasting using artificial neural networks for improved accuracy. https://www.bom.gov.au

Central Water Commission (CWC). (2022). AI-based smart irrigation system for Cauvery basin water management. https://www.cwc.gov.in

Enel Green Power. (2023). *Smart Dam Monitoring (DMS): Real-Time Dam Monitoring*. Enel Green Power Report.

Environment Agency UK. (2022). Detecting sewage leaks using machine learning for flow pattern analysis. https://www.gov.uk/government/organisations/environment-agency

FAO. (2021). *The State of the World's Land and Water Resources for Food and Agriculture*. Food and Agriculture Organization of the United Nations. Rome, Italy. https://www.fao.org

FAO. (2024). *AQUASTAT Dissemination System*. Food and Agriculture Organization of the United Nations. Rome, Italy. https://data.apps.fao.org/aquastat/?lang=en

Gebbers, R., & Adamchuk, V. I. (2010). Precision agriculture and food security. *Science*, 327(5967), 828–831. https://doi.org/10.1126/science.1183899

Gholizadeh, A., Amin, M. S. M., Anuar, A. R., Esfandiarian, V., & Saberioon, M. (2016). Soil nutrient mapping using machine learning techniques. *Geoderma*, 267, 46–56. https://doi.org/10.1016/j.geoderma.2015.12.007

Google Flood Forecasting Initiative. (2021). Google flood alerts: AI-driven flood prediction and warning system. https://sites.research.google/floods

GSDR. (2019). *The Future is Now: Science for Achieving Sustainable Development*. UNDESA. New York, USA. Retrieved https://sdgs.un.org/publications/future-now-science-achieving-sustainable-development-gsdr-2019-24576

Gupta, R., & Gupta, A. (2021). Artificial intelligence in water management: Applications and challenges. *Water Resources Management*, 35(5), 1273–1288. https://doi.org/10.1007/s11269-021-02800-6

Hashemi, S. H., Shafiei, S., & Najafi, M. (2021). SVM for water quality: River water quality monitoring. *Environmental Monitoring Journal*, 33(4), 214–220.

Huntington, T. G., Balazs, A. D., & Richardson, A. D. (2016). Climate change and its effects on water resources. *Hydrology and Earth System Sciences*, 20(8), 3219–3234. https://doi.org/10.5194/hess-20-3219-2016

ICAR-AICRP on Water Management. (2023). Crop water requirement estimation using AI for improved productivity in Maharashtra. https://www.icar.org.in

IMD, Assam Flood Report. (2022). *Flood Forecasting for Brahmaputra: AI-Driven Prediction and Early Warning System*. Indian Meteorological Department. New Delhi, India.

ITU. (2017). AI for good global summit report-2017. https://www.itu.int/en/ITU-T/AI/Documents/Report/AI_for_Good_Global_Summit_Report_2017.pdf

LAWUNA Environmental Project. (2022). Monitoring macro-pollutants with drones and smartphones. https://sodzofoundation.org/lawuna-project/

Mahlein, A. K., Kuska, M. T., Behmann, J., Polder, G., & Walter, A. (2018). Hyperspectral sensors and imaging technologies in plant phenotyping. *Sensors*, 18(2), 478. https://doi.org/10.3390/s18020478

Martinez, J. A., Gonzalez, F. J., & Lopez, M. (2020). Water demand forecasting for irrigation: Enhancing predictions with ANN, Bayesian framework, and genetic algorithms. *Agricultural Water Management,* 227, 105840.

Mishra, A., Singh, R., & Kumar, P. (2020). Role of AI in irrigation and water management. *Journal of Water Resources*, 35(4), 257–268.

Mishra, A., Yadav, R., & Sharma, M. (2022). Groundwater quality assessment using ANN and multiple linear regression. *Groundwater Hydrology Journal,* 48(2), 118–127.

Mukherjee, A., Rahman, M. M., & Sarkar, A. (2018). Flood prediction using AI and machine learning techniques. *Journal of Hydrology*, 556, 135–145. https://doi.org/10.1016/j.jhydrol.2017.11.015

Pahlevan, N., Sarkar, S., Franz, B. A., Balasubramanian, S. V., & He, J. (2019). Sentinel-2 multispectral imagery for monitoring coastal waters. *Remote Sensing of Environment*, 231, 111–126. https://doi.org/10.1016/j.rse.2019.111112

Perea, R. G., Poyato, E. C., Montesinos, P., & Díaz, J. A. R. (2019). Optimisation of water demand forecasting by artificial intelligence with short data sets. *Biosystems Engineering,* 177, 59–66.

RanMarine Technology. (2022). WasteShark: Autonomous vessel for water body cleanup and monitoring. https://www.ranmarine.io

RemixWater Project, Japan. (2021). *Seawater desalination and water reuse optimization with AI.* RemixWater Project Report, Japan.

RUBSEE Project Report. (2023). *Waste management monitoring using AI and computer vision.* RUBSEE European Project, Spain.

Sahoo, B., Jha, M. K., & Singh, A. (2020). Water resources management using machine learning techniques. *Environmental Monitoring and Assessment*, 192(7), 423. https://doi.org/10.1007/s10661-020-08396-1

The Nature Conservancy. (2023). BasinScout platform: Groundwater and watershed management using satellite data and machine learning. https://www.nature.org

UN. (2018). The sustainable development goals report 2018. https://unstats.un.org/sdgs/files/report/2018/TheSustainableDevelopmentGoalsReport2018-EN.pdf

UNICEF (2024). *Water and the global climate crisis: 10 things you should know*. https://www.unicef.org/stories/water-and-climate-change-10-things-you-should-know

Wang, H., Li, Z., Zhang, J., & Liu, X. (2019). AI applications for real-time water quality monitoring. *Water Science and Technology*, 79(6), 1124–1134. https://doi.org/10.2166/wst.2019.100

Wasko, C., & Sharma, A. (2017). Global assessment of flood and drought trends. *Journal of Climate*, 30(7), 2511–2525. https://doi.org/10.1175/JCLI-D-16-0308.1

WINT Water Intelligence. (2022). WINT – Water Intelligence: Real-time leak detection and water control. https://www.wint.ai

World Resources Institute. (2023). 25 countries, housing one-quarter of the population, face extremely high water stress. https://www.wri.org/insights/highest-water-stressed-countries

World Economic Forum. (2023). *Global freshwater demand will exceed supply 40% by 2030, expertswarn.*https://www.weforum.org/stories/2023/03/global-freshwater-demand-will-exceed-supply-40-by-2030-experts-warn/

Wu, J., Liu, H., & Zhou, W. (2020). Smart water management systems: An AI perspective. *Computers and Electronics in Agriculture*, 174, 105–120. https://doi.org/10.1016/j.compag.2020.105120

Xu, S., Wang, Y., & Li, J. (2020). U-Net in microscopic analysis for water treatment. *Journal of Water Process Engineering*, 33, 101080. https://doi.org/10.1016/j.jwpe.2020.101080

Zhang, J., Zhang, X., Chen, Y., & Liu, X. (2019). AI-driven optimization for agricultural irrigation. *Agricultural Water Management*, 216, 116–124. https://doi.org/10.1016/j.agwat.2019.01.004

Zhou, G., Chen, J., & Xu, X. (2018). Advances in AI-based water resource management. *Environmental Research Letters*, 13(9), 095010. https://doi.org/10.1088/1748-9326/aadcf8

Zhou, Q., He, Y., & Zhang, R. (2021). Smart water grids and AI applications. *Journal of Water Process Engineering*, 40, 101873. https://doi.org/10.1016/j.jwpe.2021.101873

2 Urban Water Management Challenges and Solutions

Yi Ka Fong, Hong Yee Kek, Huiyi Tan, Mohd Hafiz Dzarfan Othman, Sien Jie Wong, Mohamad Nur Hidayat Mat, Yoon Tung Chan, and Keng Yinn Wong

2.1 INTRODUCTION

Urban water management involves handling water supply, wastewater, stormwater management, and urban drainage within cities (Larsen & Gujer, 1997). Water, covering 71% of the earth's surface with approximately 3% being fresh water, is essential for life (Reidmiller et al., 2017). Rapid urbanization intensifies demand on natural resources like water, land, and air for domestic and industrial uses (Parkinson et al., 2010). This surge in urban population density significantly impacts water availability and quality across global cities (Lorenzo & Kinzig, 2020). For instance, Malaysia witnessed a dramatic increase in its urban population from 18,520,504 in 2007 to 24,124,786 in 2017, marking a 30.26% growth (Sa'adi et al., 2024). Concurrently, water availability per capita decreased by 14.1%, from 21,707 m^3 in 2007 to 18,646 m^3 in 2017, due to heightened freshwater withdrawals (Ong et al., 2023). Similarly in China, where 66.2% of the population lived in urban areas by the end of 2023, water usage in cities has surged (Peng, 2011). Despite possessing 14% of the world's freshwater resources (Harrell, 2023), the uneven distribution across regions like Shanghai, which used 7,000 million m^3 of water in 2015 for various purposes, demonstrates the complexities in managing urban water demands effectively (Wang et al., 2018). These examples underscore a troubling global trend where approximately 150 million people currently experience water scarcity in cities, a figure projected to escalate to 1 billion by 2050 (McDonald et al., 2011).

Another major challenge in urban water management was flash flooding due to climate change. For example, Ho Chi Minh City's urbanization rate of 41.5% in 2022 (Huong et al., 2022) contributed to flash floods being the predominant cause of natural disaster fatalities from 1990 to 2019 (Luu et al., 2018), affecting 22% of the population and decreasing the GDP annually by 1.2%–1.5% (Citaristi, 2022; Garschagen, 2013). A 2013 study in Quang Nam, Vietnam, reported heavy rainfall causing 47 fatalities or missing cases, damaging over 425,987 properties, and affecting 5,799 hectares of agricultural land (Evans & Falvey, 2013). Studies suggest that

 DOI: 10.1201/9781003547099-3

without improved urban planning and flood management, an additional 13%–27% of the urban population could be at risk of flooding in the future (Bangalore et al., 2019). The third significant challenge is water pollution, driven by improper waste disposal and inadequate industrial and municipal water treatment (Siddique, 2021). A notable incident occurred on March 7, 2019, when illegal chemical dumping in Malaysia's Kim Kim River affected over 2,000 people, with 143 requiring intensive treatment (Yap et al., 2019). Such incidents highlight the dire consequences of pollution and the need for robust regulatory frameworks and advanced treatment technologies.

The integration of AI technologies in water management is redefining traditional practices, shifting from reactive to proactive and predictive management. Through the application of machine learning algorithms such as long short-term memory (LSTM), convolutional neural networks (CNN), deep neural networks (DNN), and artificial neural networks (ANN), urban water systems can achieve enhanced efficiency and reliability in water distribution, filtration, monitoring, and treatment processes. Recent research reported that AI-driven technologies have optimized water distribution systems (WDS) by utilizing deep learning algorithms to identify regions with high water consumption (Nova, 2023). These AI-driven technologies also excel in detecting and locating leaks within the water distribution networks. By integrating dynamic pressure sensors and accelerometers with ANNs, these systems can model complex nonlinear relationships to accurately pinpoint leakages (Mahdi et al., 2024). Multi-Layer Perceptron ANN (MLP-ANN) has proven effective in detecting critical water quality parameters such as pH, suspended solids, and ammonium nitrogen (Najah Ahmed et al., 2019). Furthermore, the application of AI in anaerobic-oxic biological treatment processes has successfully reduced the concentration of pollutants in water bodies, enhancing overall treatment effectiveness (Shi & Xu, 2018).

This chapter discusses the myriad challenges of urban water management, drawing on global examples to highlight the urgent need for sustainable solutions. It provides a comprehensive overview of current and emerging urban water management solutions, emphasizing the integration of AI-driven technologies. This synergy aims to optimize the WDS, enhance water quality monitoring, and improve wastewater management systems, ensuring the provision of clean and high-quality water for the growing populations.

2.2 CURRENT CHALLENGES IN URBAN WATER MANAGEMENT

2.2.1 Population Growth and Urbanization

Rapid urbanization refers to the drastic growth in the proportion of the population living in urban areas (Chen, 2007) and has profound implications for water resources. Figure 2.1 shows the decline in renewable freshwater resources per capita from 2000 to 2020.

The ASEAN region witnessed an urban population increase to 326.8 million or 49.5% by 2020, with an expected rise of 70 million more by 2025 (Ong et al., 2023). From 2007 to 2017, ASEAN countries experienced a 12% average decrease in available water resources (Ong et al., 2023), attributed to increased urbanization and corresponding water demand (Lorenzo & Kinzig, 2020). In China, the world's

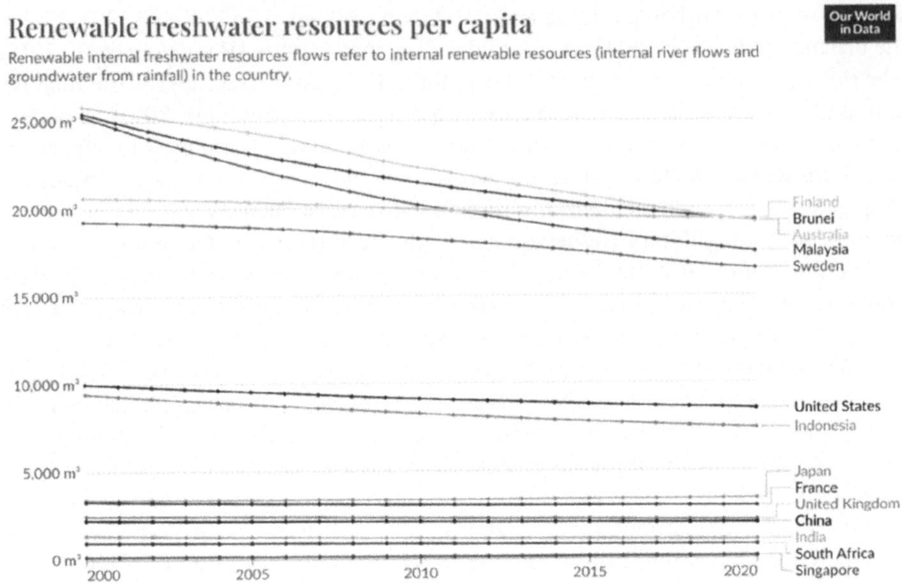

FIGURE 2.1 Comparison between the freshwater resource availability per capita for various countries between 2000 and 2020 (m³/capita/year). (Data, 2024.)

most populous country, 64% of the population lived in urban areas as of 2021, with projections suggesting this will rise to 70.12% by 2030 (Long et al., 2021; Sun et al., 2017). Dubai serves as a dramatic example, where the urban population doubled from 1.2 million to 2.4 million from 2005 to 2015, driven by rapid developments such as the Burj Khalifa (Elessawy, 2017). Such rapid urbanization often leads to water scarcity as the demand from domestic and industrial uses outpaces supply. In China, industrial water usage accounts for 50%, followed by domestic use at 33.3% and agricultural use at 16.7%, with cities like Guangzhou requiring substantial water for cooling in manufacturing industries (Wang et al., 2018). The trends in countries like Malaysia and Singapore underscore this correlation further. From 2007 to 2017, Malaysia's urban population surged from 18,520,504 to 24,124,786, marking a 30.26% increase (Sa'adi et al., 2024), which coincided with a 14.1% decrease in water availability per capita from 21,707 m³ in 2007 to 18,646 m³ in 2017 (Ong et al., 2023). Similarly, Singapore experienced an approximate 18% increase in its urban population during the same period (Guo et al., 2017; Jones, 2012), which was accompanied by an 18.3% decline in per capita water resource availability (Ong et al., 2023).

2.2.2 Water Quality and Pollution

Rapid urbanization reduces the availability of freshwater (Brar, 2013). In addition, the declining water quality further limits the amount of water suitable for daily human consumption. The characteristics of the water quality are defined by the Water Quality Index (WQI) that is widely used in evaluating surface water quality (Uddin et al., 2021). Water quality indicators are categorized into physical, chemical, biological, nutrient, and heavy metal parameters (Tomas et al., 2017). These parameters are individually assessed to create sub-indices, which are then weighted and aggregated to form the overall WQI. According to Das Kangabam et al. (2017), WQI classifications include five levels: highly polluted (0–24), polluted (25–49), fair (50–74), good (75–89), and excellent (90–100).

When the WQI falls below acceptable levels, as it often does in highly urbanized regions, the resulting water pollution can pose serious health risks. For example, water with WQI below 49 can adversely affect human health, potentially leading to diseases if used for drinking or food preparation (Das Kangabam et al., 2017). Water-related diseases remain a significant public health concern globally. Annually, diseases like diarrhoea cause approximately 1.6 million deaths, predominantly affecting children under five (Pandit & Kumar, 2015). Furthermore, contaminated water can lead to severe health issues such as liver and kidney damage from pollutants like lead and copper, neurological disorders from mercury, and gastrointestinal problems from various contaminants (Joseph et al., 2019).

The root causes of water pollution are multifaceted but often include inadequate disposal of industrial waste (Li et al., 2014). For instance, in China, rapid industrialization has led to pollution in the Yellow River Basin, where the water quality is significantly poorer than the national average due to heavy metal contamination (Du et al., 2022). Similar issues are observed in India's Vembanad Lake (Ramasamy et al., 2017) and Indonesia's Lake Metano (Melati et al., 2020), where industrial activities and improper waste management practices have led to detrimental water quality impacts. Figure 2.2 depicts the causes, effects, and primary pollutants associated with water pollution.

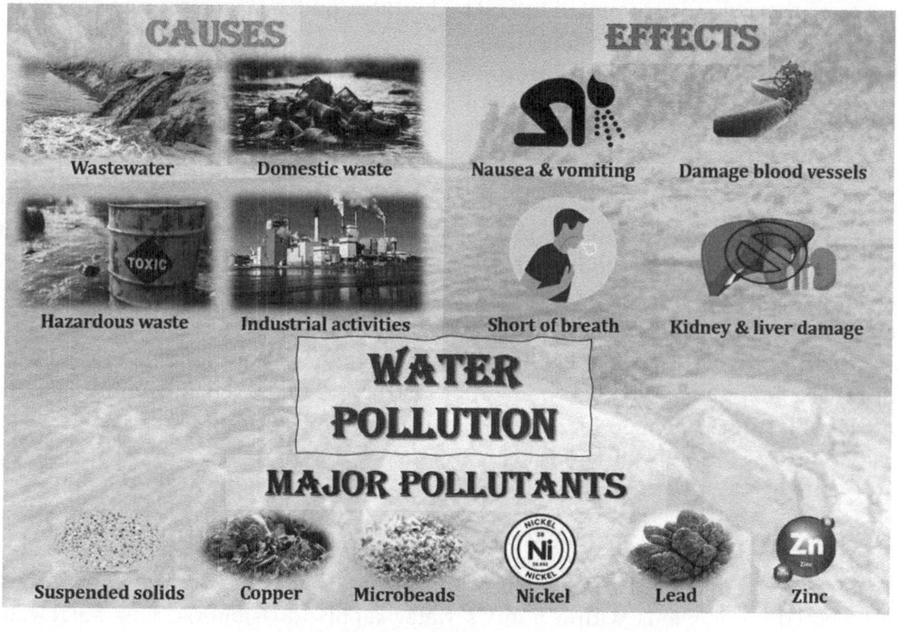

FIGURE 2.2 Causes, effects, and major pollutants of water pollution.

2.2.3 Climate Change and Its Impact

Climate change encompasses long-term alterations in temperature and weather patterns, largely driven by human activities such as deforestation for urban development (Jay et al., 1988). This phenomenon has numerous facets, including rising global temperatures, increased sea levels, and more frequent and intense weather extremes such as hurricanes, floods, droughts, and heatwaves (Dietz et al., 2020). In Europe, researchers observed that water availability is increasingly compromised by the severity and frequency of heavy rainfall events, directly impacting water management systems (Georgi et al., 2016).

Cities worldwide often struggle to cope with these climate extremes due to inadequate infrastructure (Leal Filho et al., 2019). For example, insufficient stormwater management facilities can lead to flash flooding during prolonged heavy rainfall, a common issue in wet tropical regions like Southeast Asia. Kuala Lumpur, Malaysia, experiences an average monthly precipitation of 360mm during the rainy season, dropping to 126mm in the dry months (Muhammad et al., 2020). Between 1998 and 2018, the city faced 51 major flood disasters, affecting 3 million people, causing 281 deaths, and incurring costs of approximately USD 2 billion (Zurairi, 2018). A notable mitigation measure in Kuala Lumpur is the SMART tunnel, a dual-purpose structure designed for traffic and stormwater management. It spans 4km and can handle up to 3.5 million m^3 of stormwater (Isah, 2016). However, its operation requires manual interventions, and it can take up to four days to reopen once engaged in flood mode.

In Beijing's Changping District, the cumulative rainfall reached 744.8mm in 2022–2023 (Li et al., 2023), leading to severe flooding that impacted approximately 5.5 million people and resulted in 107 fatalities or missing persons. The economic loss from these floods amounted to USD 22 billion, underscoring the acute need for effective stormwater management (Gao et al., 2024). Table 2.1 summarizes the incidents of floods due to the lack of stormwater management facilities.

Additionally, climate change exacerbates water scarcity issues. In 2015, 75% of North Africa's population, or 160 million people, faced water scarcity, a figure projected to increase to 206 million by 2050 (Gosling & Arnell, 2016). In South Asia, the situation is similarly dire, with water scarcity affecting 800 million people in 2010—a number expected to rise to 1.466 billion by 2050 due to rapid urbanization (Gosling & Arnell, 2016). Indian researchers noted that global warming is likely to increase water scarcity in India by 20%, affecting 1.2 billion residents and reducing crop yields. By 2025, it is projected that 1.8 billion people globally will face water scarcity issues (Mahato et al., 2022). These findings highlight the critical need for robust, adaptive urban planning and infrastructure improvements to mitigate the impacts of climate change on water management and to ensure sustainable water availability for future generations.

2.2.4 Ageing Water Infrastructure

Ageing water infrastructure in urban water management refers to the deterioration of essential components within a city's water supply, distribution, and wastewater treatment systems, including pipes, pumps, treatment plants, and reservoirs. After

TABLE 2.1
Incidents of Flash Floods and Current Solutions to Overcome the Floods

Location	Incidents	Current Solutions	Limitations	References
Changping District, China	Floods affected 5.5 million people and resulted in 107 fatalities and missing persons, 1.338 million of damaged property and loss of a total of USD 22 billion	Satellite imagery is used to monitor river basin levels and provide early warnings	High costs associated with system setup	Gao et al. (2024)
Kuala Lumpur, Malaysia	Seventy-nine flood locations were recorded from years 2000 to 2010	SMART tunnel constructed to divert stormwater during heavy rainfall	Potential evacuation delays due to manual vehicle removal requirements in the tunnel	Isah (2016)
Quang Nam, Vietnam	Floods caused 17 deaths, 230 injuries, and affected 91,739 properties, submerging agricultural land	Flood hazard assessment was conducted using AHP to evaluate flood probability	Inaccuracy in assessments if rainfall increases suddenly	Luu et al. (2018)
Thailand	Floods affected 13 million people in 69 provinces, with 680 deaths and a loss of USD 46.5 million	Water retention areas are created by constructing amphibious houses	Lack of ability to predict future flood trends	Phongsathorn, (2011), Poapongsakorn and Meethom (2013)
Philippines	Typhoon-related flood incidents killed 24 people	Structural equation modelling (SEM) is used to compute methods to increase flood awareness	Residents lack flood event preparation knowledge	Pascual et al. (2024)

decades of use, these components often become outdated and prone to failure (Grigg, 2019). Urban water systems, particularly in older cities, were typically installed in the early to mid-20th century and are now increasingly susceptible to leaks, water main breaks, contamination, and inefficiency without substantial upgrades or replacements.

There are several key challenges associated with ageing water infrastructure, such as leaks and water loss, reduced capacity and reliability, and significant water quality issues. In South Korea, the collapse of an old water reservoir after 60 years of operation led to a crisis in public confidence in water quality, affecting 21.3% of the population (Kang, 2019). Moreover, in 2011, a major disruption in water supply was caused by untreated sewage entering the treatment system, impacting half of the national population's access to clean water. It is projected that by 2035, 72.1% of South Korea's water infrastructure will have deteriorated significantly (Kang, 2019). In the United States, more than 73% of water pipes with a diameter of 250 mm are over 35 years old. The 2009 annual infrastructure report card issued a grade of 'D', highlighting the critical state of the water systems, which suffer from an estimated 26.5 billion litres of water leakage daily (Selvakumar & Tafuri, 2012). Similarly, in China, around 40% of dams are over 50 years old, including the Three Gorges Dam on the Yangtze River. The ageing of these dams has led to the accumulation of sediments at the basin's bottom, diminishing their operational effectiveness and impacting water management (Perera et al., 2021). These examples underline the urgent need for investment in and renewal of ageing water infrastructure to ensure reliability, efficiency, and safety in urban water management systems worldwide.

2.3 AI-DRIVEN SOLUTIONS IN URBAN WATER MANAGEMENT

2.3.1 AI in Water Distribution/Filtration Systems

Water scarcity and drought pose significant challenges to the urban water management system, necessitating efficient systems to support public health, agriculture, and environmental conservation (Gregory et al., 2011). Approximately 30% of treated water is lost during transmission (Ociepa et al., 2019), and detecting leaks within extensive water distribution networks can be challenging. Traditional methods like acoustic detection are labour-intensive, require specialized skills, and lack systematic approaches (Vanijjirattikhan et al., 2022). In contrast, AI-driven technologies offer promising solutions in WDS, particularly in detecting and managing water distribution issues. For instance, Fu et al. (2022) demonstrated that CNN could detect pipe leakages effectively using data from the pressure, flow, acoustic, and vibration signals, which has contributed to a drastic reduction in water losses. They also develop the Burst Location Identification Framework by FL-DenseNet (BLIFF), which accurately identifies burst locations by analysing pressure drops unique to burst outflows. It is a notable improvement over traditional CNN methods, which may struggle with global features like amplitude and time variations (Zhou et al., 2019). Machine learning models like ANN, DNN, clustering, and Support Vector Machines (SVM) have been widely employed to detect burst locations, yet they often fail to automatically extract and learn from complex raw data features (Vanijjirattikhan et al., 2022; Wu et al., 2016). In Cape Town, South Africa, the AI-integrated system optimizes pressure at critical network points, resulting in a 58% reduction in bursts and a 38% decrease in water lost through leaks, extending the water network asset lifespan by five years and saving approximately £300,000 annually in operations and maintenance costs (City of Cape Town Case Study, 2020). Meanwhile, in Milan, Italy, a rule-based control system optimizes hydraulic pump operations to maintain pressure within an optimal range with minimal energy use, demonstrating AI's potential to enhance efficiency in water systems (Ferrari et al., 2023). The important variables that affect the optimization model of WDS are summarized in Figure 2.3 (Parvaze et al., 2023).

AI technologies have also revolutionized the water filtration system. Amin et al. highlighted that water often contains pollutants such as heavy metals, dyes, nutrients,

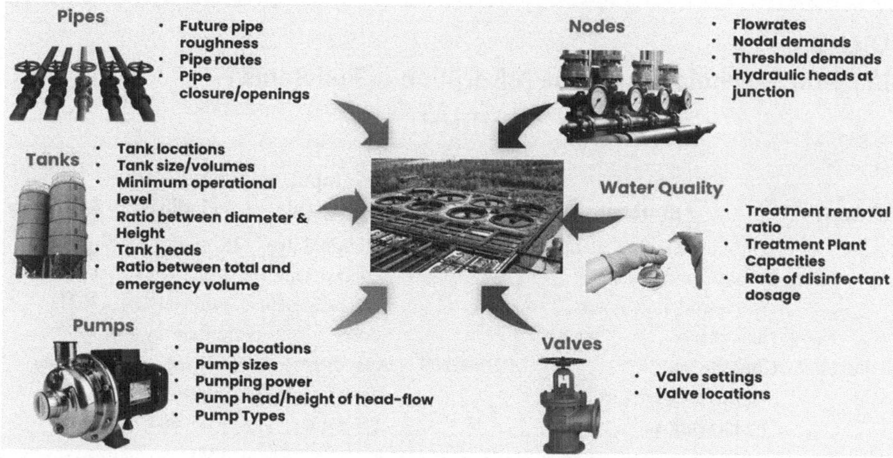

FIGURE 2.3 Decision variables on the optimization model in a WDS.

and pharmaceuticals (Amin et al., 2014). To tackle this, AI-hybrid techniques like MLP-ANN, LS-SVM, ANN-GA, ANN-PSO, and RSM-GA are employed to enhance the predictive and decision-making capabilities of filtration systems, optimizing the removal efficiency of these contaminants (Antwi et al., 2019; Man et al., 2019; Nag et al., 2018; Naghibi et al., 2021; Sabour & Amiri, 2017). The output variable expected from the system is mainly focused on the removal efficiency or the adsorption capacity of the water (Ghaedi & Vafaei, 2017). Table 2.2 summarizes the water treatment process by using AI-hybrid technologies while dealing with different types of pollutants.

AI has become integral to short-term demand forecasting in urban water management. By using machine learning (ML) algorithms like LSTM and CNN, these systems can predict future water needs based on historical demand data, weather, and demographic data (Fu et al., 2022). Nasser et al. reported successful implementations of LSTM-CNN that provided real-time monitoring and prediction insights into water consumption patterns (Nasser et al., 2020). Kavya et al. (2023) also supported LSTM-CNN as the optimal ML model for analysing water consumption patterns and optimizing WDS in urban settings. Muhammad et al. (2019) suggest that ANNs are particularly effective in forecasting scenarios where nonlinear interactions are prevalent and difficult to model with other techniques (Muhammad et al., 2019). Globally, similar AI-driven initiatives have been adopted in cities to catering unique local challenges. In Singapore, the national water authority implemented AI systems that monitor water consumption and optimize distribution, significantly reducing water losses and enhancing supply security in densely populated areas (Essamlali et al., 2024). In the Swedish cities of Gothenburg and Helsingborg, the Future City Flow project has integrated ML algorithms in pumping stations to achieve a more consistent inflow, reduced operating costs, and decreased energy consumption (Oliver Grievson, 2022). Looking forward, other cities could incorporate advanced predictive analytics and IoT-based sensors to manage water resources more efficiently. These technologies enable real-time data analysis and control, facilitating dynamic adjustments in water supply systems to meet fluctuating demands and environmental conditions effectively.

TABLE 2.2
AI-Hybrid Technologies for the Adsorption of Pollutants

Pollutants	Adsorbent	Hybrid AI Models Used	Input Variable	Findings	Reference
Mg	Chitosan/polyvinyl alcohol/ zeolite imidazolate frameworks membrane adsorbents	MLP-ANN	pH, initial dye concentration, and adsorbent dose	Increase Mg removal efficiency	Naghibi et al. (2021)
Cu, Fe, Pb	Chitosan and Chitosan-Montmorillonite nanocomposite	RBF-ANN	Adsorbent dosage, initial pH value	Increase copper, lead, and iron removal efficiency	Hamidian et al. (2019)
Zn(II)	Activated carbon derived from palm oil kernel shell	DEO Neural Network	Initial solution, pH. Adsorbent dose	Increase in Zn (II) removal efficiency	Karri and Sahu (2018)
As (III)	Cerium oxide tetraethylenepentamine	LS-SVM	Temperature, time, concentration, pH and dose	Increased adsorption capacity	Mandal et al. (2015)
BG	ZnS nanoparticles loaded activated carbon	Multi-Layer Perception ANN	Contact time, adsorbent dosage, BG concentration, EB concentration	Increased adsorption capacity	Jamshidi et al. (2016)
MB	Mesoporous rGO/Fe/Co nanohybrids	ANN-PSO, ANN-GA	Initial concentration, contact time, temperature, pH	Increased adsorption capacity	Qi et al. (2020)

Mg, Magnesium; Cu, Copper; Fe, Iron; Pb, Lead; Zn(II), Zinc II; As (III), Arsenic III; BG, Brilliant Green; MB, Methylene Blue; MLP, Multi-Layer Perceptron; ANN, Artificial Neural Network; PSO, Particle Swarm Optimization; GA, Genetic Algorithm; LS-SVM, Least Squares Support Vector Machine; DEO, Differential Evolution Optimization; RBF, Radial Basis Function.

2.3.2 AI in Water Quality Monitoring

Environmental quality has become a vital aspect of human well-being, influencing daily life and health. In alignment with Sustainable Development Goals (SDGs) 6, which emphasizes the importance of clean water and sanitation, the expectation is for all individuals to have access to safe, pollutant-free water. Achieving this goal is essential not only for safeguarding physical and mental health but also for preserving ecosystem integrity. Addressing pollutants that jeopardize water quality requires a comprehensive approach involving pollution prevention, advanced water treatment technologies, and public awareness efforts to ensure sustainable water resources for current and future generations (Vrana et al., 2005). Sousa et al. (2018) reported that the detected water pollutants include organic pollutants, microbeads, malachite green, etc. All these primary pollutants and their descriptions are provided in "Section 3.3 Wastewater Treatment System". In recent years, AI has played a significant role in water quality monitoring by analysing parameter data and predicting water quality trends based on the concentration patterns of key physicochemical parameters, such as COD, BOD, and pH (Hayder et al., 2020). Machine learning, a branch of AI technology, uses past data to learn patterns and predict future trends (Mahesh, 2020). Several algorithms are commonly used in ML, including ANN, Deep Learning Neural Network (DLNN) (Kek et al., 2024), and SVM (Abobakr Yahya et al., 2019; Jamshidzadeh et al., 2024).

One of the most widely used AI ML models in water quality monitoring is ANN. ANN employs nonlinear input-output mapping methods that can learn from data observation without relying on specific mechanics understanding (Mustafa et al., 2020). This capability enables ANN to predict complex hydrological processes without understanding the relationship between input and output (Ayer et al., 2013). The ANN model can be used to predict the physical characteristics of the basin, which represent the physicochemical parameters and pollutants in freshwater as defined by Mustafa et al. (2020)., pollutants in the freshwater. Guo et al. (2015) conducted a study comparing two AI ML models, ANN and SVM, to monitor nitrogen parameters. While the SVM model showed high-performance accuracy, the ANN model demonstrated superior precision in estimating nitrogen pollutant concentrations on water surfaces. Further research by Azrour et al. (2022) tested both ANN and SVM models, concluding that ANN achieved an accuracy of 85% and a precision of 90%, compared to SVM's 78% accuracy and 72% precision when detecting parameters like pH and turbidity. Table 2.3 summarizes different types of AI ML models and their application in monitoring various parameters, including pH, Suspended Solids (SS), Ammonia Nitrogen (AN), and Dissolved Oxygen (DO).

AI-driven technologies in water quality monitoring have been widely implemented across various locations worldwide to enhance real-time data collection, analysis, and predictive insights for better management of water resources. For instance, in China, rivers such as Huaxi, Wubu, and Dahan have adopted AI to monitor water quality parameters continuously and address pollution more proactively. Similarly, in Germany, the Mar Frisio River has integrated AI systems to track pollutants and provide early warnings of water quality degradation. In Southeast Asia, Indonesia's Surabaya River and Malaysia's Johor River are utilizing AI-based monitoring technologies, helping local authorities respond promptly to changes in water quality by predicting pollution patterns and supporting more effective resource management strategies (Mustafa et al., 2021). A recent study revealed that this AI system could further differentiate between contaminated, clean, and UV-disinfected water samples (Durgun, 2024). This advancement highlights how integrating AI with spectral analysis offers a significant breakthrough in ensuring a safe water supply and enhancing water resource management.

TABLE 2.3
Different Types of AI Models Used in Water Quality Monitoring from Various Sources

Source/Country	AI Models	Detected Parameters	Findings	References
River/Johor, Malaysia	Multi-Layer Perceptron ANN	pH, SS, AN	Data set = 110 pH = 8 SS = 165 mg/L AN = 1.1 mg/L	Najah Ahmed et al. (2019)
River/Johor, Malaysia	ANFIS	pH, SS, AN	Data set = 110 pH = 7.8 SS = 165 mg/L AN = 1.1 mg/L	Najah Ahmed et al. (2019)
River/Johor, Malaysia	WDT-ANFIS	pH, SS, AN	Data set = 110 pH = 7.9 SS = 80 mg/L AN = 0.1 mg/L	Najah Ahmed et al. (2019)
River/Kelantan, Malaysia	ANN	DO, BOD, COD, pH, AN and SS	Data set = 250 DO = 83.65 % Sat BOD = 2.75 mg/L COD = 17.26 mg/L pH = 6.03 AN = 0.11 mg/L SS = 144.2 mg/L	Hayder et al. (2020)
Lake/Turkey	SVM	pH, BOD, WQI	pH = 7.18 BOD = 3.13 mg/L WQI = 68.15	Nasir et al. (2022)
Coastal Water/ China	ANN	SS, turbidity	Sample size = 42 SS = 3.3 mg/L Turbidity = 4 NTU	Hafeez et al. (2019)
Rivers & Lakes/ India	ANFIS, FFNN	WQI	WQI = 7	Hmoud Al-Adhaileh and Waselallah Alsaade (2021)

ANFIS, Adaptive Neuro-Fuzzy Inference System; FFNN, Feed-Forward Neural Network; SVM, Support Vector Machine; AN, Artificial Neural Network; WDT-ANFIS, Wavelet Decomposition Technique-ANFIS.

2.3.3 AI in Wastewater Treatment

Wastewater treatment (WWT) faces increasing challenges due to growing urban populations and industrial activities, which intensify the presence of pollutants like heavy metals and organic compounds in water bodies (Zhao et al., 2020). These challenges necessitate more efficient and effective treatment processes to ensure water safety and sustainability. To address these issues, ML models such as recurrent neural networks (RNN) and CNN play important roles in the modelling, prediction, and optimization of WWT processes (Wang et al., 2023). AI-driven tools significantly improve the pretreatment stages by integrating with urban drainage systems to remove solids such as water bottles, plastic bags, and other debris before the water enters the main treatment stages (Albannay et al., 2021; Altowayti et al., 2022; Fan et al., 2022; Kwon & Kim, 2021). This pretreatment helps in reducing the load on subsequent processing stages like sedimentation, which separates sludge from the water over several hours (Kalfa et al., 2020). The real power of AI comes into play during the monitoring phase, where it can continuously assess water quality parameters, including BOD, pH, COD, turbidity, and microbial activity (Saboe et al., 2021). This is facilitated by integrating various sensors such as pH sensors, potentiometric sensors for NH_4^+ detection, FIA-thermal sensors for COD, and biosensors based on bioluminescent bacteria (Banna et al., 2014; Kruse, 2018; Liu et al., 2020; Yao et al., 2015). Real-time data provided by these AI systems enables operation teams to react swiftly to changes in water quality, with user-friendly interfaces offering customizable threshold alerts for pollutant levels (Zhang et al., 2023). This proactive approach ensures that treatment processes are initiated precisely when needed, enhancing efficiency and reducing labour costs. Figure 2.4 shows the integration of AI-driven technologies in wastewater treatment plant.

The AI-assisted WWT plant will reduce the concentration of the contaminants in the wastewater by using various treatment process techniques such as aeration diffusion, adsorption, and photocatalysis, before discharging the water for the purpose of recycling water for domestic, industrial, and agricultural purposes. The table below summarizes the types of AI ML models and the treatment process used for different types of pollutants.

Practical applications of these technologies have shown promising results. For example, the use of genetic algorithms to optimize COD concentration in treatment plants has led to a 25% reduction in energy consumption and a 7% improvement in

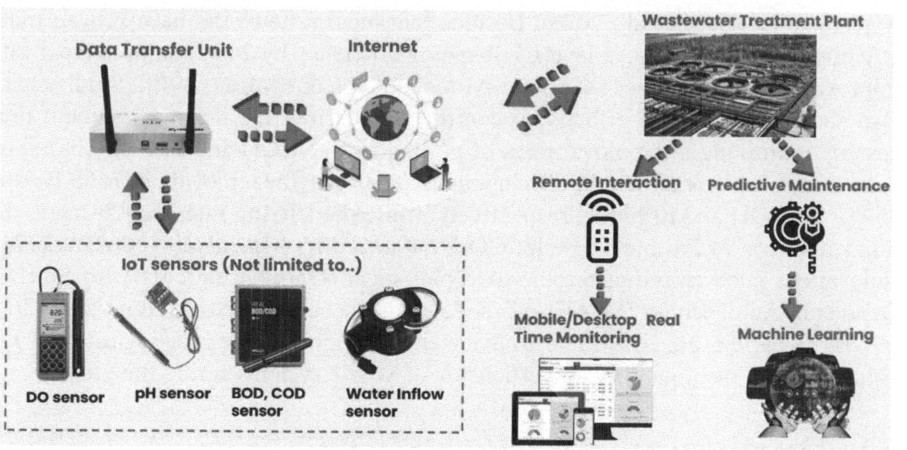

FIGURE 2.4 Integration of AI-driven technologies in wastewater treatment plant. (Miao et al., 2021.)

TABLE 2.4
Application of AI Technologies in Treating Different Pollutants during WWT

Parameter	Recommended Threshold Value	AI Machine Learning Model	Treatment Process	References
COD	21 mg/L	ANN + Genetic Algorithms	Aeration diffusion	Alipour et al., (2017), Man et al. (2019)
Pb^{2+}	30 µg/100 mL	Back- Propagation ANN	Adsorption on thiosemicarbazide chitosan	Nagpal et al. (2024), Reynel-Avila et al. (2015)
BOD	5–10 mg/L	Feed-Forward Neural Networks + Adaptive Network-based Fuzzy Inference System + Support Vector Machine	Activated sludge bioreactor	Lotfi et al. (2019)
Methylene Blue	2 mg/kg	Radial Basis function—ANN	Photocatalytic	Asfaram et al. (2017), Ginimuge andJyothi (2010)
NH_4^+	0.01 ppm	SDAE	Anaerobic-oxic biological	Shi and Xu (2018)
Microbeads	1–1,000 µm	Convolutional Neural Network	Physico-chemical	Alex et al. (2024), Yurtsever and Yurtsever (2019)
Malachite green	0.5–20 µg/kg	Multi-layer ANN	Adsorption	Dil et al. (2017), Verdon and Andersen (2016)

water quality (Nam et al., 2023). Besides that, neural networks have proven more effective in predicting BOD levels, enhancing efficiency by 24%, compared to a 14% improvement with simpler averaging AI tools (Abba & Elkiran, 2019). Studies have also shown that AI can effectively control the stirring rate during treatment processes, optimizing the removal rates of pollutants like COD and turbidity based on their concentrations. When the stirring rate is at 50 rpm, the removal rate is 63.1% and 35.3% for COD and turbidity, respectively, while the stirring rate is at 200 rpm, the removal rate is 74.2% and 37.3% for COD and turbidity, respectively (Hassan, 2024). Innovations such as the anaerobic-oxic biological treatment process with Stacked Denoising Autoencoder (SDAE) AI models have successfully treated up to 93.36% of NH_4^+ (Khalid et al., 2019). These advancements highlight the growing role of AI in enhancing the sustainability and efficiency of WWT systems across the globe.

2.4 CONCLUSION

In conclusion, this chapter has explored the profound impact of AI on urban water management, transforming traditional systems into more efficient, responsive, and sustainable networks through advanced ML algorithms like LSTM, CNN, DNN, and ANN. These technologies have enhanced predictive maintenance, real-time monitoring, and operational optimization, contributing to smarter water distribution, filtration, and wastewater treatment. Moreover, the practical implementations across various cities have showcased their versatility in addressing unique local challenges, from reducing energy consumption and operational costs to enhancing the overall security of water supply systems. These case studies provide valuable insights into the potential for AI to revolutionize urban water management globally. As urban populations continue to grow and environmental challenges become more complex, the integration of AI in urban water systems presents a crucial opportunity for sustainable development. However, challenges such as data privacy, technological disparities, and high initial costs must be addressed to fully realize AI's potential in ensuring water resilience and sustainability. Addressing these challenges requires a concerted effort from policymakers, engineers, and urban planners. This chapter underscores the transformative potential of AI, suggesting a future where intelligent water management systems are not just ideal but essential for the survival and thriving of urban landscapes worldwide.

ACKNOWLEDGEMENT

The authors would like to acknowledge the financial support from Asia Technological University (ATU) Network under ATU-Net Young Researcher Grant (YRG) under Vot. No. R.J130000.7724.4J714.

DECLARATION OF GENERATIVE AI IN SCIENTIFIC WRITING

I hereby declare that no generative AI tools were used in the preparation, writing, or editing of this manuscript, except for grammar and English language checks.

REFERENCES

Abba, S. I., & Elkiran, G. (2019). *Wastewater Treatment Plant Performance Analysis Using Artificial Intelligence–An Ensemble Approach*. Near East University.

Abobakr Yahya, A. S., Ahmed, A. N., Binti Othman, F., Ibrahim, R. K., Afan, H. A., El-Shafie, A., . . . Elshafie, A. (2019). Water quality prediction model based support vector machine model for ungauged river catchment under dual scenarios. *Water*, 11(6), 1231.

Albannay, S., Kazama, S., Oguma, K., Hashimoto, T., & Takizawa, S. (2021). Water demand management based on water consumption data analysis in the emirate of Abu Dhabi. *Water*, 13(20), 2827.

Alex, R. K., Maes, T., & Devipriya, S. P. (2024). Clean, but not green: emission assessment, forecast modelling and policy solutions for plastic microbeads from personal care products in India. *Emerging Contaminants*, 10(3), 100326. https://doi.org/10.1016/j.emcon.2024.100326

Alipour, V., Moein, F., & Rezaei, L. (2017). Determining the salt tolerance threshold for biological treatment of salty wastewater. *Health Scope*, 6(1), e36425.

Altowayti, W. A. H., Shahir, S., Othman, N., Eisa, T. A. E., Yafooz, W. M., Al-Dhaqm, A., . . . Abaker, M. (2022). The role of conventional methods and artificial intelligence in the wastewater treatment: a comprehensive review. *Processes*, 10(9), 1832.

Amin, M. T., Alazba, A. A., & Manzoor, U. (2014). A review of removal of pollutants from water/wastewater using different types of nanomaterials. *Advances in Materials Science and Engineering*, 2014(1), 825910.

Antwi, P., Zhang, D., Xiao, L., Kabutey, F. T., Quashie, F. K., Luo, W., Meng, J., & Li, J. (2019). Modeling the performance of Single-stage Nitrogen removal using Anammox and Partial nitritation (SNAP) process with backpropagation neural network and response surface methodology. *Science of the Total Environment*, 690, 108–120.

Asfaram, A., Ghaedi, M., Hajati, S., Goudarzi, A., & Dil, E. A. (2017). Screening and optimization of highly effective ultrasound-assisted simultaneous adsorption of cationic dyes onto Mn-doped Fe_3O_4-nanoparticle-loaded activated carbon. *Ultrasonics Sonochemistry*, 34, 1–12.

Ayer, T., Chen, Q., & Burnside, E. S. (2013). Artificial neural networks in mammography interpretation and diagnostic decision making. *Computational and Mathematical Methods in Medicine*, 2013(1), 832509.

Azrour, M., Mabrouki, J., Fattah, G., Guezzaz, A., & Aziz, F. (2022). Machine learning algorithms for efficient water quality prediction. *Modeling Earth Systems and Environment*, 8(2), 2793–2801.

Bangalore, M., Smith, A., & Veldkamp, T. (2019). Exposure to floods, climate change, and poverty in Vietnam. *Economics of Disasters and Climate Change*, 3, 79–99.

Banna, M. H., Najjaran, H., Sadiq, R., Imran, S. A., Rodriguez, M. J., & Hoorfar, M. (2014). Miniaturized water quality monitoring pH and conductivity sensors. *Sensors and Actuators B: Chemical*, 193, 434–441.

Brar, A. S. (2013). *Consumer Behaviour and Perception for Efficient Water Use in Urban Punjab*. Punjab, India: Punjab Technical University.

Chen, J. (2007). Rapid urbanization in China: a real challenge to soil protection and food security. *CATENA*, 69(1), 1–15. https://doi.org/10.1016/j.catena.2006.04.019

Citaristi, I. (2022). Asian Development Bank—ADB. In: *The Europa Directory of International Organizations* (pp. 453–457). Routledge.

City of Cape Town Case Study, i-2O Water. (2020).

Das Kangabam, R., Bhoominathan, S. D., Kanagaraj, S., & Govindaraju, M. (2017). Development of a water quality index (WQI) for the Loktak Lake in India. *Applied Water Science*, 7, 2907–2918.

Data, O. W. I. (2024). *Data Page: Renewable Freshwater Resources Per Capita*. Food and Agriculture Organization of the United Nations (via World Bank). https://ourworldindata.org/grapher/renewable-water-resources-per-capita

Dietz, T., Shwom, R. L., & Whitley, C. T. (2020). Climate change and society. *Annual Review of Sociology*, 46(1), 135–158.

Dil, E. A., Ghaedi, M., Asfaram, A., Hajati, S., Mehrabi, F., & Goudarzi, A. (2017). Preparation of nanomaterials for the ultrasound-enhanced removal of Pb^{2+} ions and malachite green dye: chemometric optimization and modeling. *Ultrasonics Sonochemistry*, 34, 677–691.

Du, H., Ji, X., & Chuai, X. (2022). Spatial differentiation and influencing factors of water pollution-intensive industries in the Yellow River Basin, China. *International Journal of Environmental Research and Public Health*, 19(1), 497.

Durgun, Y. (2024). Real-time water quality monitoring using AI-enabled sensors: detection of contaminants and UV disinfection analysis in smart urban water systems. *Journal of King Saud University – Science*, 36(9), 103409. https://doi.org/10.1016/j.jksus.2024.103409

Elessawy, F. (2017). The boom: population and urban growth of Dubai City. *Horizons Humanities and Social Sciences*, 2, 26–41.

Essamlali, I., Nhaila, H., & El Khaili, M. (2024). Advances in machine learning and IoT for water quality monitoring: a comprehensive review. *Heliyon*, 10(6), e27920. https://doi.org/10.1016/j.heliyon.2024.e27920

Evans, A., & Falvey, R. (2013). *Annual Tropical Cyclone Report*. The Joint Typhoon Warning Center (JTWC).

Fan, C., Lv, C., Wang, Z., Wu, S., Jin, Z., Bei, K., . . . Zhao, M. (2022). Influence of regular addition of ore on treatment efficiency and aquatic organisms in living machine system for black water treatment. *Journal of Cleaner Production*, 341, 130928.

Ferrari, E., Verda, D., Pinna, N., & Muselli, M. (2023). Optimizing water distribution through explainable AI and rule-based control. *Computers*, 12(6), 123.

Fu, G., Jin, Y., Sun, S., Yuan, Z., & Butler, D. (2022). The role of deep learning in urban water management: a critical review. *Water Research*, 223, 118973. https://doi.org/10.1016/j.watres.2022.118973

Gao, H., Xu, C., Xie, C., Ma, J., & Xiao, Z. (2024). *Landslides Triggered by the July 2023 Extreme Rainstorm in the Haihe River Basin*. China. Springer.

Garschagen, M. (2013). Resilience and organisational institutionalism from a cross-cultural perspective: an exploration based on urban climate change adaptation in Vietnam. *Natural Hazards*, 67(1), 25–46.

Georgi, B., Isoard, S., Asquith, M., Garzillo, C., Swart, R., & Timmerman, J. (2016). *Urban Adaptation to Climate Change in Europe* 2016*: Transforming Cities in a Changing Climate*. European Environment Agency

Ghaedi, A. M., & Vafaei, A. (2017). Applications of artificial neural networks for adsorption removal of dyes from aqueous solution: a review. *Advances in Colloid and Interface Science*, 245, 20–39.

Ginimuge, P. R., & Jyothi, S. D. (2010). Methylene blue: revisited. *Journal of Anaesthesiology Clinical Pharmacology*, 26(4), 517–520.

Gosling, S. N., & Arnell, N. W. (2016). A global assessment of the impact of climate change on water scarcity. *Climatic Change*, 134(3), 371–385. https://doi.org/10.1007/s10584-013-0853-x

Gregory, K. B., Vidic, R. D., & Dzombak, D. A. (2011). Water management challenges associated with the production of shale gas by hydraulic fracturing. *Elements*, 7(3), 181–186.

Grigg, N. (2019). *Aging Water Infrastructure*. IWA Publishing.

Guo, H., Cao, K., & Wang, P. (2017). Population estimation in Singapore based on remote sensing and open data. *The International Archives of the Photogrammetry, Remote Sensing and Spatial Information Sciences*, 42, 1181–1187.

Guo, H., Jeong, K., Lim, J., Jo, J., Kim, Y. M., Park, J.-P., . . . Cho, K. H. (2015). Prediction of effluent concentration in a wastewater treatment plant using machine learning models. *Journal of Environmental Sciences*, 32, 90–101.

Hafeez, S., Wong, M. S., Ho, H. C., Nazeer, M., Nichol, J., Abbas, S., . . . Pun, L. (2019). Comparison of machine learning algorithms for retrieval of water quality indicators in case-II waters: a case study of Hong Kong. *Remote Sensing*, 11(6), 617. https://www.mdpi.com/2072-4292/11/6/617

Hamidian, A. H., Esfandeh, S., Zhang, Y., & Yang, M. (2019). Simulation and optimization of nanomaterials application for heavy metal removal from aqueous solutions. *Inorganic and Nano-Metal Chemistry*, 49(7), 217–230.

Harrell, S. (2023). *An Ecological History of Modern China*. University of Washington Press.

Hassan, R. (2024). Efficient pulp and paper wastewater treatment using green tea-synthesized magnetite: artificial intelligence and kinetic analysis. *Desalination and Water Treatment*, 320, 100804. https://doi.org/10.1016/j.dwt.2024.100804

Hayder, G., Kurniawan, I., & Mustafa, H. M. (2020). Implementation of machine learning methods for monitoring and predicting water quality parameters. *Biointerface Research in Applied Chemistry*, 11, 9285–9295.

Hmoud Al-Adhaileh, M., & Waselallah Alsaade, F. (2021). Modelling and Prediction of Water Quality by Using Artificial Intelligence. *Sustainability*, 13(8), 4259. https://www.mdpi.com/2071-1050/13/8/4259

Huong, T. T. L., Van Anh, D. T., Dat, T. T., Truong, D. D., & Tam, D. D. (2022). Disaster risk management system in Vietnam: progress and challenges. *Heliyon*, 8(10), e10701.

Isah, N. (2016). Flood occurrence, smart tunnel operating system and traffic flow: a case of Kuala Lumpur smart tunnel Malaysia Universiti Tun Hussein Onn Malaysia].

Jamshidi, M., Ghaedi, M., Dashtian, K., Ghaedi, A., Hajati, S., Goudarzi, A., & Alipanahpour, E. (2016). Highly efficient simultaneous ultrasonic assisted adsorption of brilliant green and eosin B onto ZnS nanoparticles loaded activated carbon: artificial neural network modeling and central composite design optimization. *Spectrochimica Acta Part A: Molecular and Biomolecular Spectroscopy*, 153, 257–267.

Jamshidzadeh, Z., Ehteram, M., & Shabanian, H. (2024). Bidirectional long short-term memory (BILSTM) – Support vector machine: a new machine learning model for predicting water quality parameters. *Ain Shams Engineering Journal*, 15(3), 102510. https://doi.org/10.1016/j.asej.2023.102510

Jay, M. S., Bridges, C. E., Gottlieb, A. A., & DuRant, R. H. (1988). Adolescent contraception: an overview. *Adolescent and Pediatric Gynecology*, 1(2), 83–95.

Jones, G. W. (2012). Population policy in a prosperous city-state: dilemmas for Singapore. *Population and Development Review*, 38(2), 311–336.

Joseph, L., Jun, B.-M., Flora, J. R. V., Park, C. M., & Yoon, Y. (2019). Removal of heavy metals from water sources in the developing world using low-cost materials: a review. *Chemosphere*, 229, 142–159. https://doi.org/10.1016/j.chemosphere.2019.04.198

Kalfa, A., Shapira, B., Shopin, A., Cohen, I., Avraham, E., & Aurbach, D. (2020). Capacitive deionization for wastewater treatment: opportunities and challenges. *Chemosphere*, 241, 125003.

Kang, H. (2019). Challenges for water infrastructure asset management in South Korea. *Water Policy*, 21(5), 934–944.

Karri, R. R., & Sahu, J. (2018). Process optimization and adsorption modeling using activated carbon derived from palm oil kernel shell for Zn (II) disposal from the aqueous environment using differential evolution embedded neural network. *Journal of Molecular Liquids*, 265, 592–602.

Kavya, M., Mathew, A., Shekar, P. R., & Sarwesh, P. (2023). Short term water demand forecast modelling using artificial intelligence for smart water management. *Sustainable Cities and Society*, 95, 104610.

Kek, H. Y., Bazgir, A., Tan, H., Lee, C. T., Hong, T., Othman, M. H. D., . . . Wong, K. Y. (2024). Particle dispersion for indoor air quality control considering air change approach: a novel accelerated CFD-DNN prediction. *Energy and Buildings*, 306, 113938. https://doi.org/10.1016/j.enbuild.2024.113938

Khalid, A. A. H., Yaakob, Z., Abdullah, S. R. S., & Takriff, M. S. (2019). Analysis of the elemental composition and uptake mechanism of Chlorella sorokiniana for nutrient removal in agricultural wastewater under optimized response surface methodology (RSM) conditions. *Journal of Cleaner Production*, 210, 673–686.

Kruse, P. (2018). Review on water quality sensors. *Journal of Physics D: Applied Physics*, 51(20), 203002.

Kwon, S. H., & Kim, J. H. (2021). Machine learning and urban drainage systems: state-of-the-art review. *Water*, 13(24), 3545.

Larsen, T. A., & Gujer, W. (1997). The concept of sustainable Urban Water Management. *Water Science and Technology*, 35(9), 3–10. https://doi.org/10.1016/S0273-1223(97)00179-0

Leal Filho, W., Balogun, A.-L., Olayide, O. E., Azeiteiro, U. M., Ayal, D. Y., Muñoz, P. D. C., . . . Toamukum, N. Y. (2019). Assessing the impacts of climate change in cities and their adaptive capacity: towards transformative approaches to climate change adaptation and poverty reduction in urban areas in a set of developing countries. *Science of the Total Environment*, 692, 1175–1190.

Li, X., Zhang, Y., Li, W., & Chen, Y. (2023). Extreme characteristics of "23· 7" heavy rain in Beijing-Tianjin-Hebei and its implications for urban flood control in China. *China Flood & Drought Management*, 33(11), 13–18.

Li, Z., Ma, Z., van der Kuijp, T. J., Yuan, Z., & Huang, L. (2014). A review of soil heavy metal pollution from mines in China: pollution and health risk assessment. *Science of the Total Environment*, 468, 843–853.

Liu, Y., Li, J., Wan, N., Fu, T., Wang, L., Li, C., . . . Zhu, A. (2020). A current sensing biosensor for BOD rapid measurement. *Archaea*, 2020(1), 8894925.

Long, H., Kong, X., Hu, S., & Li, Y. (2021). Land use transitions under rapid urbanization: a perspective from developing China. *Land*, 10, 935.

Lorenzo, T. E., & Kinzig, A. P. (2020). Double exposures: future water security across urban Southeast Asia. *Water*, 12(1), 116.

Lotfi, K., Bonakdari, H., Ebtehaj, I., Mjalli, F. S., Zeynoddin, M., Delatolla, R., & Gharabaghi, B. (2019). Predicting wastewater treatment plant quality parameters using a novel hybrid linear-nonlinear methodology. *Journal of Environmental Management*, 240, 463–474.

Luu, C., Von Meding, J., & Kanjanabootra, S. (2018). Assessing flood hazard using flood marks and analytic hierarchy process approach: a case study for the 2013 flood event in Quang Nam, Vietnam. *Natural Hazards*, 90, 1031–1050.

Mahato, A., Upadhyay, S., & Sharma, D. (2022). Global water scarcity due to climate change and its conservation strategies with special reference to India: a review. *Plant Archives*, 22(1), 64–69.

Mahdi, N. M., Jassim, A. H., Abulqasim, S. A., Basem, A., Ogaili, A. A. F., & Al-Haddad, L. A. (2024). Leak detection and localization in water distribution systems using advanced feature analysis and an Artificial Neural Network. *Desalination and Water Treatment*, 320, 100685. https://doi.org/10.1016/j.dwt.2024.100685

Mahesh, B. (2020). Machine learning algorithms-a review. *International Journal of Science and Research (IJSR)*, 9(1), 381–386.

Man, Y., Hu, Y., & Ren, J. (2019). Forecasting COD load in municipal sewage based on ARMA and VAR algorithms. *Resources, Conservation and Recycling*, 144, 56–64.

Mandal, S., Mahapatra, S., Adhikari, S., & Patel, R. (2015). Modeling of arsenic (III) removal by evolutionary genetic programming and least square support vector machine models. *Environmental Processes*, 2, 145–172.

McDonald, R. I., Green, P., Balk, D., Fekete, B. M., Revenga, C., Todd, M., & Montgomery, M. (2011). Urban growth, climate change, and freshwater availability. *Proceedings of the National Academy of Sciences*, 108(15), 6312–6317.

Melati, I., Maghfiroh, M., & Jasalesmana, T. (2020). Public perception and contribution of domestic waste to water pollution of Lake Matano, South Sulawesi-Indonesia. *IOP Conference Series: Earth and Environmental Science*, 535, 012065.

Miao, S., Zhou, C., AlQahtani, S. A., Alrashoud, M., Ghoneim, A., & Lv, Z. (2021). Applying machine learning in intelligent sewage treatment: a case study of chemical plant in sustainable cities. *Sustainable Cities and Society*, 72, 103009. https://doi.org/10.1016/j.scs.2021.103009

Muhammad, A. U., Li, X., & Feng, J. (2019). Artificial intelligence approaches for urban water demand forecasting: a review. *Machine Learning and Intelligent Communications: 4th International Conference, MLICOM 2019*, Nanjing, China, August 24–25, 2019, Proceedings 4. Springer, pp. 595–622.

Muhammad, N., Abdullah, J., & Julien, P. (2020). Characteristics of rainfall in peninsular Malaysia. *Journal of Physics: Conference Series*, 1529, 052014.

Mustafa, H. M., Hayder, G., Mustapha, A., Kushiar, K. F., Tahir, S. M., & Elsadig, E. H. O. (2020). Bibliometric analysis of sediment concentration prediction using artificial neural network. *Journal of Energy and Environment*, 12(2)

Mustafa, H. M., Mustapha, A., Hayder, G., & Salisu, A. (2021). Applications of IoT and artificial intelligence in water quality monitoring and prediction: a review. *2021 6th International Conference on Inventive Computation Technologies (ICICT)*, 20–22 January, Coimbatore, Inda. IEEE.

Nag, S., Mondal, A., Roy, D. N., Bar, N., & Das, S. K. (2018). Sustainable bioremediation of Cd (II) from aqueous solution using natural waste materials: kinetics, equilibrium, thermodynamics, toxicity studies and GA-ANN hybrid modelling. *Environmental Technology & Innovation*, 11, 83–104.

Naghibi, S. A., Salehi, E., Khajavian, M., Vatanpour, V., & Sillanpää, M. (2021). Multivariate data-based optimization of membrane adsorption process for wastewater treatment: multi-layer perceptron adaptive neural network versus adaptive neural fuzzy inference system. *Chemosphere*, 267, 129268.

Nagpal, M., Siddique, M. A., Sharma, K., Sharma, N., & Mittal, A. (2024). Optimizing wastewater treatment through artificial intelligence: recent advances and future prospects. *Water Science and Technology*, 90(3), 731–757. https://doi.org/10.2166/wst.2024.259

Najah Ahmed, A., Binti Othman, F., Abdulmohsin Afan, H., Khaleel Ibrahim, R., Ming Fai, C., Shabbir Hossain, M., . . . Elshafie, A. (2019). Machine learning methods for better water quality prediction. *Journal of Hydrology*, 578, 124084. https://doi.org/10.1016/j.jhydrol.2019.124084

Nam, K., Heo, S., Kim, S., & Yoo, C. (2023). A multi-agent AI reinforcement-based digital multi-solution for optimal operation of a full-scale wastewater treatment plant under various influent conditions. *Journal of Water Process Engineering*, 52, 103533. https://doi.org/10.1016/j.jwpe.2023.103533

Nasir, N., Kansal, A., Alshaltone, O., Barneih, F., Sameer, M., Shanableh, A., & Al-Shamma'a, A. (2022). Water quality classification using machine learning algorithms. *Journal of Water Process Engineering*, 48, 102920. https://doi.org/10.1016/j.jwpe.2022.102920

Nasser, A. A., Rashad, M. Z., & Hussein, S. E. (2020). A two-layer water demand prediction system in urban areas based on micro-services and LSTM neural networks. *IEEE Access*, 8, 147647–147661.

Nova, K. (2023). AI-enabled water management systems: an analysis of system components and interdependencies for water conservation. *Eigenpub Review of Science and Technology*, 7(1), 105–124.

Ociepa, E., Mrowiec, M., & Deska, I. (2019). Analysis of water losses and assessment of initiatives aimed at their reduction in selected water supply systems. *Water*, 11(5), 1037. https://www.mdpi.com/2073-4441/11/5/1037

Oliver Grievson, T. H., & Johnson, B. (2022). *A Strategic Digital Transformation for the Water Industry*. IWA Publishing.

Ong, C., Tortajada, C., & Arora, O. (2023). Key Challenges to Urban Water Management in ASEAN. In Ong, C., Tortajada, C., & Arora, O. (Eds.), *Urban Water Demand Management: A Guidebook for ASEAN* (pp. 3–14). Springer Nature Singapore. https://doi.org/10.1007/978-981-19-8677-2_2

Pandit, A. B., & Kumar, J. K. (2015). Clean water for developing countries. *Annual Review of Chemical and Biomolecular Engineering*, 6(1), 217–246.

Parkinson, J., Goldenfum, J., & Tucci, C. (2010). Integrated urban water management: humid tropics. https://doi.org/10.1201/b10534

Parvaze, S., Kumar, R., Khan, J. N., Al-Ansari, N., Parvaze, S., Vishwakarma, D. K., . . . Kuriqi, A. (2023). Optimization of water distribution systems using genetic algorithms: a review. *Archives of Computational Methods in Engineering*, 30(7), 4209–4244.

Pascual, L. A. C. A., Ong, A. K. S., Briggs, C. M., Diaz, J. F. T., & German, J. D. (2024). Factors affecting the intention to prepare for flash floods in the Philippines. *International Journal of Disaster Risk Reduction*, 112, 104794. https://doi.org/10.1016/j.ijdrr.2024.104794

Peng, X. (2011). China's demographic history and future challenges. *Science*, 333(6042), 581–587.

Perera, D., Smakhtin, V., Williams, S., North, T., & Curry, A. (2021). Ageing water storage infrastructure: an emerging global risk. *UNU-INWEH Report Series*, 11, 25.

Phongsathorn, P. (2011). Environment and migration: the 2011 floods in Thailand. In: Gemenne, F., Brückner, P., & Ionesco D. (Eds), *The State of Environmental Migration.*

Poapongsakorn, N., & Meethom, P. (2013). Impact of the 2011 floods, and flood management in Thailand. *ERIA Discussion Paper Series*, 34, 2013.

Qi, J., Hou, Y., Hu, J., Ruan, W., Xiang, Y., & Wei, X. (2020). Decontamination of methylene Blue from simulated wastewater by the mesoporous rGO/Fe/Co nanohybrids: artificial intelligence modeling and optimization. *Materials Today Communications*, 24, 100709.

Ramasamy, E., Jayasooryan, K., Chandran, M. S., & Mohan, M. (2017). Total and methyl mercury in the water, sediment, and fishes of Vembanad, a tropical backwater system in India. *Environmental Monitoring and Assessment*, 189, 1–19.

Reidmiller, D. R., Avery, C. W., Easterling, D. R., Kunkel, K. E., Lewis, K. L., Maycock, T. K., & Stewart, B. C. (2017). *Impacts, Risks, and Adaptation in the United States: Fourth National Climate Assessment* (vol. II). Washington, DC: U.S. Global Change Research Program.

Reynel-Avila, H. E., Bonilla-Petriciolet, A., & de la Rosa, G. (2015). Analysis and modeling of multicomponent sorption of heavy metals on chicken feathers using Taguchi's experimental designs and artificial neural networks. *Desalination and Water Treatment*, 55(7), 1885–1899.

Sa'adi, Z., Ramli, M. W. A., Wan Ahmad Tajuddin, W. A. N., Arman, N. Z., Che Hassan, C. H., Ramzan, M. A., . . . Alias, N. E. (2024). Evaluating flood early warning system and public preparedness and knowledge in urban and semi-urban areas of Johor, Malaysia: challenges and opportunities. *International Journal of Disaster Risk Reduction*, 113, 104870. https://doi.org/10.1016/j.ijdrr.2024.104870

Saboe, D., Ghasemi, H., Gao, M. M., Samardzic, M., Hristovski, K. D., Boscovic, D., . . . Hoffman, D. A. (2021). Real-time monitoring and prediction of water quality parameters and algae concentrations using microbial potentiometric sensor signals and machine learning tools. *Science of the Total Environment*, 764, 142876.

Sabour, M. R., & Amiri, A. (2017). Comparative study of ANN and RSM for simultaneous optimization of multiple targets in Fenton treatment of landfill leachate. *Waste Management*, 65, 54–62.

Selvakumar, A., & Tafuri, A. N. (2012). Rehabilitation of aging water infrastructure systems: key challenges and issues. *Journal of Infrastructure Systems*, 18(3), 202–209.

Shi, S., & Xu, G. (2018). Novel performance prediction model of a biofilm system treating domestic wastewater based on stacked denoising auto-encoders deep learning network. *Chemical Engineering Journal*, 347, 280–290.

Siddique, I. (2021). Sustainable water management in urban areas: integrating innovative technologies and practices to address water scarcity and pollution. *The Pharmaceutical and Chemical Journal*, 8(1), 172–178.

Sousa, J. C. G., Ribeiro, A. R., Barbosa, M. O., Pereira, M. F. R., & Silva, A. M. T. (2018). A review on environmental monitoring of water organic pollutants identified by EU guidelines. *Journal of Hazardous Materials*, 344, 146–162. https://doi.org/10.1016/j.jhazmat.2017.09.058

Sun, D., Zhou, L., Li, Y., Liu, H., Shen, X., Wang, Z., & Wang, X. (2017). New-type urbanization in China: predicted trends and investment demand for 2015–2030. *Journal of Geographical Sciences*, 27, 943–966.

Tomas, D., Čurlin, M., & Marić, A. S. (2017). Assessing the surface water status in Pannonian ecoregion by the water quality index model. *Ecological Indicators*, 79, 182–190.

Uddin, M. G., Nash, S., & Olbert, A. I. (2021). A review of water quality index models and their use for assessing surface water quality. *Ecological Indicators*, 122, 107218. https://doi.org/10.1016/j.ecolind.2020.107218

Vanijjirattikhan, R., Khomsay, S., Kitbutrawat, N., Khomsay, K., Supakchukul, U., Udomsuk, S., . . . Anusart, K. (2022). AI-based acoustic leak detection in water distribution systems. *Results in Engineering*, 15, 100557. https://doi.org/10.1016/j.rineng.2022.100557

Verdon, E., & Andersen, W. C. (2016). Certain dyes as pharmacologically active substances in fish farming and other aquaculture products. In: *Chemical Analysis of Non-Antimicrobial Veterinary Drug Residues in Food* (pp. 497–548).

Vrana, B., Allan, I. J., Greenwood, R., Mills, G. A., Dominiak, E., Svensson, K., . . . Morrison, G. (2005). Passive sampling techniques for monitoring pollutants in water. *TrAC – Trends in Analytical Chemistry*, 24(10), 845–868. https://doi.org/10.1016/j.trac.2005.06.006

Wang, Q., Jiang, R., & Li, R. (2018). Decoupling analysis of economic growth from water use in City: a case study of Beijing, Shanghai, and Guangzhou of China. *Sustainable Cities and Society*, 41, 86–94. https://doi.org/10.1016/j.scs.2018.05.010

Wang, Y., Cheng, Y., Liu, H., Guo, Q., Dai, C., Zhao, M., & Liu, D. (2023). A review on applications of artificial intelligence in wastewater treatment. *Sustainability*, 15(18), 13557.

Wu, Y., Liu, S., Wu, X., Liu, Y., & Guan, Y. (2016). Burst detection in district metering areas using a data driven clustering algorithm. *Water Research*, 100, 28–37.

Yao, N., Liu, Z., Chen, Y., Zhou, Y., & Xie, B. (2015). A novel thermal sensor for the sensitive measurement of chemical oxygen demand. *Sensors*, 15(8), 20501–20510.

Yap, C. K., Peng, S. H. T., & Leow, C. S. (2019). Contamination in Pasir Gudang area, peninsular Malaysia: what can we learn from Kim Kim River chemical waste contamination. *Journal of Humanities and Education Development*, 1(2), 82–87.

Yurtsever, M., & Yurtsever, U. (2019). Use of a convolutional neural network for the classification of microbeads in urban wastewater. *Chemosphere*, 216, 271–280.

Zhang, S., Jin, Y., Chen, W., Wang, J., Wang, Y., & Ren, H. (2023). Artificial intelligence in wastewater treatment: a data-driven analysis of status and trends. *Chemosphere*, 336, 139163.

Zhao, L., Dai, T., Qiao, Z., Sun, P., Hao, J., & Yang, Y. (2020). Application of artificial intelligence to wastewater treatment: a bibliometric analysis and systematic review of technology, economy, management, and wastewater reuse. *Process Safety and Environmental Protection*, 133, 169–182.

Zhou, X., Tang, Z., Xu, W., Meng, F., Chu, X., Xin, K., & Fu, G. (2019). Deep learning identifies accurate burst locations in water distribution networks. *Water Research*, 166, 115058. https://doi.org/10.1016/j.watres.2019.115058

Zurairi, A. (2018). *Climate-Related Natural Disasters Cost Malaysia RM8b in the Last 20 Years*. Malay Mail.

3 Machine Learning, IoT, and GIS for Water Quality Monitoring

Huiyi Tan, Mohd Hafiz Dzarfan Othman, Fong Chi Chan, Hong Yee Kek, Kok Sin Woon, Ruzairi Abdul Rahim, Sien Jie Wong, Elmy Johana Mohamad, and Keng Yinn Wong

3.1 INTRODUCTION

Water quality monitoring is crucial for safeguarding public health, promoting environmental sustainability, and supporting industrial applications. Contaminated water containing pathogenic microorganisms, such as bacteria, protozoa, and viruses, which can lead to the spread of waterborne diseases such as cholera, dysentery, and typhoid fever (Shayo et al., 2023). Delays in treatment have been associated with increased risks of disability, illness, disorders, or even death due to prolonged exposure to contaminated water (Demeneix & Hahn, 2020). United Nations reported that approximately 2.1 billion people do not have access to clean and safe drinking water, leading to 2.2 million deaths annually from waterborne diseases (Uhlenbrook & Connor, 2019). To combat waterborne diseases, governments and relevant authorities have implemented various water treatment procedures. For instance, ceramic-based water purification systems have been widely used in developing countries, effectively reducing waterborne diseases by 50%–90% due to their affordability and cost-effectiveness (Shayo et al., 2023). By continuously assessing water bodies, harmful pathogens, heavy metals, and chemical pollutants can be detected early, allowing for timely interventions to prevent health risks. Contaminants commonly found in water supplies can be categorized into non-biological (such as ammonia, arsenic, barium, boron, chlorine, cadmium, lead, mercury, nickel, nitrate, and sodium) and biological (including *Cryptosporidium*, *Escherichia coli*, *Giardia*, *Legionella*, pesticides, and *Pseudomonas*) (Zulkifli et al., 2018). Prolonged exposure to these contaminants can increase the risk of cancer, kidney failure, diarrhoea, toxic accumulation, and other health issues (Zulkifli et al., 2018). Hence, water quality monitoring is crucial for maintaining the health of human beings, aquatic ecosystems, which are essential for biodiversity, regulating climate, and supporting natural processes (Kamali et al., 2019).

In industrial sectors, high-quality water is essential for efficient and sustainable operations across agriculture, manufacturing, energy production, and pharmaceuticals

DOI: 10.1201/9781003547099-4

(Strade et al., 2020). Monitoring water quality ensures that the water used in industrial processes meets necessary standards, helping to prevent equipment damage and operational inefficiencies. For instance, water contamination can compromise pharmaceutical products due to stringent requirements for chemical and microbiological purity as well as Good Manufacturing Practice (GMP) regulations (Strade et al., 2020). Industries are often major contributors to water pollution, making monitoring vital for regulating the discharge of industrial effluents and ensuring compliance with environmental standards. As environmental challenges like climate change and pollution continue to escalate, water quality monitoring becomes crucial for managing water resources responsibly, safeguarding public health, protecting ecosystems, and promoting sustainable industrial practices (Awewomom et al., 2024). While wastewater treatment has garnered considerable research attention, its connection to sustainable resource management has not been adequately addressed (Silva, 2023). Previous studies have examined various wastewater treatment techniques, including chemical treatment, physical treatment, biological processes, and sludge management. Common steps in wastewater treatment highlighted in the literature include screening and pumping, primary treatment, secondary treatment, disinfection, and sludge handling (Nair K et al., 2021). Additionally, innovative solutions for water treatment and filtration have emerged, utilizing materials such as empty fruit bunch biowaste (Tan et al., 2022), pre-treated biowaste (Tan et al., 2023), and ceramic membrane (Omar et al., 2024).

This chapter highlights the potential of AI-powered techniques for water quality monitoring and early detection of contaminants, surpassing the limitations of traditional methods. Past studies have focused on conventional approaches, such as manual sampling and statistical models, which are often constrained by delayed data collection, limited real-time analysis, and reduced predictive capabilities. In contrast, AI-driven solutions, which integrate machine learning (ML) and IoT approaches, offer advanced data processing, real-time monitoring, and the ability to predict contaminant spread in water before it significantly affects human health and environmental issues. This chapter emphasizes AI's ability to manage complex, high-dimensional datasets, automate data collection, as well as enhance early detection frameworks. It also emphasizes AI's vital role in promoting sustainable water management, aligning with SDG 6 (Clean Water and Sanitation), and enhancing Environmental, Social, and Governance (ESG) practices. AI is set to transform water quality monitoring by enabling timely interventions, ultimately protecting public health and ecosystems.

3.2 OVERVIEW OF AI IN ENVIRONMENTAL MONITORING

AI has advanced environmental monitoring by enabling more efficient and proactive management of ecosystems and natural resources (Bibri et al., 2024). Traditional environmental monitoring methods often rely on manual onsite data sampling and analysis. These approaches can be time-consuming, costly, and prone to human errors (Azha et al., 2023). AI—through advanced algorithms, computational resources, and big data analytics—provides tools to automate these processes, analyse complex datasets, and generate real-time insights that would be challenging to achieve with conventional approaches. AI is being applied in diverse areas such as air quality control, water quality monitoring, biodiversity conservation, and climate change modelling. Often integrated with IoT sensors, AI facilitates the collection and processing of large volumes of environmental data, including pollutant levels, temperature, humidity, and biological indicators. This integration enables real-time monitoring and early detection of anomalies. For instance, AI in recent environmental monitoring has been extended to marine litter detection and classification (Politikos et al., 2023), air pollution in industrial environments (Ramadan et al., 2024), environmental quality in urban areas (Bainomugisha et al., 2024), air quality in healthcare facilities (Tan et al., 2024), sea ice thickness (Wu et al., 2024), climate change modelling (Secci et al., 2023), wildlife and biodiversity (August et al., 2020), etc. Figure 3.1 shows the overview of AI applications for environmental monitoring in recent years.

AI is extensively employed in environmental monitoring to enhance data management, analysis, and interpretation of environmental data across various fields. In air quality monitoring, AI models predict pollution levels in real-time, track emissions, and identify pollution sources (Imam et al., 2024). For water quality monitoring, AI-powered sensors detect contaminants and forecast changes in water availability or scarcity (Jayaraman et al., 2024). AI also supports wildlife and biodiversity monitoring by using drones and remote sensing for species tracking and poaching prevention (York et al., 2023). In deforestation and land use monitoring, AI analyses satellite images to detect forest cover changes and predict land degradation (Solórzano et al., 2023). Marine pollution tracking leverages AI to detect microplastics and monitor ocean health (Jin et al., 2024). Climate change modelling uses AI to monitor carbon emissions, forecast extreme weather events, and assess mitigation strategies (Amiri et al., 2024). In agriculture, AI optimizes precision farming practices and pest control, while in energy efficiency, AI enhances smart grids and building sustainability (Mana et al., 2024). AI also aids in disaster management by predicting natural events

FIGURE 3.1 Applications of AI in environmental monitoring.

and assessing damage, supporting waste management through smart sorting systems, and tracking plastic waste in both terrestrial and marine environments (Bari et al., 2023). Also, AI contributes to ecosystem restoration by assisting with reforestation and monitoring the health of wetlands and coral reefs (Suggett et al., 2023).

3.3 AI-POWERED TECHNIQUES FOR MONITORING WATER QUALITY

AI-powered techniques for monitoring water quality can be categorized into two main approaches: ML and the Internet of Things (IoT). Both of these are crucial for advancing real-time assessment and early detection of contaminants. In ML, key methods include (i) neural networks, (ii) support vector machines, and (iii) random forests. On the IoT side, the main approaches are (i) sensor networks, (ii) AI decision support systems, and (iii) cloud-based platforms. Figure 3.2 depicts the overview of AI-powered techniques for monitoring water quality.

3.3.1 ML Models

ML models have emerged as powerful tools for predicting water quality, offering improved accuracy and efficiency compared to traditional statistical methods. The key parameters for water quality monitoring and prediction include total dissolved solids prediction (estimating the concentration of dissolved particles in water), dissolved oxygen prediction (measuring the oxygen levels essential for aquatic life), pH prediction (assessing water acidity or alkalinity), salinity prediction (tracking salt concentration), anomaly detection (identifying unusual patterns or deviations in water quality), and dimension reduction (simplifying large datasets while retaining critical information) (Kalaivanan & Vellingiri, 2022). Several ML algorithms are particularly effective for modelling complex, nonlinear relationships within environmental datasets, making them ideal for monitoring and predicting water quality parameters

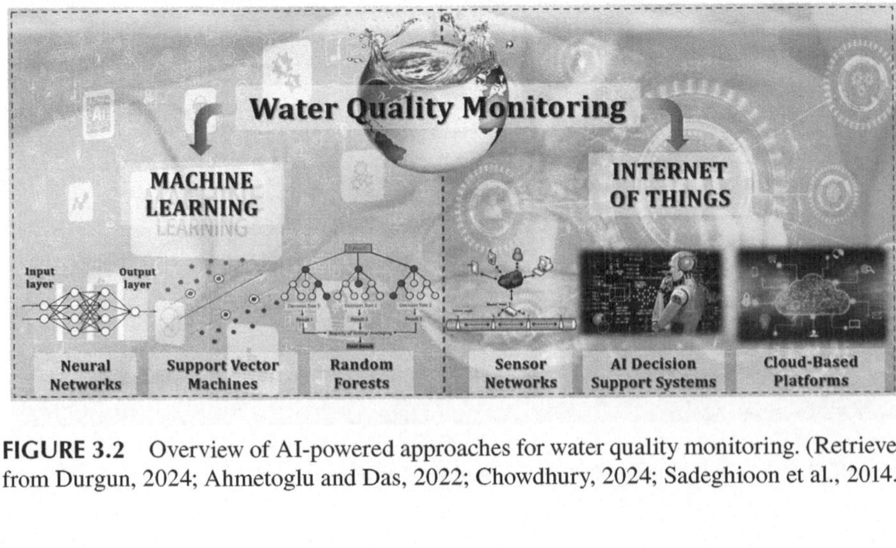

FIGURE 3.2 Overview of AI-powered approaches for water quality monitoring. (Retrieved from Durgun, 2024; Ahmetoglu and Das, 2022; Chowdhury, 2024; Sadeghioon et al., 2014.)

(Najah Ahmed et al., 2019). Among these, support vector machines (SVMs), Random Forests, and neural networks are widely used for their unique capabilities.

SVM is a supervised learning model well-suited for classification and regression tasks. It operates by identifying the optimal hyperplane to distinguish between different classes of data points. For instance, calculating the Water Quality Index (WQI) typically involves labour-intensive sub-index calculations like biochemical oxygen demand (BOD) and chemical oxygen demand (COD) (Leong et al., 2021). However, with SVM, WQI can be instantly predicted using directly measured physical data, eliminating the need for these sub-index calculations. The SVM model, particularly with a polynomial kernel function, demonstrated strong predictive performance, achieving an R^2 of 0.8796, while the LS-SVM model showed even higher accuracy with an R^2 of 0.9227 (Leong et al., 2021). A recent innovation combined Bidirectional Long Short-Term Memory (BILSTM) and SVM, using the Ali Baba and Forty Thieves (ATF) optimization algorithm, to improve the predictive accuracy of various water quality parameters, such as pH, calcium, magnesium, sodium, potassium, bicarbonate, sulphate, and chlorine. This hybrid model, tested on real-world datasets, proved reliable in predicting total dissolved solids and electrical conductivity, with a mean absolute percentage error (MAPE) of 2% and an R^2 of 0.94, respectively (Jamshidzadeh et al., 2024). To further enhance contaminant detection (including water temperature, pH, turbidity, and conductivity) in water distribution systems, a Least-Squares SVM (LS-SVM) model was developed. This model enables real-time water quality monitoring and facilitates timely interventions when contaminants exceed allowable thresholds. The LS-SVM achieved a root mean square error of prediction (RMSEP) of 1.3158 and an R^2 of 0.9712, outperforming multiple linear regression models, which had an RMSEP of 2.5327 and an R^2 of 0.8531 (Wang et al., 2018).

Random Forest is an ensemble learning technique that constructs multiple decision trees and combines their outputs to enhance predictive accuracy. In the context of water quality, Random Forest models can analyse large datasets with numerous variables, such as dissolved oxygen, nitrates, and temperature, offering insights into how these factors interact and influence overall water quality. Random Forest is also highly effective in identifying the key variables impacting water quality, making it ideal for both prediction and feature selection. For instance, a study develops a Random Forest modelling framework to capture the spatial variability of water quality parameters. Socioeconomic variables, land use, and impervious surface cover were found to have significant correlations with stream water quality, emphasizing the influence of human activities on aquatic ecosystems (Wang et al., 2021). The insights from this approach provide valuable information for policymakers and environmental managers, suggesting that integrating land use planning with water quality monitoring could improve pollution control and resource management efforts (Wang et al., 2021). Interestingly, integrating the Random Forest algorithm with blockchain technology could address data integrity concerns while also improving prediction reliability, as verified using the Secure Hash Algorithm-256 (Geetha Jenifel, 2024). The Random Forest model demonstrated a mean absolute error of 0.56, a mean square error of 0.33, and a root mean square error of 0.56 (Geetha Jenifel, 2024). Also, the Random Forest algorithm has been extended to assess the impact of water quality parameters on the biological status of surface waters. For example, researchers used Random Forest to classify the biological status of phytoplankton, phytobenthos, and macrophytes in surface waters, achieving a median error rate of 33.3%, with individual errors ranging from 8% to 60% (Szomolányi & Clement, 2023). Although Random Forest cannot fully replace biological assessments due to notable prediction errors, it remains a promising tool for enhancing water quality monitoring and understanding ecological dynamics in surface waters (Szomolányi & Clement, 2023).

Neural networks, particularly deep learning models, are highly adaptable and capable of capturing nonlinear and complex relationships within data. In water quality monitoring, these models can process large volumes of continuous data from sensors—such as chemical concentrations and temperature changes—and predict future water quality conditions based on historical data patterns. Traditional methods often struggle to model the intricate interactions between various water quality parameters, whereas deep learning algorithms can learn hierarchical data representations, enabling them to detect subtle trends and correlations that conventional approaches might overlook. ANNs, CNNs, and RNNs are especially effective at identifying complex patterns in high-dimensional water quality data. For example, a convolutional neural network and gated recurrent unit-based water quality monitoring model (CNGRU-WQM) was used to monitor a river in India, achieving a validation accuracy of 97.86% (Gandhimathi et al., 2024). In this model, CNNs perform feature extraction, identifying spatial patterns in the data, whereas GRUs handle temporal analysis, capturing time-series trends and fluctuations in water quality. This integration showcases how advanced computational methods can significantly enhance environmental monitoring efforts and be adapted to other river systems facing similar challenges (Gandhimathi et al., 2024). Also, Radial Basis Function Neural Networks (RBF-ANN) and Multi-Layer Perceptron Neural Networks (MLP-ANN) have been implemented for water quality monitoring. More recently, an augmented wavelet de-noising technique, integrated with a Neuro-Fuzzy Inference System (WDT-ANFIS), has been proposed to enhance the predictive accuracy of key water quality parameters, such as ammoniacal nitrogen, suspended solids, and pH levels (Najah Ahmed et al., 2019).

In 2023, a model combining CNNs with remote sensing was developed to quantitatively predict water quality parameters like chlorophyll-a, phycocyanin, turbidity, and dissolved oxygen (Yang et al., 2023). This model can analyse images from remote sensing or multispectral data to detect changes in water quality due to pollution or seasonal variations, with significant implications for environmental management in areas like Guilin's wetlands (Yang et al., 2023). RNNs, with their ability to retain memory, are well-suited for tracking temporal changes in water quality, making them effective at predicting future conditions based on historical trends (Waqas & Humphries, 2024). This capability enhances the accuracy and reliability of water quality assessments, supporting proactive management strategies. However, RNNs can face challenges like vanishing and exploding gradients, limiting their ability to learn long-term dependencies, which is critical in hydrology due to long-term temporal correlations. Long short-term memory (LSTM) networks were developed to overcome these issues by incorporating memory cells that allow them to retain information over extended periods (Waqas & Humphries, 2024). Recently, CNNs have also been used to analyse satellite imagery for water quality monitoring, identifying harmful phenomena such as algal blooms. By training models on labelled datasets, researchers can develop systems that automatically flag abnormal conditions, potentially mitigating the risks posed to coastal ecosystems by algal blooms (Ding & Li, 2024). While ANN remains the most widely used ML algorithm in water quality monitoring (Pérez-Beltrán et al., 2024), no single technique has emerged as the most accurate. Each approach has its limitations. For instance, although the WDT-ANFIS method can quickly identify input data, it does not significantly reduce the time needed for water quality predictions (Kalaivanan & Vellingiri, 2022). Similarly, while the ANFIS model can easily predict dissolved solids in aquifers, it does not improve output accuracy (Kalaivanan & Vellingiri, 2022). Further exploration and development are still required to optimize the integration of ML in water quality monitoring.

3.3.2 AI-Driven IoT for Water Quality Monitoring

The integration of AI with IoT has significantly advanced water quality monitoring systems, transforming the way data is collected, analysed, and applied. AI enhances the efficiency and effectiveness of IoT-based monitoring by enabling intelligent data processing and decision-making. This synergy allows for the automatic detection of contaminants, pattern recognition, and predictions about future water quality conditions, supporting proactive water resource management. With AI, these systems can not only identify current issues but also anticipate potential problems before they worsen, making water quality monitoring more responsive and preventive. For instance, drinking water parameters such as pH (6.5–8.5), turbidity (<5 NTU), conductivity (200–800 μS/cm), and carbon dioxide levels (<2.0 mg/L) must be within acceptable ranges (Lakshmikantha et al., 2021). Recently, researchers have developed an affordable IoT-based water quality monitoring system connected to a cloud

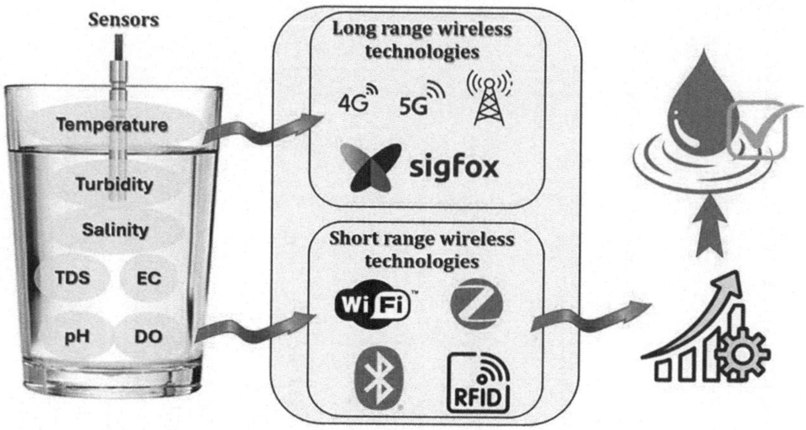

FIGURE 3.3 Water quality monitoring using IoT sensors and data analysis. (Retrieved from Essamlali et al., 2024.)

TABLE 3.1
Key Parameters Measured by IoT Sensors in Water Quality Monitoring

Processor	Parameters									Connection	Reference
	pH	Turbidity	Temperature	Humidity	Ultrasonic	Carbon dioxide	Total dissolved solids	Electrical conductivity	Oxidation-reduction potential		
Microcontroller Unit (Model: Espressif Systems-ESP8266)	✓	✓	✓	✓	✓					Short range: Wi-Fi module	Pasika and Gandla (2020)
Microcontroller Unit (Model: Microchip ATmega328)	✓	✓		✓		✓		✓		Short range: Wi-Fi module	Lakshmikantha et al. (2021)
Microcontroller Unit (Model: Arduino UNO)	✓	✓	✓				✓			Short range: Bluetooth module HC-05	Nishan et al. (2024)
Microcontroller Unit (Model: Arduino Mega 2560 & Arduino UNO R3)	✓	✓	✓							Short range: Wi-Fi module	Chowdury et al. (2019)
Microcontroller Unit (Model: Microchip ATmega328P-AU)	✓	✓								Long range: Transceiver-LoRa RFM 95	Murti et al. (2024)
Microcontroller Unit (Model: Texas Instruments CC3200)	✓	✓						✓	✓	Short range: ZigBee protocol & Wi-Fi module	Geetha and Gouthami (2017)

server to assess the suitability of water for consumption across various water sources (Lakshmikantha et al., 2021). The IoT water quality monitoring system could achieve 95% accuracy in measuring pH, turbidity, total dissolved solids, and water temperature, compared to the traditional method's 85% accuracy (Jayaraman et al., 2024).

The most commonly used IoT wireless technologies include low-power wide area networks (LPWAN), Wi-Fi, ZigBee, radio frequency identification (RFID), cellular networks, and Bluetooth (Essamlali et al., 2024). LPWAN stands out for its energy efficiency and extended range, making it suitable for water quality monitoring, while cellular networks offer reliable, cost-effective, real-time monitoring. Depending on the application's specific needs, cellular networks can range from 2G, 3G, and 4G to 5G. However, recent studies have primarily focused on 4G/5G and 4G-LTE for supervisory control and data acquisition (SCADA) (Saravanan et al., 2018) and unmanned aerial vehicle (UAV)-based (Esakki et al., 2018) water quality monitoring purposes. These studies monitor parameters such as temperature, colour, turbidity, pH, dissolved oxygen, and electrical conductivity. Figure 3.3 presents an overview of the IoT sensor for water quality monitoring.

Referring to Figure 3.3, the AI-driven IoT systems for water quality monitoring integrated with IoT sensors enable real-time data collection and analysis. The data collected by these sensors is then processed using AI algorithms, which can analyse vast amounts of information to identify trends, anomalies, and correlations. This integration enables water quality managers to make informed decisions based on accurate, real-time insights. Based on recent studies, commonly measured parameters using low-cost IoT sensors are pH, turbidity, temperature, humidity, and ultrasonic paired up with a microcontroller. Table 3.1 tabulates the parameters monitored by IoT sensors in water quality assessment.

3.4 DETECTION OF CONTAMINANTS AND POLLUTANTS USING GIS-AI

Integrating AI with Geographic Information Systems (GIS) enhances the effectiveness of water monitoring systems by incorporating spatial analysis and mapping capabilities. GIS allows for the visualization of water quality data across geographical landscapes, enabling the identification of contamination hotspots and the assessment of potential sources of pollution. By combining spatial data with AI analytics, water managers can gain insights into how environmental factors, such as land use and weather patterns, influence water quality. This integration supports more informed decision-making and strategic planning for water resource management, helping to identify areas that may require targeted interventions or monitoring efforts to prevent contamination (Figure 3.4).

Several examples of real-time alert systems illustrate the successful application of these technologies across various sectors, including municipal water supply (Athanasiou et al., 2018; Pandey et al., 2024; Saravani et al., 2024), agriculture (Kourgialas et al., 2022; Yousaf et al., 2021), and industrial operations. In municipal water supply systems, real-time monitoring can detect changes in water quality that may signal contamination from external sources, allowing for immediate corrective actions to safeguard public health. In agriculture, AI-driven systems can monitor

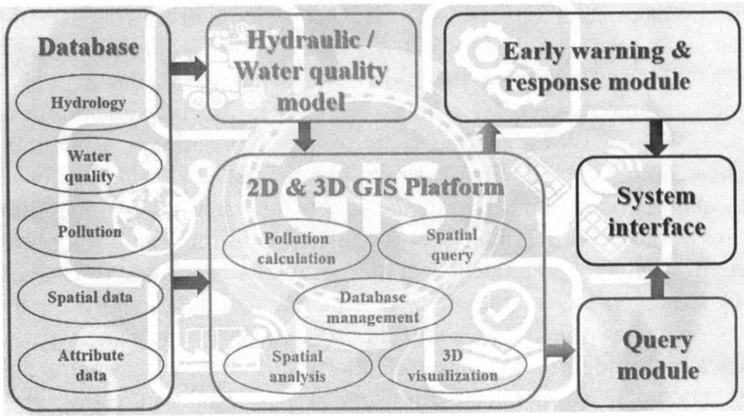

FIGURE 3.4 Overview of GIS system for contaminant detection, warning, and response. (Retrieved from Rui et al., 2015.)

irrigation water quality, alerting farmers to harmful pollutants that could affect crop health and safety. Additionally, industries that utilize large volumes of water can implement real-time monitoring systems to ensure compliance with environmental regulations and minimize the risk of chemical spills. These applications demonstrate the versatility and effectiveness of AI-driven real-time monitoring and alert systems in promoting sustainable water management practices across different sectors.

GIS fuses vast amounts of data, from natural topography to anthropogenic activities to global hydrological cycle in a geospatial context, allowing the analysis and visualization of the water quality application from the perspective of geographical reference (Habeeb & Weli, 2021). Multiple sources of Earth observation data that are archived in GIS can provide many of information, such as the distribution of the inland water bodies tells the water flow patterns, spatial layers about water discharge, and land use/cover, aiding the selection of the dam site (Al-Ruzouq et al., 2019). Meanwhile, the implementation of ML is offering a strong analytical and predictive tool in handling nonlinear relationships between the different types of datasets from different sources (Al-Ruzouq et al., 2019; Ekeanyanwu et al., 2022). The integration has made the data-driven ML model feasible, with the support of massive amounts of data archived in GIS. On the other hand, the analytical capabilities of ML investigate the correlation between the environment parameters and water quality through regression, while identifying the pattern, a similar trend through pattern recognition abilities, yet, the ML predictive capabilities support the estimation of water quality in the region the temporal period that experiences data scarcity.

The integration of GIS-AI enriches each phase of the water quality monitoring from pre-processing, data analysis and predictive modelling, anomaly detection, and even in visualization (Habeeb & Weli, 2021; Ramadas & Samantaray, 2018; Yang et al., 2022). GIS aids in the exploratory data analysis, such as data cleaning and data alignment, before the in-depth analysis. Next, GIS can provide insights into water quality spatially, such as identifying contamination hotspots based on land use and

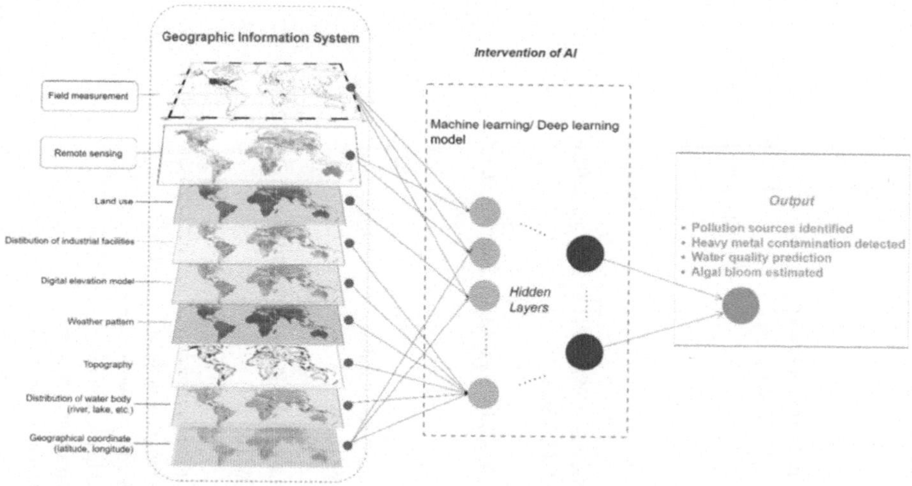

FIGURE 3.5 Illustration of the integration of GIS and machine learning in water quality monitoring and prediction.

proximity of the industrial facilities while the capability of ML in identifying the underlying pattern, trends, seasonal cycles, and anomalies is able to provide insights into the temporal dynamics of the water quality. The synergy of GIS-AI allowing the spatio-temporal analysis which may result in a map of water quality over time, and proceed to predictive ML model development by analysing patterns and trends in the historical data and real-time data (Figure 3.5)

In GIS field, remote sensing (RS) data is one of the critical layers which providing long-term Earth observation data that acquired via a range of payloads i.e., sensor such as camera or detector onboard the space-borne platform (satellite) or short-term observation through the sensors onboard the air-borne platforms (e.g., unmanned aerial vehicle, aircraft). In the context of water quality monitoring, it is able to monitor remote area on a large scale where such as rivers and oceans, where human access is challenging. For example, Soil Moisture and Ocean Salinity (SMOS) is a satellite mission that is dedicated for the global observation moisture over land and salinity over the ocean. The other satellite missions, like Aqua satellite, Landsat-8, Sentinel-2, and Sentinel-3 satellites are able to monitor the ocean colour, chlorophyll levels, which are essential indicators of water quality. Extensive research has been conducted to investigate the performance of ML in elevating the RS data in water quality management (Arias-Rodriguez et al., 2021; Chebud et al., 2012; Kim et al., 2014; Li et al., 2022; Najafzadeh & Basirian, 2023; Ramadas & Samantaray, 2018; Villota-González et al., 2023; Zhang et al., 2021). The ML is trained to recognize the deviations in water quality parameters, detecting patterns of the RS data, and seeking correlation between anomalies of the water quality measurements with the RS data. Then, GIS software is employed to place the anomalies and RS data in a geographical context, enabling the insights to be visualized for the decision-makers and stakeholders (Eid et al., 2023; El-Rawy et al., 2023; Gaagai et al., 2023; Solangi

TABLE 3.2
Integration of GIS Data and Machine Learning Techniques to Study Water Quality

Water Body	Application	Machine Learning Technique	Implementation of GIS or Remote Sensing Data	Indicator of Water Quality	Reference
Water reservoir	To identify the most appropriate location to construct a new dam	ER, GBM, SVM	Precipitation, drainage stream density, geomorphology, geology, curve number, total dissolved solid elevation, slope, and major fracture thematic data archived in GIS	–	Al-Ruzouq et al. (2019)
Ground water	To forecast irrigation water quality indices (IWQIs)	SVM	GIS to determine the spatial distribution of the IWQIs	SAR, Na%, SSP, PS, KI, PI, RSC	Eid et al. (2023)
Ground water	To assess and forecast the quality of groundwater used for irrigation	ANN, GBM	GIS for mapping purpose (maps presenting the geographical distribution of groundwater quality indicators)	SAR, KI, PI, Na%, MH	Gaagai et al. (2023)
Ground water	To assess the suitability of groundwater for drinking purposes	LR, DTC, XGB, RF, KNN	GIS for mapping purposes	Water Quality Index (WQI)	Solangi et al. (2024)
Ground water	To predict the WQI	LR, SVM, GPR, ANN, and Fit binary tree	GIS for mapping purpose (mapping of the spatial distribution of groundwater quality metrics)	Water Quality Index (WQI)	El-Rawy et al. (2023)
River water	To develop empirical relationships based on AI models and remote sensing data in order to estimate WQI	MT, MARS, GEP, EPR	RS data from Landsat-8	Turbidity, sulphate, sodium, potassium, hardness, fluoride, DO, chloride, arsenic, alkalinity, pH, nitrate, and magnesium	Najafzadeh and Basirian (2023)

(*Continued*)

TABLE 3.2 (*Continued*)
Integration of GIS Data and Machine Learning Techniques to Study Water Quality

Water Body	Application	Machine Learning Technique	Implementation of GIS or Remote Sensing Data	Indicator of Water Quality	Reference
River water	To develop retrieval models of WQPs based on RS data and ML algorithm	SVM, RF, ANN, RT, GBM	RS data from Landsat-8	TN, AN, TP	Li et al. (2022)
River water	To develop technique coupling RS and ANN to monitor WQPs	ANN	RS data from Landsat-8	Chl-a, turbidity, and TP	Chebud et al. (2012)
Lake	To construct a RS spectral index and establish a WQI model based on ML for monitoring water quality	SVM	RS data from Sentinel-2 and Sentinel-3	Water Quality Index (WQI)	Zhang et al. (2021)
Lake	To develop ML model for estimating WQPs with the integration of field measurement and RS data	SVM, ELM, LR	RS data from Landsat-8, Sentinel-3, and Sentinel-2	Chl-a, Turbidity, TSM and SDD	Arias-Rodriguez et al. (2021)
Lake	To train and evaluate ML algorithms in estimating WQ parameters	GBM, RF, KNN, DT, SVM, SLA, MLP, Ridge,	RS data from Landsat-8, and Sentinel-2	Chl-a, Turbidity, TSS	Villota-González et al. (2023)

(*Continued*)

TABLE 3.2 (*Continued*)
Integration of GIS Data and Machine Learning Techniques to Study Water Quality

Water Body	Application	Machine Learning Technique	Implementation of GIS or Remote Sensing Data	Indicator of Water Quality	Reference
Coastal water	To evaluate coastal water quality	RF, Cubist, SVM	RS data from GOCI	Chl-a, SPM	Kim et al. (2014)

Machine learning models: Artificial Neural Network (ANN), Decision Tree (DT), Evolutionary Polynomial Regression (EPR), Extreme Gradient Boosting (XGB), Extreme Learning Machine (ELM), Fit Binary Tree, Gaussian Process Regression (GPR), Gene Expression Programming (GEP), Gradient Boost Machine (GBM), K-Nearest Neighbours (KNN), Linear Regression (LR), Logistic Regression (LR), M5 Model Tree (MT), Multilayer Perceptron (MLP), Multivariate Adaptive Regression Spline (MARS), Random Forest (RF), Regression Tree (RT), Ridge, Super Learner Algorithm (SLA), and Support Vector Machine (SVM).

The Irrigation Water Quality Index (IWQI) includes Sodium Adsorption Ratio (SAR), Sodium Percentage (Na%), Soluble Sodium Percentage (SSP), Kelly Index (KI), Permeability Index (PI), Potential Salinity (PS), Residual Sodium Carbonate (RSC), and Magnesium Hazard (MH).

Water quality parameters (WQPs) include Alkalinity, Ammoniacal Nitrogen (AN), Arsenic, Biological Oxygen Demand (BOD), Chloride, Chlorophyll-a (Chl-a), Conductivity, Dissolved Oxygen (DO), Fluoride, Hardness, Magnesium, Nitrate, pH, Potassium, Secchi Disk Depth (SDD), Sodium, Sulphate, Suspended Particulate Matter (SPM) concentrations, Total Nitrogen (TN), Total Phosphorus (TP), Total Suspended Solids (TSS) / Total Suspended Matter (TSM), and Turbidity.

Satellite mission: Landsat-8, Sentinel-3, and Sentinel-2, GeostationaryOceanColor Imager (GOCI) et al., 2024). Table 3.2 tabulates the ML models and the input from GIS and RS in water quality monitoring.

3.5 CHALLENGES AND FUTURE DIRECTIONS

Utilizing AI for water quality monitoring presents several main challenges related to data quality, sensor limitations, computational requirements, and model interpretability. AI models rely on large, high-quality datasets. However, water quality data is often inconsistent or incomplete, particularly in remote or resource-limited areas. Environmental factors, sensor calibration issues, and noise in the data can reduce the accuracy of AI predictions (Chen et al., 2024). Also, sensors used to monitor water parameters often face durability issues, require frequent maintenance, or fail to provide accurate readings in harsh conditions. These limitations hinder the effectiveness of AI models, which depend on continuous, high-resolution data for optimal performance. Beyond data and sensor challenges, computational requirements and model interpretability pose further challenges. Advanced AI models, such as neural networks, demand substantial computational power for training and real-time prediction, which can be difficult to implement in regions with limited infrastructure. Also, many AI models, especially deep learning, operate as "black boxes," making it difficult for decision-makers and regulatory bodies to interpret how predictions are made (Herm et al., 2023). This lack of interpretability can limit the adoption of AI-driven insights in water management.

Future research should address these challenges by focusing on advancements in AI algorithms, sensor technology, and data integration. For instance, data obtained from monitoring stations could possibly be polluted by noise signals as a result of random and systematic errors (Najah Ahmed et al., 2019). Developing more sophisticated AI models, such as ensemble methods, transfer learning, and explainable AI, can enhance prediction accuracy while improving model interpretability. Hybrid approaches that combine traditional statistical methods with ML can also provide more robust insights into complex environmental interactions. Improving sensor technologies is crucial for collecting reliable, high-quality data (Okafor et al., 2020). Researchers should prioritize the development of durable, sensitive, and versatile sensors capable of operating in diverse conditions and combining multiple sensing technologies. Effective data integration frameworks that unify information from IoT sensors, satellite imagery, and historical datasets will further enhance monitoring capabilities and the accuracy of AI-driven insights. Another promising future direction is the development of AI-powered digital twin frameworks for water quality monitoring that incorporate AI-driven analytics (Brahmbhatt et al., 2023). Researchers should explore the integration of diverse data sources, such as IoT sensors, satellite imagery, and historical datasets, to create more accurate and comprehensive digital twins of water systems. Also, efforts should be made to enhance the interoperability of digital twins with existing water management systems and regulatory frameworks. Collaborations among various stakeholders, including technology developers, environmental scientists, and policymakers, will be essential to advancing this approach.

3.6 CONCLUSION

This chapter highlighted the transformative potential of AI in enhancing water quality monitoring, focusing on ML models such as SVM, Random Forests, and neural networks, as well as AI-driven IoT and GIS-AI approaches. These AI-powered techniques significantly improve accuracy, enabling early detection of pollution and facilitating real-time monitoring of key parameters such as pH, dissolved oxygen, turbidity, and chemical concentrations. By streamlining data analysis and providing critical insights, AI enhances water resource management. However, realizing AI's full potential for sustainable water management requires interdisciplinary collaboration among researchers, engineers, policymakers, and industry stakeholders. Such efforts will drive the development of innovative AI-based solutions, address existing challenges in data quality and sensor technology, and safeguard vital water resources for future generations.

ACKNOWLEDGEMENT

The authors would like to acknowledge the financial support from Asia Technological University (ATU) Network under ATU-Net Young Researcher Grant (YRG) under Vot. No. R.J130000.7724.4J714.

DECLARATION OF GENERATIVE AI IN SCIENTIFIC WRITING

I hereby declare that no generative AI tools were used in the preparation, writing, or editing of this manuscript, except for grammar and English language checks.

REFERENCES

Ahmetoglu, H., & Das, R. (2022). A comprehensive review on detection of cyber-attacks: data sets, methods, challenges, and future research directions. *Internet of Things*, *20*, 100615. https://doi.org/https://doi.org/10.1016/j.iot.2022.100615

Al-Ruzouq, R., Shanableh, A., Yilmaz, A. G., Idris, A., Mukherjee, S., Khalil, M. A., & Gibril, M. B. A. (2019). Dam site suitability mapping and analysis using an integrated GIS and machine learning approach. *Water*, *11*(9). https://doi.org/10.3390/w11091880

Amiri, Z., Heidari, A., & Navimipour, N. J. (2024). Comprehensive survey of artificial intelligence techniques and strategies for climate change mitigation. *Energy*, *308*, 132827. https://doi.org/https://doi.org/10.1016/j.energy.2024.132827

Arias-Rodriguez, L. F., Duan, Z., Diaz-Torres, J. J., Basilio Hazas, M., Huang, J., Kumar, B. U., . . . Disse, M. (2021). Integration of remote sensing and Mexican water quality monitoring system using an extreme learning machine. Sensors, *21*(12). https://doi.org/10.3390/s21124118

Athanasiou, T., Salmas, D., Karvelis, P., Angelis, I., Andrea, V., Schismenos, P., . . . Stylios, C. (2018). A Web-geographical information system for real time monitoring of Arachthos River, Greece. *IFAC-PapersOnLine*, *51*(30), 384–389. https://doi.org/https://doi.org/10.1016/j.ifacol.2018.11.335

August, T. A., Pescott, O. L., Joly, A., & Bonnet, P. (2020). AI naturalists might hold the key to unlocking biodiversity data in social media imagery. *Patterns*, *1*(7), 100116. https://doi.org/https://doi.org/10.1016/j.patter.2020.100116

Awewomom, J., Dzeble, F., Takyi, Y. D., Ashie, W. B., Ettey, E. N. Y. O., Afua, P. E., . . . Akoto, O. (2024). Addressing global environmental pollution using environmental control techniques: a focus on environmental policy and preventive environmental management. *Discover Environment*, *2*(1), 8. https://doi.org/10.1007/s44274-024-00033-5

Azha, S. F., Sidek, L. M., Ahmad, Z., Zhang, J., Basri, H., Zawawi, M. H., . . . Ahmed, A. N. (2023). Enhancing river health monitoring: developing a reliable predictive model and mitigation plan. *Ecological Indicators*, *156*, 111190. https://doi.org/https://doi.org/10.1016/j.ecolind.2023.111190

Bainomugisha, E., Adrine Warigo, P., Busigu Daka, F., Nshimye, A., Birungi, M., & Okure, D. (2024). AI-driven environmental sensor networks and digital platforms for urban air pollution monitoring and modelling. *Societal Impacts*, *3*, 100044. https://doi.org/https://doi.org/10.1016/j.socimp.2024.100044

Bari, L. F., Ahmed, I., Ahamed, R., Zihan, T. A., Sharmin, S., Pranto, A. H., & Islam, M. R. (2023). Potential use of artificial intelligence (AI) in disaster risk and emergency health management: a critical appraisal on environmental health. *Environ Health Insights*, *17*, 11786302231217808. https://doi.org/10.1177/11786302231217808

Bibri, S. E., Huang, J., & Krogstie, J. (2024). Artificial intelligence of things for synergizing smarter eco-city brain, metabolism, and platform: pioneering data-driven environmental governance. *Sustainable Cities and Society*, *108*, 105516. https://doi.org/https://doi.org/10.1016/j.scs.2024.105516

Brahmbhatt, P., Maheshwari, A., & Gudi, R. D. (2023). Digital twin assisted decision support system for quality regulation and leak localization task in large-scale water distribution networks. *Digital Chemical Engineering*, *9*, 100127. https://doi.org/https://doi.org/10.1016/j.dche.2023.100127

Chebud, Y., Naja, G. M., Rivero, R. G., & Melesse, A. M. (2012). Water quality monitoring using remote sensing and an artificial neural network. *Water, Air, & Soil Pollution*, *223*(8), 4875–4887. https://doi.org/10.1007/s11270-012-1243-0

Chen, L., Xia, C., Zhao, Z., Fu, H., & Chen, Y. (2024). AI-driven sensing technology: review. *Sensors*, *24*(10), 2958.

Chowdhury, M. S. (2024). Comparison of accuracy and reliability of random forest, support vector machine, artificial neural network and maximum likelihood method in land use/cover classification of urban setting. *Environmental Challenges*, *14*, 100800. https://doi.org/https://doi.org/10.1016/j.envc.2023.100800

Chowdury, M. S. U., Emran, T. B., Ghosh, S., Pathak, A., Alam, M. M., Absar, N., . . . Hossain, M. S. (2019). IoT based real-time river water quality monitoring system. *Procedia Computer Science*, *155*, 161–168. https://doi.org/https://doi.org/10.1016/j.procs.2019.08.025

Demeneix, B., & Hahn, M. E. (2020). Human health and ocean pollution. *Annals of Global Health*, *86*(1), 151.

Ding, W., & Li, C. (2024). Algal blooms forecasting with hybrid deep learning models from satellite data in the Zhoushan fishery. *Ecological Informatics*, *82*, 102664. https://doi.org/https://doi.org/10.1016/j.ecoinf.2024.102664

Durgun, Y. (2024). Real-time water quality monitoring using AI-enabled sensors: detection of contaminants and UV disinfection analysis in smart urban water systems. *Journal of King Saud University – Science*, *36*(9), 103409. https://doi.org/https://doi.org/10.1016/j.jksus.2024.103409

Eid, M. H., Elbagory, M., Tamma, A. A., Gad, M., Elsayed, S., Hussein, H., . . . Péter, S. (2023). Evaluation of groundwater quality for irrigation in deep aquifers using multiple graphical and indexing approaches supported with machine learning models and GIS techniques, Souf Valley, Algeria. *Water*, *15*(1). https://doi.org/10.3390/w15010182

Ekeanyanwu, C. V., Obisakin, I. F., Aduwenye, P., & Dede-Bamfo, N. (2022). Merging GIS and machine learning techniques: a paper review. *Journal of Geoscience and Environment Protection, 10*(09), 61–83. https://doi.org/10.4236/gep.2022.109004

El-Rawy, M., Batelaan, O., Alshehri, F., Almadani, S., Ahmed, M. S., & Elbeltagi, A. (2023). An Integrated GIS and machine-learning technique for groundwater quality assessment and prediction in Southern Saudi Arabia. *Water, 15*(13). https://doi.org/10.3390/w15132448

Esakki, B., Ganesan, S., Mathiyazhagan, S., Ramasubramanian, K., Gnanasekaran, B., Son, B., . . . Choi, J. S. (2018). Design of amphibious vehicle for unmanned mission in water quality monitoring using internet of things. *Sensors, 18*(10). https://doi.org/10.3390/s18103318

Essamlali, I., Nhaila, H., & El Khaili, M. (2024). Advances in machine learning and IoT for water quality monitoring: a comprehensive review. *Heliyon, 10*(6). https://doi.org/10.1016/j.heliyon.2024.e27920

Gaagai, A., Aouissi, H., Bencedira, S., Hinge, G., Athamena, A., Heddam, S., . . . Ibrahim, H. (2023). Application of water quality indices, machine learning approaches, and GIS to identify groundwater quality for irrigation purposes: a case study of Sahara Aquifer, Doucen Plain, Algeria. *Water, 15*(2). https://doi.org/10.3390/w15020289

Gandhimathi, G., Chellaswamy, C., & Thiruvalar Selvan, P. (2024). Comprehensive river water quality monitoring using convolutional neural networks and gated recurrent units: a case study along the Vaigai River. *Journal of Environmental Management, 365*, 121567. https://doi.org/https://doi.org/10.1016/j.jenvman.2024.121567

Geetha, S., & Gouthami, S. (2017). Internet of things enabled real time water quality monitoring system. *Smart Water, 2*(1), 1. https://doi.org/10.1186/s40713-017-0005-y

Geetha Jenifel, M. (2024). Secure water quality prediction system using machine learning and blockchain technologies. *Journal of Environmental Management, 350*, 119357. https://doi.org/https://doi.org/10.1016/j.jenvman.2023.119357

Habeeb, N. J., & Weli, S. T. (2021). Combination of GIS with different technologies for water quality: an overview. *HighTech and Innovation Journal, 2*(3), 262–272. https://hightech-journal.org/index.php/HIJ/article/view/127

Herm, L.-V., Heinrich, K., Wanner, J., & Janiesch, C. (2023). Stop ordering machine learning algorithms by their explainability! A user-centered investigation of performance and explainability. *International Journal of Information Management, 69*, 102538. https://doi.org/https://doi.org/10.1016/j.ijinfomgt.2022.102538

Imam, M., Adam, S., Dev, S., & Nesa, N. (2024). Air quality monitoring using statistical learning models for sustainable environment. *Intelligent Systems with Applications, 22*, 200333. https://doi.org/https://doi.org/10.1016/j.iswa.2024.200333

Jamshidzadeh, Z., Ehteram, M., & Shabanian, H. (2024). Bidirectional Long Short-Term Memory (BILSTM) – support vector machine: a new machine learning model for predicting water quality parameters. *Ain Shams Engineering Journal, 15*(3), 102510. https://doi.org/https://doi.org/10.1016/j.asej.2023.102510

Jayaraman, P., Nagarajan, K. K., Partheeban, P., & Krishnamurthy, V. (2024). Critical review on water quality analysis using IoT and machine learning models. *International Journal of Information Management Data Insights, 4*(1), 100210. https://doi.org/https://doi.org/10.1016/j.jjimei.2023.100210

Jin, H., Kong, F., Li, X., & Shen, J. (2024). Artificial intelligence in microplastic detection and pollution control. *Environmental Research, 262*, 119812. https://doi.org/https://doi.org/10.1016/j.envres.2024.119812

Kalaivanan, K., & Vellingiri, J. (2022). Survival study on different water quality prediction methods using machine learning. *Nature Environment & Pollution Technology, 21*(3), 1259–1267.

Kamali, M., Persson, K. M., Costa, M. E., & Capela, I. (2019). Sustainability criteria for assessing nanotechnology applicability in industrial wastewater treatment: current status and future outlook. *Environment International*, *125*, 261–276. https://doi.org/https://doi.org/10.1016/j.envint.2019.01.055

Kim, Y. H., Im, J., Ha, H. K., Choi, J.-K., & Ha, S. (2014). Machine learning approaches to coastal water quality monitoring using GOCI satellite data. *GIScience & Remote Sensing*, *51*(2), 158–174. https://doi.org/10.1080/15481603.2014.900983

Kourgialas, N. N., Hliaoutakis, A., Argyriou, A. V., Morianou, G., Voulgarakis, A. E., Kokinou, E., . . . Soupios, P. (2022). A web-based GIS platform supporting innovative irrigation management techniques at farm-scale for the Mediterranean island of Crete. *Science of the Total Environment*, *842*, 156918. https://doi.org/https://doi.org/10.1016/j.scitotenv.2022.156918

Lakshmikantha, V., Hiriyannagowda, A., Manjunath, A., Patted, A., Basavaiah, J., & Anthony, A. A. (2021). IoT based smart water quality monitoring system. *Global Transitions Proceedings*, *2*(2), 181–186. https://doi.org/https://doi.org/10.1016/j.gltp.2021.08.062

Leong, W. C., Bahadori, A., Zhang, J., & Ahmad, Z. (2021). Prediction of water quality index (WQI) using support vector machine (SVM) and least square-support vector machine (LS-SVM). *International Journal of River Basin Management*, *19*(2), 149–156. https://doi.org/10.1080/15715124.2019.1628030

Li, N., Ning, Z., Chen, M., Wu, D., Hao, C., Zhang, D., . . . Zhang, L. (2022). Satellite and machine learning monitoring of optically inactive water quality variability in a tropical river. Remote Sensing, *14*(21). https://doi.org/10.3390/rs14215466

Mana, A. A., Allouhi, A., Hamrani, A., Rehman, S., el Jamaoui, I., & Jayachandran, K. (2024). Sustainable AI-based production agriculture: exploring AI applications and implications in agricultural practices. *Smart Agricultural Technology*, *7*, 100416. https://doi.org/https://doi.org/10.1016/j.atech.2024.100416

Murti, M. A., Saputra, A. R. A., Alinursafa, I., Ahmed, A. N., Yafouz, A., & El-Shafie, A. (2024). Smart system for water quality monitoring utilizing long-range-based Internet of Things. *Applied Water Science*, *14*(4), 69. https://doi.org/10.1007/s13201-024-02128-z

Nair K, S., Manu, B., & Azhoni, A. (2021). Sustainable treatment of paint industry wastewater: current techniques and challenges. *Journal of Environmental Management*, *296*, 113105. https://doi.org/https://doi.org/10.1016/j.jenvman.2021.113105

Najafzadeh, M., & Basirian, S. (2023). Evaluation of river water quality index using remote sensing and artificial intelligence models. *Remote Sensing*, *15*(9). https://doi.org/10.3390/rs15092359

Najah Ahmed, A., Binti Othman, F., Abdulmohsin Afan, H., Khaleel Ibrahim, R., Ming Fai, C., Shabbir Hossain, M., . . . Elshafie, A. (2019). Machine learning methods for better water quality prediction. *Journal of Hydrology*, *578*, 124084. https://doi.org/https://doi.org/10.1016/j.jhydrol.2019.124084

Nishan, R. K., Akter, S., Sony, R. I., Hoque, M. M., Anee, M. J., & Hossain, A. (2024). Development of an IoT-based multi-level system for real-time water quality monitoring in industrial wastewater. *Discover Water*, *4*(1), 43. https://doi.org/10.1007/s43832-024-00092-y

Okafor, N. U., Alghorani, Y., & Delaney, D. T. (2020). Improving data quality of low-cost IoT sensors in environmental monitoring networks using data fusion and machine learning approach. *ICT Express*, *6*(3), 220–228. https://doi.org/https://doi.org/10.1016/j.icte.2020.06.004

Omar, N. M. A., Othman, M. H. D., Tai, Z. S., Kurniawan, T. A., Puteh, M. H., Jaafar, J., . . . Wong, K. Y. (2024). Recent strategies for enhancing the performance and lifespan of low-cost ceramic membranes in water filtration and treatment processes: a review. *Journal of Water Process Engineering*, *62*, 105399. https://doi.org/https://doi.org/10.1016/j.jwpe.2024.105399

Pandey, M., Senapati, S., & Bhunia, G. S. (2024). Chapter 27- GIS-based modeling for water resource monitoring and management: a critical review. In Chandra Pal, S., Roy, S. S., Saha, A., & Abioui, M. (Eds.), *Developments in Environmental Science* (Vol. 16, pp. 621–636). Elsevier. https://doi.org/https://doi.org/10.1016/B978-0-443-23665-5.00027-2

Pasika, S., & Gandla, S. T. (2020). Smart water quality monitoring system with cost-effective using IoT. *Heliyon*, *6*(7), e04096. https://doi.org/https://doi.org/10.1016/j.heliyon.2020.e04096

Pérez-Beltrán, C. H., Robles, A. D., Rodriguez, N. A., Ortega-Gavilán, F., & Jiménez-Carvelo, A. M. (2024). Artificial intelligence and water quality: from drinking water to wastewater. *TrAC Trends in Analytical Chemistry*, *172*, 117597. https://doi.org/https://doi.org/10.1016/j.trac.2024.117597

Politikos, D. V., Adamopoulou, A., Petasis, G., & Galgani, F. (2023). Using artificial intelligence to support marine macrolitter research: a content analysis and an online database. *Ocean & Coastal Management*, *233*, 106466. https://doi.org/https://doi.org/10.1016/j.ocecoaman.2022.106466

Ramadan, M. N. A., Ali, M. A. H., Khoo, S. Y., Alkhedher, M., & Alherbawi, M. (2024). Real-time IoT-powered AI system for monitoring and forecasting of air pollution in industrial environment. *Ecotoxicology and Environmental Safety*, *283*, 116856. https://doi.org/https://doi.org/10.1016/j.ecoenv.2024.116856

Ramadas, M., & Samantaray, A. (2018). Applications of remote sensing and GIS in water quality monitoring and remediation: a state-of-the-art review. In *Water Remediation. Energy, Environment, and Sustainability* (pp. 225–246). https://doi.org/10.1007/978-981-10-7551-3_13

Rui, Y., Shen, D., Khalid, S., Yang, Z., & Wang, J. (2015). GIS-based emergency response system for sudden water pollution accidents. *Physics and Chemistry of the Earth, Parts A/B/C*, 79–82, 115–121. https://doi.org/https://doi.org/10.1016/j.pce.2015.03.001

Sadeghioon, A. M., Metje, N., Chapman, D. N., & Anthony, C. J. (2014). SmartPipes: smart wireless sensor networks for leak detection in water pipelines. *Journal of Sensor and Actuator Networks*, *3*(1), 64–78.

Saravanan, K., Anusuya, E., Kumar, R., & Son, L. H. (2018). Real-time water quality monitoring using Internet of Things in SCADA. *Environmental Monitoring and Assessment*, *190*(9), 556. https://doi.org/10.1007/s10661-018-6914-x

Saravani, M. J., Saadatpour, M., & Shahvaran, A. R. (2024). A web GIS based integrated water resources assessment tool for Javeh Reservoir. *Expert Systems with Applications*, *252*, 124198. https://doi.org/https://doi.org/10.1016/j.eswa.2024.124198

Secci, D., Giovanna Tanda, M., D'Oria, M., & Todaro, V. (2023). Artificial intelligence models to evaluate the impact of climate change on groundwater resources. *Journal of Hydrology*, *627*, 130359. https://doi.org/https://doi.org/10.1016/j.jhydrol.2023.130359

Shayo, G. M., Elimbinzi, E., Shao, G. N., & Fabian, C. (2023). Severity of waterborne diseases in developing countries and the effectiveness of ceramic filters for improving water quality. *Bulletin of the National Research Centre*, *47*(1), 113. https://doi.org/10.1186/s42269-023-01088-9

Silva, J. A. (2023). Wastewater treatment and reuse for sustainable water resources management: a systematic literature review. *Sustainability*, *15*(14), 10940.

Solangi, G. S., Ali, Z., Bilal, M., Junaid, M., Panhwar, S., Keerio, H. A., . . . Zaman, N. (2024). Machine learning, Water Quality Index, and GIS-based analysis of groundwater quality. *Water Practice & Technology*, *19*(2), 384–400. https://doi.org/10.2166/wpt.2024.014

Solórzano, J. V., Mas, J. F., Gallardo-Cruz, J. A., Gao, Y., & Fernández-Montes de Oca, A. (2023). Deforestation detection using a spatio-temporal deep learning approach with synthetic aperture radar and multispectral images. *ISPRS Journal of Photogrammetry and Remote Sensing*, *199*, 87–101. https://doi.org/https://doi.org/10.1016/j.isprsjprs.2023.03.017

Strade, E., Kalnina, D., & Kulczycka, J. (2020). Water efficiency and safe re-use of different grades of water – Topical issues for the pharmaceutical industry. *Water Resources and Industry*, *24*, 100132. https://doi.org/https://doi.org/10.1016/j.wri.2020.100132

Suggett, D. J., Edwards, M., Cotton, D., Hein, M., & Camp, E. F. (2023). An integrative framework for sustainable coral reef restoration. *One Earth*, *6*(6), 666–681. https://doi.org/https://doi.org/10.1016/j.oneear.2023.05.007

Szomolányi, O., & Clement, A. (2023). Use of random forest for assessing the effect of water quality parameters on the biological status of surface waters. *GEM – International Journal on Geomathematics*, *14*(1), 20. https://doi.org/10.1007/s13137-023-00229-6

Tan, H., Ong, P. Y., Abdul Wahab, R., Goh, P. S., Wong, K. Y., Klemeš, J. J., . . . Lee, C. T. (2023). Mechanistic insight and optimisation of hydrothermally pre-treated biowaste-derived biochar for saline water treatment. *Journal of Cleaner Production*, *421*, 138465. https://doi.org/https://doi.org/10.1016/j.jclepro.2023.138465

Tan, H., Othman, M. H. D., Kek, H. Y., Chong, W. T., Nyakuma, B. B., Wahab, R. A., . . . Wong, K. Y. (2024). Revolutionizing indoor air quality monitoring through IoT innovations: a comprehensive systematic review and bibliometric analysis. *Environmental Science and Pollution Research*, *31*(32), 44463–44488. https://doi.org/10.1007/s11356-024-34075-2

Tan, H., Wahab, R. A., Ong, P. Y., Goh, P. S., Wong, K. Y., Klemeš, J. J., & Lee, C. T. (2022). Functionalisation of Biowaste-Derived Biochar Via Accelerated Hydrothermal-Assisted Pre-Treatment for Enhanced Sodium Ion Adsorption. *Available at SSRN 4257964*.

Uhlenbrook, S., & Connor, R. (2019). *The United Nations World Water Development Report 2019: Leaving No One Behind.* United Nations. https://www.unwater.org/publications/un-world-water-development-report-2019

Villota-González, F. H., Sulbarán-Rangel, B., Zurita-Martínez, F., Gurubel-Tun, K. J., & Zúñiga-Grajeda, V. (2023). Assessment of machine learning models for remote sensing of water quality in Lakes Cajititlán and Zapotlán, Jalisco—Mexico. *Remote Sensing*, *15*(23). https://doi.org/10.3390/rs15235505

Wang, F., Wang, Y., Zhang, K., Hu, M., Weng, Q., & Zhang, H. (2021). Spatial heterogeneity modeling of water quality based on random forest regression and model interpretation. *Environmental Research*, *202*, 111660. https://doi.org/https://doi.org/10.1016/j.envres.2021.111660

Wang, K., Wen, X., Hou, D., Tu, D., Zhu, N., Huang, P., . . . Zhang, H. (2018). Application of least-squares support vector machines for quantitative evaluation of known contaminant in water distribution system using online water quality parameters. *Sensors*, *18*(4), 938.

Waqas, M., & Humphries, U. W. (2024). A critical review of RNN and LSTM variants in hydrological time series predictions. *MethodsX*, *13*, 102946. https://doi.org/https://doi.org/10.1016/j.mex.2024.102946

Wu, H., Wang, Y., Zhang, R., Yan, H., & Hong, M. (2024). Bias correction of Arctic sea ice thickness products based on factor selection and machine learning methods. *Applied Ocean Research*, *149*, 104069. https://doi.org/https://doi.org/10.1016/j.apor.2024.104069

Yang, L., Driscol, J., Sarigai, S., Wu, Q., Lippitt, C. D., & Morgan, M. (2022). Towards synoptic water monitoring systems: a review of AI methods for automating water body detection and water quality monitoring using remote sensing. *Sensors*, *22*(6). https://doi.org/10.3390/s22062416

Yang, W., Fu, B., Li, S., Lao, Z., Deng, T., He, W., . . . Chen, Z. (2023). Monitoring multi-water quality of internationally important karst wetland through deep learning, multi-sensor and multi-platform remote sensing images: a case study of Guilin, China. *Ecological Indicators*, *154*, 110755. https://doi.org/https://doi.org/10.1016/j.ecolind.2023.110755

York, N. D. L., Pritchard, R., Sauls, L. A., Enns, C., & Foster, T. (2023). Justice and ethics in conservation remote sensing: current discourses and research needs. *Biological Conservation*, *287*, 110319. https://doi.org/https://doi.org/10.1016/j.biocon.2023.110319

Yousaf, W., Awan, W. K., Kamran, M., Ahmad, S. R., Bodla, H. U., Riaz, M., . . . Chohan, K. (2021). A paradigm of GIS and remote sensing for crop water deficit assessment in near real time to improve irrigation distribution plan. *Agricultural Water Management, 243*, 106443. https://doi.org/https://doi.org/10.1016/j.agwat.2020.106443

Zhang, F., Chan, N. W., Liu, C., Wang, X., Shi, J., Kung, H.-T., . . . Cao, N. (2021). Water Quality Index (WQI) as a potential proxy for remote sensing evaluation of water quality in arid areas. *Water, 13*(22). https://doi.org/10.3390/w13223250

Zulkifli, S. N., Rahim, H. A., & Lau, W.-J. (2018). Detection of contaminants in water supply: a review on state-of-the-art monitoring technologies and their applications. *Sensors and Actuators B: Chemical, 255*, 2657–2689. https://doi.org/https://doi.org/10.1016/j.snb.2017.09.078

4 Advantages and Challenges of AI-Driven Water Quality Monitoring

Jerome P. Romero, Liliane M. Suarez, Joan Cecilia C. Casila, and Eric B. Casila

4.1 INTRODUCTION

Due to urbanization and industrialization caused by a rapid increase in population, water quality degradation has become a major environmental issue globally for both developed and developing countries (Lin et al., 2022) Water quality refers to the chemical, physical, and biological characteristics of water, usually determined depending on suitability for various purposes (Arora, 2017). The primary cause of water pollutants are the organic and inorganic substances, toxic solvents and volatile organic chemicals discharged during industrial and agricultural processes (Zhang et al., 2023). Increased usage of chemical fertilizers and pesticides, and untreated wastewater in irrigation, pollutes groundwater and surface water.

Various methods were used to assess the water quality and determine its characteristics. Sample collection and laboratory analysis are the traditional practices in water quality assessment. These traditional methods demand careful handling and measurements. The raw, filtered, or treated water samples were assessed according to different water quality parameters, like turbidity, pH, and dissolved oxygen (DO), using laboratory equipment. Despite the intensive sample handling, the results of the analysis are prone to errors due to the complicated process and other causes brought by equipment miscalibration and field sampling practices (Zainurin et al., 2022). Moreover, the traditional method is inefficient as human energy is used to handle work, and the testing frequency can be very low. The laboratory facilities can also be accompanied by high cost and maintenance. Overall, the traditional water quality assessment methods require high cost and longer periods of analysis resulting in lack of real-time data availability. These factors limit the immediate detection of contaminants and increase the risks in the environment and human health sectors. Accurate real-time monitoring of water quality must be executed to implement better measures and policies.

The World Health Organization (WHO, 2023) reported that as of 2022, around 2.2 billion people lack access to safely managed drinking water, while over 2 billion people live in water-stressed countries, which is expected to be exacerbated in some regions because of climate change and population growth. Industrial activities, agricultural runoff, and inadequate waste management are among the primary

DOI: 10.1201/9781003547099-5

contributors to water pollution (Russo et al., 2020). Contaminated water is linked to the transmission of various diseases such as cholera, diarrhoea, dysentery, hepatitis, and others. Water management in cities is being seriously neglected and poses a major issue, especially in the cities. Household effluent, human waste, toxic chemicals, and medical waste are exposed to the air, contaminating the environment in often densely populated residential zones. On the other hand, in poorer urban areas, a large proportion of untreated wastewater is discharged directly into the closest drainage channel or water bodies (WHO, 2023).

According to the United Nations Development Programme (2025), the 17 Sustainable Development Goals (SDGs) address the global challenges, such as problems in water management and quality, to achieve a better and more sustainable future for all by 2030. Among the 17 SDGs, SDG 6 focuses on Clean Water and Sanitation, which aims to secure and ensure the availability and sustainable management of water and sanitation for all. In particular, the goal of SDG 6.3 is to reduce water pollution by eliminating dumps and discharge of hazardous chemicals and materials, halving the percentage of untreated wastewater, increasing global water recycling and reuse, and all lead to improved water quality by 2030.

Advantageously, the world has experienced significant shifts and changes due to the emergence of modern technologies. These shifts are necessary to navigate the numerous challenges of ensuring excellent water quality. The recent development of sensors and automated water quality monitoring systems can greatly improve the assessment of water quality.

By practicing the triangulation approach by integrating various data sources from remote sensing, earth observation, modelling, and the actual onsite data, a more comprehensive global water quality assessment can be obtained (Mohan et al., 2025). This approach has more benefits than traditional methods since modern methods can produce output results and analyse the water quality parameters in detail and in real time. After an immediate identification of water quality, faster action can be taken to handle any poor water quality indicators, including pH, temperature, turbidity, and pollutant levels (Zainurin et al., 2022). The emerging trends in artificial intelligence (AI) have also provided a concrete path to make better predictions in a variety of fields, including water quality assessment, compared to the conventional techniques, because of its enhanced accuracy, scalability, and efficiency.

4.2 ADVANTAGES OF AI IN WATER QUALITY MONITORING

The AI-driven systems have been steadily growing, especially in water resource management systems, as they provide proactive supervision of water resources, simplifying early detection of possible hazards and consequently decreasing risks to the environment and society. The incorporation of AI into the monitoring and treatment of water quality serves as a remarkable demonstration of technological progress.

In the following sections, the implications of AI on monitoring water quality are discussed, signifying the dedication of the scientific community to promoting environmental sustainability and protecting public health through recent technology development.

4.3 ARTIFICIAL INTELLIGENCE METHODS FOR MONITORING WATER QUALITY

The application of AI in water quality monitoring has revolutionized the approach to managing water resources, enabling faster, more accurate, and scalable solutions to detect contaminants and optimize water management practices. This section delves into various AI techniques such as machine learning (ML), deep learning (DL), and predictive analytics, integrating AI in Internet of Things (IoT).

4.3.1 Machine Learning

Machine learning (ML) is a core AI technology focused on developing algorithms that enable systems to learn and improve from experience without explicit programming. In the context of water quality monitoring, ML models play a crucial role by analysing large datasets, detecting anomalies, and predicting contamination events before they become critical. This proactive approach can prevent significant water quality issues, ensuring better public health and environmental sustainability. Various ML algorithms, such as Decision Trees, Support Vector Machines (SVMs), Random Forests, and Artificial Neural Networks (ANNs), have been widely applied in water systems to enhance water quality prediction, assess potability, and manage water resources. These algorithms offer substantial potential for improving the accuracy and efficiency of water monitoring systems, which is essential for achieving sustainable water management practices.

For instance, Random Forests and Decision Trees have been effectively used to predict water quality by analysing physicochemical properties such as pH, hardness, and turbidity. Random Forest model outperformed the Support Vector Machine model with an accuracy of 70%, a precision of 72%, and an ROC-AUC score of 75% in predicting water potability (Rachid et al., 2025). While Decision Trees are simpler, they provide clear visualizations of the decision-making process, making them useful for initial water quality assessments (Iyer et al., 2023). SVMs, on the other hand, have been employed in water quality classification, often combined with feature selection techniques like Principal Component Analysis (PCA) to enhance accuracy. A study that integrated SVM with PCA reported an impressive accuracy of 99.24% in classifying water quality, highlighting the algorithm's robustness in handling complex datasets (Ladjal et al., 2021). ML algorithms have also proven valuable in predicting water potability and contaminants. ANNs, for example, have been used to classify water potability by analysing datasets that include parameters such as pH, hardness, and total dissolved solids. While ANNs achieved moderate accuracy of 70%, this result suggests that the performance could improve with larger and more diverse datasets (Darmawan et al., 2024). Meanwhile, more advanced models like XGBoost have outperformed traditional models in water quality prediction, demonstrating the potential of ensemble methods in managing large datasets with multiple features (Makumbura et al., 2024).

In water resource management, ML algorithms, including ANNs, have been employed to model and predict water demand, aiding in the efficient management of water supply systems. These models help utilities make informed decisions by forecasting demand and detecting anomalies, which in turn optimize resource allocation and ensure efficient water usage (Zanfei et al., 2023). Additionally, ML techniques have been integrated into public water management systems to improve monitoring and strategic planning for water resources. This integration is essential in developing sustainable water management strategies, particularly in the face of challenges such as climate change and population growth (Oğuz & Ertuğrul, 2023).

4.3.2 Deep Learning

DL, a subset of ML, leverages multi-layered neural networks to model complex patterns in data, making it a powerful tool for water quality monitoring. By utilizing advanced architectures such as convolutional neural networks (CNNs) and recurrent neural networks (RNNs), DL addresses the nonlinear relationships inherent in water quality data, offering enhanced accuracy and efficiency. The integration of DL models with real-time data collection systems has dramatically improved the ability to monitor, predict, and manage water quality, ensuring the safety and sustainability of water resources. This response explores the application of various DL models in water quality monitoring, discussing their methodologies, advantages, and specific use cases.

CNNs have been highly effective in classifying water quality images, achieving impressive accuracy rates. For instance, one study reported an average accuracy of 98.6% using CNNs for water quality image classification, showcasing the model's strength in processing visual data for water assessments (Jin et al., 2024). Another approach integrated CNNs with feature selection techniques like AdaBoost, further improving the precision of water quality evaluations by refining feature extraction, which is crucial for optimizing model performance (Dave et al., 2024).

RNNs, particularly Gated Recurrent Units (GRUs), have been employed in real-time river water quality monitoring, with one system, CNNs and Gated Recurrent Unit-based water quality monitoring (CNGRU-WQM), achieving a validation accuracy of 97.86%. This demonstrates RNNs' effectiveness in providing timely warnings, thereby aiding environmental conservation efforts (Geetha et al., 2024). Furthermore, integrating RNNs with biosensors has been proposed to enhance water quality monitoring. This system enables early detection of issues by analysing data from multiple sensors, providing adaptive responses to environmental changes (Pokkuluri et al., 2024).

In addition to CNNs and RNNs, ensemble and hybrid models have been developed to boost water quality prediction. Models like TNX (with temporal attention) and STNX (with spatio-temporal attention) incorporate temporal and spatio-temporal attention mechanisms, significantly improving short-step and long-step predictions by capturing variations in water quality parameters (Zheng et al., 2024). Hybrid models, such as Hierarchical Multimodal Long Short-Term Memory (HM-LSTM), combine different neural networks to detect anomalies in water data, enhancing the ability to identify deviations from normal patterns and providing detailed insights into the causes of these anomalies (Tharayil et al., 2024).

Advanced DL techniques, such as Bi-directional Long Short-Term Memory (Bi-LSTM) networks combined with computational fluid dynamics (CFD), have been proposed to model contaminant movement in water bodies. This framework delivers superior predictive accuracy and efficiency, addressing the limitations of traditional methods (Alfwzan et al., 2024). Additionally, interpretable DL models like the Gated Residual Variable Selection (GRVS) model have been employed to estimate key water quality parameters. These models use techniques like SHapley Additive exPlanations (SHAP) to identify influential factors, improving transparency and trust in predictions (Majnooni et al., 2024). By employing various neural network architectures and advanced techniques, DL ensures more efficient, reliable, and transparent water quality management, which is critical for maintaining safe and sustainable water resources.

4.3.3 Predictive Analytics

Predictive analytics for water quality forecasting utilizes ML and DL techniques to predict key water quality parameters and indices, playing a critical role in public health, environmental sustainability, and resource management. By analysing parameters such as pH, turbidity, and dissolved oxygen, these methods help ensure timely interventions to maintain water quality. Advanced ML models, including Decision Trees, Random Forests, and gradient boosting, have been integrated with data preprocessing techniques, improving prediction accuracy and reliability. Predictive analytics for water quality forecasting involves a multi-step process that includes data preprocessing, model optimization, and validation. These steps are crucial for developing accurate and reliable models that can predict water quality indices (WQI) and classifications (WQC). The integration of ML techniques has significantly enhanced the ability to forecast water quality by leveraging historical data and optimizing model parameters. This approach not only aids in environmental monitoring but also supports sustainable water management practices.

Several ML models have been applied for water quality prediction. Decision tree models, such as J48, Logistic Model Tree (LMT), and Random Forest, have been implemented using Waikato Environment for Knowledge Analysis (WEKA) software to classify water samples into clean and contaminated categories, streamlining laboratory analysis (Suma et al., 2023). Gradient boosting methods, like eXtreme Gradient Boost (XGBoost), have demonstrated superior accuracy in predicting WQI, with XGBoost achieving 97.06% accuracy in some studies (Elvin & Wibowo, 2024), and gradient boosting models reaching 99.50% accuracy in classification tasks (Nand Jha et al., 2024). DL techniques such as LSTM networks have been employed alongside causal inference and time-frequency analysis to forecast parameters like total nitrogen, significantly improving prediction accuracy (Zhang et al., 2024). RNNs, often combined with LSTM, have also been shown to deliver more precise predictions compared to other techniques, particularly in time-series analysis (Amireddy & Dileep, 2024). These ML and DL models highlight the transformative potential of predictive analytics in ensuring effective water quality management and resource sustainability.

4.3.4 AI in IoT-Based Systems

The integration of AI and IoT in water quality monitoring represents a significant advancement in environmental management, offering real-time, accurate, and efficient solutions for assessing water conditions. This combination leverages IoT's ability to collect data from various sensors and AI's analytical power to predict and manage water quality issues. These technologies are essential in addressing challenges posed by traditional water monitoring methods, which are often slow and resource-intensive.

IoT systems enable continuous monitoring of water quality by deploying sensors that measure parameters such as pH, turbidity, temperature, and dissolved oxygen (Leonila et al., 2024; Nutipalli et al., 2024). These systems provide real-time data, which is crucial for timely interventions and management of water resources (Puneeth et al., 2024). IoT-based frameworks are cost-effective and scalable, making them suitable for widespread deployment in various environments, including industrial and domestic settings (Londhe et al., 2024).

AI techniques, particularly ML algorithms, are used to analyse the data collected by IoT sensors, enabling the prediction of water quality trends and the detection of anomalies (Leonila et al., 2024; Shashank et al., 2024). AI models such as LSTM and AdaBoost regressor have been effectively used to forecast water quality parameters and predict concentrations of pollutants like *E. coli* (Nguyen et al., 2024). The integration of AI allows for the development of smart systems that can operate autonomously, providing insights and alerts for potential water quality issues (Singh & Walingo, 2024)

4.4 EARLY DETECTION OF CONTAMINANTS

Toxic chemicals, fertilizers, heavy metals, salts, minerals, and other substances that are not naturally present in water are referred to as water pollutants or contaminants and can cause detrimental effects on the environment and human well-being (LG Sonic, 2024). These substances are often invisible and do not always change the colour of the water. Small amounts of water and aquatic organisms are needed in determining the water quality (Lakshmikantha et al., 2021).

According to EPA Victoria State (2021), water pollutants generally come from two main sources: point sources and nonpoint sources. From the term itself, point source water pollution is easy to identify because it enters the water environment from an easily identified place. The three sources of point source pollution are the factories, power plants, and wastewater treatment plants. The nonpoint source pollution, on the other hand, came from non-discernible sources and cannot be easily attributed to a single point, as it was a combination of other contaminant sources. This also commonly originated from larger areas (Kuzniewski, 2023).

Contaminants coming from these sources degrade the water quality of aquifers, streams, lakes, oceans, and other bodies of water, which has a huge impact on aquatic life and wildlife (Mishra et al., 2023). When nutrients wash into waterways through storm runoff, they deplete oxygen in the water that fish need to survive (Mustafa et al., 2024). The nutrient build-up in the body of water promotes eutrophication, which lowers the oxygen levels in the water (Yang et al., 2008). The heavy metals and pesticides in the water also affect the ability of the fish to locate food and their ability to protect themselves from predators. Moreover, based on the report of NOAA, when contaminants enter the water body, they harm the aquatic habitat of marine species, and as the contaminants are consumed or absorbed by fish, they become unsafe to consume by people. Additionally, it could also affect several recreational activities such as fishing and swimming (LG Sonic, 2024).

As previously mentioned, contaminants have a very serious impact on human health. The US EPA has set standards for more than 80 contaminants in drinking water. According to the WHO, 80% of global diseases and 50% of child deaths are caused by poor quality of drinking water. The most highly vulnerable are the low-income households with close proximity to the sources of water pollution such as fecal coliforms (Berihun et al., 2023).

According to Sharma & Bhattachary (2016), there are four general categories of contaminants in drinking water, which are the following: physical, chemical, biological, and radiological. Aside from these contaminants, there are various emerging contaminants in water bodies, including per- and polyfluoroalkyl substances (PFAS), methane, and microplastics. The PFAS are a broad and complex class of synthetic chemicals that have been commonly used since the 1950s as components of consumer products such as non-stick pots and pans, food packaging, cleaning products, textiles, and industrial processes, due to their unique chemical and physical properties (Prevedouros et al., 2006). Since it has been found in various commonly used materials, PFAS are widespread in the environment, contaminating both surface water and groundwater, including sources of drinking water, causing serious detrimental health effects on humans. In the case of methane in water, it is produced naturally by the decay of organic matter and in the process of digestion of various organisms (Illinois Department of Public Health, 2018). Methanogens are methane-producing bacteria and are a primary source of methane in aquatic environments. Emissions from natural gas industries, as well as oil spills, can elevate the concentration of methane in water. It alters the chemical composition of water, making it unsuitable for drinking. It also facilitates the release of iron and manganese, which affects the water's taste (Almomani, 2022). Lastly, the microplastics in water are a broad variety of materials made up of diverse substances, each with varying densities, chemical compositions, shapes, and sizes. The potential risks posed by microplastics occur in three forms: physical particles, chemical substances, and microbial pathogens present in biofilms (Marsden et al., 2019). The primary way humans are exposed to microplastics is through the consumption of food, which includes seafood contaminated with microplastics, commercially processed fish, sea salt, honey, beer, and various food ingredients (Ziani et al., 2023).

People can suffer from acute and chronic health effects from exposure to almost any contaminant. When certain levels of contaminants are reached in the body, they can pose serious risks to vital organs and can be deadly for individuals with weakened immune systems, such as children and infants. With the increasing number of people exposed to unsafe drinking water, the need for early detection of contaminants has grown. Significant time and resources have been invested in research and development over recent years to enable efficient, cost-effective, and energy-efficient monitoring and regulation of water contaminants. This section will discuss these advancements in water quality monitoring, highlighting the capability of AI technologies in the early detection of heavy metals, chemical and biological contaminants, and emerging pollutants in water bodies.

4.4.1 AI in Heavy Metals Detection

AI technologies have revolutionized the detection of heavy metals in water, offering enhanced sensitivity, accuracy, and efficiency. These innovations combine AI with traditional methods to create systems that are both effective and cost-efficient. One approach is the Neuro-Particle Swarm Optimization (NN-PSO) model, which predicts heavy metal concentrations in water with greater accuracy than traditional models, achieving a Kling–Gupta efficiency of over 0.95. The NN-PSO model effectively incorporates physicochemical parameters like pH, making it highly reliable (De Jesus et al., 2022). Optical and spectroscopic techniques, such as Laser-induced breakdown spectroscopy (LIBS) has become a promising technique compared with the currently available water quality control methods, offering advantages such as minimal or no sample pre-treatment, fast and easy operation, and a chemical-free process, which are crucial for detecting heavy metals in water samples (Goncharova et al., 2024). Furthermore, a multi-channel optical sensing system using a BP neural network improves detection limits for metals like Cd, Cu, and Cr, demonstrating enhanced performance compared to conventional methods (Tang et al., 2022)

Electrochemical sensors, like those using square wave voltammetry with ML, offer high classification accuracy (98.31%) for detecting heavy metals such as As, Pb, and Cd. These sensors allow online monitoring, reducing the need for extensive sample preprocessing (Leon-Medina et al., 2022). Similarly, fluorescent sensors, including DNA-based sensors, provide a simple and cost-effective method for detecting metal ions in water (Liu et al., 2024). Portable solutions such as smartphone-based spectrometers offer an accessible and cost-effective option for onsite heavy metal detection, integrating chemical kits and spectral sensors to measure concentrations accurately (Srivastava & Sharma, 2021). Lastly, nanoparticle-based optical techniques enable ultrasensitive detection of pollutants like mercury and lead, demonstrating great potential for widespread use in water quality monitoring (Singh et al., 2020).

4.4.2 AI in Chemical Contaminants Detection

AI technologies have demonstrated significant promise in the early detection of chemical pollutants, including pesticides, fertilizers, and volatile organic compounds, which are crucial for environmental monitoring and food safety. Various AI-driven methodologies have been developed to enhance the sensitivity and accuracy of pollutant detection, leveraging ML, neural networks, and sensor technologies. These approaches not only improve detection capabilities but also provide rapid and cost-effective solutions for real-time monitoring.

One notable study utilized an ML-assisted approach with a surface-enhanced Raman scattering (SERS) substrate to detect pesticides in water. This SERS platform, prepared using silver nanoparticles, achieved nanomolar sensitivity for detecting pesticides like myclobutanil and phosmet. ML was employed to accurately distinguish between complex Raman spectra of different pesticides, showcasing a promising application for environmental monitoring (Sahin et al., 2022).

In agriculture, the Hybrid Chronic Multi-Residual Framework employs Q-Learning-based RNN to predict pesticide contamination in fruits and vegetables, achieving a 98.57% accuracy rate. This model highlights its potential in minimizing pesticide use and ensuring consumer safety by providing early warnings about harmful pesticide levels (Sellamuthu & Kaliappan, 2023). Electronic nose (e-nose) systems were developed to predict pesticide presence in groundwater, utilizing Random Forest and support vector machine algorithms, achieving 100% accuracy in pollution detection and 98.08% in pesticide type prediction (Wang et al., 2024). Additionally, conducting polymer e-nose technology has been utilized to detect organophosphate insecticides, which are highly toxic and pose significant health risks (Wilson, 2014). Finally, a DL approach combined with SERS was used to detect mixed pesticides on fruit surfaces. The multi-channel convolutional neural networks-gate recurrent unit model achieved 99% accuracy, demonstrating the effectiveness of integrating AI with spectroscopy for rapid and accurate pesticide detection (Dong et al., 2023).

4.4.3 AI in Biological Contaminants Detection

The early detection of biological contaminants such as bacteria, viruses, and parasites in water is crucial for preventing waterborne diseases and ensuring public health safety. AI and ML are increasingly being used to monitor and predict the presence of pathogens like *E. coli*, *Cryptosporidium*, and *Giardia* in water sources. These technologies enable real-time monitoring systems and predictive models that significantly enhance the detection and management of waterborne disease outbreaks. Various ML models, such as Random Forest and eXtreme Gradient Boosting, have been employed to predict the presence and intensity of *Cryptosporidium* and *Giardia*, using meteorological, physicochemical, and microbiological parameters for high prediction accuracy (Ligda et al., 2024). Additionally, Bayesian Belief Networks are used to predict concentrations of Fecal Indicator Bacteria (FIB) and *E. coli*, providing probabilistic predictions that improve microbial water quality assessments (Aliashrafi & Peleato, 2023). AI-enhanced microfluidic devices offer portable and efficient solutions for real-time water quality analysis, overcoming the limitations of traditional lab-based methods (Zhang et al., 2024). In the realm of wastewater surveillance, biosensors play a key role in real-time pathogen monitoring, although challenges remain in processing complex wastewater matrices (Singh et al., 2024). Techniques like TaqMan Array Cards allow for the simultaneous detection of multiple pathogens in wastewater, broadening the scope and utility of wastewater-based epidemiology (Rao et al., 2024).

4.4.4 Detection of Emerging Contaminants

4.4.4.1 PFAS (Per- and Polyfluoroalkyl Substances)

Various ML models are being developed to track, detect, and remediate PFAS, addressing the limitations of traditional methods that are often expensive and time-consuming (Andraju et al., 2023). For instance, a multilabel semi-supervised ML approach has been applied to predict PFAS concentrations in California groundwater, achieving high accuracy with techniques like Random Forest and XGBoost (Dong et al., 2023)

In terms of dual-mode detection techniques, researchers have combined electrochemical impedance spectroscopy and SERS to create a portable method for detecting PFAS, utilizing substrates with a plasmon-active Au layer (Kukralova et al., 2024). When paired with ML, SERS offers ultrasensitive detection capabilities, making it a promising approach for environmental analysis (Rothstein et al., 2024).

ML techniques have shown promise in enhancing PFAS-controlling systems, achieving over 80% prediction accuracy in treatment performance and resource identification (Hosseinzadeh et al., 2023). Furthermore, the integration of eco-friendly technologies, such as advanced sorbents and electrocoagulation, can complement these ML approaches, offering cost-effective and sustainable solutions for PFAS removal (Ezeorba et al., 2024). The combination of these methodologies not only improves treatment efficacy but also aligns with sustainability goals, addressing both environmental and public health concerns (Tushar et al., 2024).

4.4.4.2 Methane

AI systems utilizing satellite data and IoT sensors are increasingly effective in detecting methane leaks and emissions in real time, which is crucial for preventing water contamination near agricultural or industrial operations. These systems leverage advanced technologies to provide high-resolution, global coverage of methane emissions, enabling timely alerts and mitigation efforts.

For satellite-based detection, vision transformers applied to multispectral satellite imagery can identify methane emissions with high temporal and spatial resolution, detecting plumes as small as 0.01 km², corresponding to sources emitting 200–300 kg CH_4 per hour (Rouet-Leduc & Hulbert, 2024). The United Nations Environment Programme's Methane Alert and Response System (MARS) utilizes satellite data to detect and notify stakeholders of significant methane emissions, focusing on strong point sources from sectors like oil and gas, with plans to expand to smaller sources and other sectors (Irakulis-Loitxate et al., 2024).

In terms of IoT and sensor technologies, high-precision sensors combined with IoT capabilities can monitor methane emissions from point sources such as abandoned wells in real time, even under varying environmental conditions. These systems can detect ultra-low leak rates, providing a scalable and cost-effective solution for remote sensing and real-time notification of methane leaks (Daneshvarjejad et al., 2024).

Moreover, integrating diverse data streams, such as satellite observations with socio-economic and environmental data, enhances the accuracy of methane emission estimates, aiding in climate change mitigation efforts (Maliehe et al., 2024). IMEO's efforts to create a centralized public record of methane emissions data facilitate global climate action by providing reliable, policy-relevant information (Zavala-Ariza et al., 2024).

4.4.4.3 Microplastics

AI-based image recognition techniques are increasingly utilized to monitor microplastic concentrations in oceans and rivers, offering a promising solution to the environmental hazard posed by microplastics. These techniques leverage advanced ML models to classify different types of plastics and predict their movement patterns, enabling targeted clean-up efforts and facilitating more efficient environmental monitoring practices. DL models, particularly those based on image segmentation, have demonstrated impressive performance, achieving a 96% accuracy in identifying microplastic particles and providing a robust alternative to manual counting methods (Royer et al., 2024). Additionally, ML technologies automate the detection of floating plastics using satellite imagery, aiding in identifying pollution sources and tracking waste movement dynamics (Danilov & Serdiukova, 2024). Holographic imaging offers a three-dimensional representation of microplastics, enhancing classification accuracy and efficiency (Munagala et al., 2024). CNNs are among the most used models for plastic waste detection and classification, frequently achieving accuracy levels exceeding 80% (Ramos et al., 2024). Moreover, AI-empowered frameworks streamline the collection and characterization of microplastics, integrating robotics and ML to improve environmental monitoring (Guo et al., 2024). Through these advanced methodologies, AI technologies play a crucial role in addressing the pressing challenge of microplastic pollution in aquatic ecosystems.

4.5 CHALLENGES AND LIMITATIONS

While the integration of AI in water quality monitoring systems offers numerous advantages, it also presents several challenges and limitations that must be addressed for successful implementation.

4.5.1 Data Availability and Quality

Data availability and quality are significant challenges in implementing AI-driven water quality systems, especially in developing countries. These issues arise from factors such as inadequate infrastructure, high costs, and the complexity of water quality data. The lack of reliable and comprehensive datasets limits the effectiveness of ML models for water quality prediction and monitoring. In countries like Nigeria, for example, insufficient environmental monitoring data restricts the application of ML (Omeka, 2024). High costs associated with IoT-based systems for real-time monitoring further hinder large-scale implementation in developing regions (Addow & Jimale, 2023). Additionally, manual data collection is often impractical and costly in rural areas, leading to limited data availability (Alipio, 2020)

Data quality is affected by technical issues, such as sensor failures, calibration difficulties, and connectivity problems, making it challenging to maintain reliable, data-driven systems (William et al., 2023). High levels of missing data, as seen in studies from Uruguay, require imputation methods, which can sometimes yield inaccurate results (Rodríguez et al., 2021). Moreover, the process of imputing missing data itself can introduce errors, affecting the overall reliability of AI models used for water quality assessment.

Technological solutions offer some hope for addressing these challenges. The integration of IoT and AI technologies can enable real-time data collection and analysis, potentially improving water quality monitoring in developing areas (Addow & Jimale, 2023). Hybrid ML models, like the Wavelet-Adaptive Neural Fuzzy Interference System (W-ANFIS), have also shown potential in enhancing prediction accuracy for water quality parameters (Omeka, 2024). Additionally, intelligent data validation techniques, such as anomaly detection and data reconciliation, can help improve data quality and support more effective water management (William et al., 2023).

4.5.2 Financial Constraints

Financial constraints pose significant challenges in the implementation of AI-driven water quality monitoring systems. These systems, while promising in their ability to provide real-time and accurate data, often require substantial initial investments in technology, infrastructure, and skilled personnel. The high costs associated with traditional water quality monitoring methods, such as laboratory analyses and field sampling, are a primary motivator for adopting AI and ML technologies. However, the transition to AI-driven systems is not without its financial hurdles.

Implementing AI-driven systems necessitates investment in advanced technologies, such as sensors and IoT devices, which can be costly (Madhavireddy & Koteswarrao, 2018; Manjakkal et al., 2021). The integration of AI with existing infrastructure can also incur additional costs, as seen in the development of real-time monitoring systems using AI for anomaly detection (Czyczula Rudjord et al., 2022). Effective AI systems require robust data management and processing capabilities, which can be financially demanding due to the need for high-quality data and sophisticated algorithms (Lowe et al., 2022). The use of microbial fuel cells (MFCs) as a low-cost, energy-autonomous solution for water quality monitoring highlights the potential for reducing operational costs, but the technology is still in its nascent stages and requires further development (Shabani et al., 2021).

Despite these challenges, AI and ML technologies have the potential to reduce long-term costs by optimizing water quality monitoring processes and decreasing the need for manual interventions (Lowe et al., 2022; Miller et al., 2023). Low-cost AI systems, such as the proposed buoy system, demonstrate the feasibility of reducing costs through innovative design and technology integration (Lu et al., 2022)

4.5.3 Public Perception and Acceptance

Public perception and acceptance are significant challenges in the implementation of AI-driven water quality monitoring systems. These challenges stem from a lack of public understanding, trust, and engagement with AI technologies, as well as the complexities involved in effectively communicating water quality data. The integration of AI in water quality monitoring offers numerous benefits, such as enhanced data resolution and operational efficiency, but these are often overshadowed by public scepticism and the need for improved communication strategies.

The public's comprehension of water quality issues is often limited due to deficiencies in communication processes. Effective communication strategies are essential to convey complex water quality data to the public in an understandable manner (Amador-Castro et al., 2024). The use of social media and AI-driven natural language processing frameworks can enhance public engagement by analysing water-related posts and providing data-driven insights (Auyb et al., 2024). Developing user-friendly interfaces, such as conversational AI agents, can bridge the gap between data collection and interpretation, making water quality data more accessible to diverse user groups (Vald et al., 2024).

Trust in AI systems is crucial for public acceptance. Explainable AI (XAI) can help build user confidence by providing transparent and understandable AI processes (Rane et al., 2024). Trust in information sources is a critical factor influencing public perception of water quality. Sources like universities and scientific organizations are often more trusted than governmental bodies due to their perceived objectivity and expertise (Wilkins et al., 2018). In the Netherlands, people tend to trust drinking water companies more than other institutions, suggesting that direct communication from these companies can enhance trust (Brouwer et al., 2020). A study in the Midwest found that familiarity with an organization increases trust, particularly among agricultural respondents (Mase et al., 2015).

Engaging stakeholders and the public in the development of water quality monitoring systems can enhance acceptance. This involves mobilizing local knowledge to identify key performance indicators (KPIs) and support effective strategies (Kokkinos et al., 2019). Policy frameworks should incorporate sustainability indicators and consider the socio-economic impacts of AI technologies to ensure equitable and sustainable water management (Kokkinos et al., 2019)

4.5.4 Data Security, Integrity, and Accuracy

The application of AI emerged as one of the important tools in water quality monitoring (Olawade et al., 2024). However, this presents various information security challenges (Jada & Mayayise, 2024) that need to be addressed for effective and reliable implementation. ANNs rely on a large amount of datasets to make predictions and decisions (Chawishborwornworng et al., 2024). The accuracy and integrity of these datasets are critical because inaccurate, incomplete or tampered data can lead to erroneous predictions with potential health risks, environmental hazards or economic losses. Securing the compliance and data collection process is vital to ensure the effectiveness of AI-driven solutions.

4.5.5 Cyber and Information Security Risks

The AI technology has been rapidly growing and being widely implemented by various sectors (Safitra et al., 2024). The AI systems have diverse applications to water quality monitoring and management (Kamyab et al., 2023), and a typical stand-alone system becomes more interconnected with other infrastructure systems. With the growing number of applications, they become vulnerable to cyberattacks. The malicious threat actors could target these systems to disrupt operations, manipulate data or completely compromise the entire systems (Ramos-Cruz et al., 2024). Establishing and strengthening cybersecurity measures, such as authentication, authorization, encryption, and intrusion detection systems (Hernandez-Jaimes et al., 2023), are critical to protect AI-powered monitoring systems from these known threats.

4.5.6 AI Model Vulnerabilities

AI models are susceptible to using artificial inputs produced by an adversary (Sharma et al., 2024). An adversarial attack is one of the security challenges in implementing an AI, which could lead to inefficiency (Jada & Mayayise, 2024). AI models used for a water quality management system may have vulnerabilities that can be exploited by adversaries. The adversarial attack can cause the system to produce incorrect results. For example, by altering sensor data slightly, an attacker could deceive the water quality system into reporting safe water when the reality it is already contaminated. In order to counter this, robust AI models need to be developed with built-in defences against such attacks (Li et al., 2024), enhancing data protection.

4.5.7 Data Privacy Issues and Concerns

Water quality monitoring systems using AI often include collections of environmental and industrial data that may contain sensitive information such as pollutant sources, water usage, and industrial waste management, among others (Zhao et al., 2024). Unauthorized sharing and access to such information could lead to privacy violations, particularly for industries and communities concerned with personal and sensitive data. Aside from the technological controls, it is highly recommended to establish and implement a data privacy protocol and ensure compliance with external regulations such as General Data Protection Regulation (GDPR). Further, the world's first major act to regulate AI, the AI Act, was passed by the European Union on 13 March 2024. With GDPR compliance, providers of AI-based systems could prevent penalties and ensure compliance with data processing (Zhao et al., 2024).

4.5.8 System Resilience and Reliability

An AI-powered water quality monitoring system needs to be highly resilient, reliable, and available, particularly in critical situations like detecting toxic contaminants or predicting pollution spikes. However, AI models, due to dependence on high-quality data, may face challenges in adapting to unexpected environmental changes or disruptions in data flow (Jada & Mayayise, 2024). Hence, the system must be regularly updated and should have a backup protocol, such as redundancy or cloud backup, which can help maintain its functionality during changes in environmental conditions or cyberattacks and calamities.

4.5.9 Ethical Standards and Accountability

AI-driven decision-making in water quality monitoring can have far-reaching implications for public health and the environment. However, the potential for AI system failures, such as undetected contamination or misclassified water quality levels, raises critical questions of accountability. In order to mitigate these risks and build public trust, it is essential to develop guidelines for AI system accountability policy and ensure transparency in their decision-making processes (Oduro et al., 2022).

4.6 FUTURE PROSPECTS

The future of AI in water quality monitoring is poised for significant advancements, driven by the integration of ML, IoT, and innovative sensor technologies. AI's ability to process large datasets and predict water quality parameters offers a transformative potential for real-time monitoring and management of water resources. This evolution is crucial for addressing the growing challenges of water pollution and resource management in a sustainable manner. The following sections explore key aspects of AI's future role in water quality monitoring.

4.6.1 Advancements in AI Technologies

Future advancements in AI technologies for water quality monitoring are poised to revolutionize the field by enhancing data analysis, prediction accuracy, and real-time monitoring capabilities. These advancements are driven by the integration of AI with various technologies such as conversational agents, DL models, and microfluidic platforms. The development of these technologies aims to address the challenges of data complexity, real-time analysis, and user accessibility in water quality management.

The AI Data Expert (AI-DE) system utilizes conversational AI such as ChatGPT for natural language processing, enabling users to interact with the system through conversational queries. By integrating AI technologies, the system supports participatory approaches that involve diverse stakeholders in decision-making processes, thereby fostering collaboration and adaptive management strategies essential for sustainable development (Santos & Carvalho, 2025).

DL models, such as KNN, CNNs, and Inception V3, are being employed to improve the precision of water quality assessments. These models, when combined with feature selection techniques like Random Forest and AdaBoost, have shown significant improvements in prediction accuracy, highlighting the importance of feature extraction in enhancing model performance (Dave et al., 2024). The use of hybrid ML models, such as the Adaptive Neuro-Fuzzy Inference System (ANFIS), has been identified as particularly effective in predicting water quality parameters like temperature and dissolved oxygen (Omeka, 2024).

AI-based microfluidic platforms are emerging as efficient tools for real-time water quality analysis. These platforms leverage AI algorithms to recognize and classify contaminants, offering a portable and convenient alternative to traditional laboratory methods (Zhang et al., 2024). The integration of IoT with AI models facilitates remote monitoring of water quality parameters. IoT-based systems have demonstrated high accuracy in measuring parameters such as pH and turbidity, significantly outperforming traditional methods (Jayaraman et al., 2024). Real-time monitoring systems using ANN and Kalman filters have been developed to enhance the accuracy and efficiency of wastewater management, achieving high recognition rates and low false alarm rates (Essahlaoui et al., 2024).

Advanced predictive models, such as those using polynomial regression and light gradient boosting (LGB), are being developed to forecast water quality changes and assess risks associated with seasonal fluctuations. These models provide valuable insights for ecological evaluations and management strategies (Lin et al., 2022; Vaivude, 2023). Additionally, data-centric intelligent systems, leveraging AI, further enhance water quality management by tracking and predicting pollutants, thus improving monitoring, assessment, and control processes (Sahin et al., 2022). Finally, the integration of AI with geographic information systems (GIS) in water quality monitoring represents a significant advancement in environmental management. GIS tools are essential for mapping and monitoring water quality over large areas. They provide spatial distribution maps that help visualize the results of AI models, enhancing the understanding of water quality dynamics (Kumari et al., 2023). This integration leverages the strengths of AI in data analysis and prediction with the spatial visualization capabilities of GIS, offering a comprehensive approach to monitoring and managing water resources.

4.6.2 AI for Climate Adaptation and Resilience

AI technologies are increasingly being leveraged to address climate adaptation and resilience, particularly in the context of water quality issues exacerbated by climate change. These technologies offer innovative solutions for monitoring, predicting, and managing the impacts of extreme weather events such as flooding and droughts, which can lead to pollutant runoff and deteriorate water quality. AI's ability to process large datasets and model complex interactions makes it a valuable tool in developing adaptive strategies for climate resilience. This response explores the role of AI in climate adaptation, focusing on its applications in water quality management and multi-risk assessment.

This helps in detecting changes in water quality due to climate-induced events like flooding and droughts (Yang et al., 2022). ML models are applied to predict and enhance water resource management, ensuring sustainability and resilience in the face of climate change (Bhadauria et al., 2024). AI-driven synoptic water monitoring systems facilitate the intelligent extraction of water information, crucial for maintaining water quality and ecosystem health (Yang et al., 2022). AI supports multi-risk assessment by modelling complex interactions between various climate hazards that affect water quality. The INTELLIGENT multi-risk (i.e., InNovaTive machinE Learning methodoLogy to assess multi-rIsk dynamics under climate chanGe futurE coNdiTions) methodology uses ML to evaluate the impacts of multi-risk events and predict future scenarios, aiding in the development of adaptation strategies (Dal Barco et al., 2024). AI frameworks, like those applied in the Veneto Region, integrate data from diverse sources to assess and predict the impacts of extreme climate events, enhancing preparedness and resilience (Ferrario et al., 2024).

AI technologies are pivotal in climate modelling and disaster prediction, providing insights into climate patterns and ecosystem dynamics (Moley, 2023). By optimizing resource allocation and infrastructure resilience, AI aids in mitigating the adverse effects of climate change on water quality (Arya, 2024). AI's capacity to model nonlinear relationships and process big data supports the development of sophisticated climate-adaptive measures (Gatla, 2019).

4.6.3 AI towards Net-Zero Emissions

AI and early prediction of pollutants play a crucial role in supporting net-zero emissions by optimizing water treatment processes, preventing large-scale contamination events, and reducing energy consumption. Early detection allows for precise interventions, reducing the need for energy-intensive treatments and excessive chemical use, which are significant contributors to greenhouse gas emissions. Ucun Ozel et al. (2020), utilize AI models to predict heavy metal contamination, efficiently monitor and manage water quality, thereby reducing the need for energy-intensive monitoring methods, leading to sustainable water resource management. Additionally, AI's detection of methane leaks, as demonstrated in the study of Rouet-Leduc & Hulbert (2024), helps to contain one of the most potent greenhouse gases by providing high-resolution, global monitoring capabilities, thereby supporting global climate targets through the reduction of emissions directly at the source.

By enabling real-time monitoring, AI-driven systems ensure more efficient water management, leading to reduced energy use, minimized waste, and better resource allocation. This combination not only improves the sustainability of water systems but also directly contributes to achieving net-zero emissions goals by lowering the carbon footprint of water treatment operations.

4.7 CONCLUSION

AI plays a crucial role in transforming water quality monitoring practices. This tool not only enhances the accuracy and speed of contaminant detection but also paves the way towards achieving net-zero emissions in water management. Integration of ML, DL, and predictive analytics in the system enables a proactive approach in water quality monitoring that traditional methods often struggle with. AI-driven systems optimize treatment processes, reducing energy consumption and minimizing the overall carbon footprint, aligning with global sustainability goals.

AI is increasingly being adopted for comprehensive water quality monitoring systems due to its significant benefits. However, as AI systems expand, they bring new challenges such as data availability, technological limitations, ethical concerns, and cyber and information security. Addressing both current and potential risks is essential to ensure data integrity, accuracy, privacy, resilience, reliability, and ethical accountability. With the right technological controls in place, AI has the potential to greatly enhance global water management and contribute to safer water resources for everyone.

The future of AI in water quality monitoring is set for significant advancements, particularly through the integration of IoT and enhanced sensor technologies. These technologies, combined with AI's predictive capabilities, will provide comprehensive, real-time water quality assessments that adapt to environmental changes and support efficient resource management. Prioritizing data security, enhancing public acceptance, and developing ethical frameworks, AI's role in achieving sustainable, resilient water systems will be instrumental, supporting not only the protection of water resources but also the broader objective of environmental sustainability and climate adaptation.

REFERENCES

Addow, M. A., & Jimale, A. D. (2023). IoT-based real-time water quality monitoring for sustainable water management: a case study in Somalia. *SSRG International Journal of Electrical and Electronics Engineering*. https://doi.org/10.14445/23488379/IJEEE-V10I8P116

Alfwzan, W. F., Selim, M. M., Almalki, A. S., & Alharbi, I. S. (2024). Water quality assessment using Bi-LSTM and computational fluid dynamics (CFD) techniques. *Alexandria Engineering Journal*, 97, 346–359.

Aliashrafi, A., & Peleato, N. M. (2023). Monitoring microbial quality of source waters using Bayesian belief networks. *Lecture Notes in Civil Engineering*. https://doi.org/10.1007/978-981-19-1061-6_25

Alipio, M. I. (2020). Data-driven IoT-based water quality monitoring and potability classification system in rural areas. In: *International Conference on ICT Convergence*. https://doi.org/10.1109/ICTC49870.2020.9289505

Almomani, S. (2022). *Microplastics: Modern Innovations to Reduce Their Presence in Water*. Save the WaterTM.

Amador-Castro, F., González-López, M. E., Lopez-Gonzalez, G., Garcia-Gonzalez, A., Díaz-Torres, O., Carbajal-Espinosa, O., & Gradilla-Hernández, M. S. (2024). Internet of Things and citizen science as alternative water quality monitoring approaches and the importance of effective water quality communication. *Journal of Environmental Management*. https://doi.org/10.1016/j.jenvman.2023.119959

Amireddy, R., & Dileep, P. (2024). A comparative study on water quality prediction using machine learning and deep learning techniques. *Third IEEE International Conference on Distributed Computing and Electrical Circuits and Electronics (ICDCECE)*, (pp. 1–5), Ballari, India, doi: 10.1109/ICDCECE60827.2024.10548555.

Andraju, N., Curtzwiler, G., Ji, Y., Kozliak, E. I., & Ranganathan, P. (2023). Machine Learning Models for PFAS Tracking, Detection and Remediation: A Review, *2023 IEEE International Conference on Electro Information Technology (eIT)*, Romeoville, IL, USA, 2023, pp. 137–142, doi: 10.1109/eIT57321.2023.10187

Arora, P. (2017). *Physical, Chemical and Biological Characteristics of Water*. E-PG Pathshala.

Arya, S. (2024). Harnessing Big Data and Machine Learning for Climate Change Predictions and Mitigation. In B. Singh, C. Kaunert, K. Vig, & S. Dutta (Eds.), *Maintaining a Sustainable World in the Nexus of Environmental Science and AI* (pp. 383–408). IGI Global Scientific Publishing. https://doi.org/10.4018/979-8-3693-6336-2.ch015

Auyb, M. A., Zamir, M. T., Khan, I., Naseem, H., Ahmad, N., & Ahmad, K. (2024). *Social Media and Artificial Intelligence for Sustainable Cities and Societies: A Water Quality Analysis Use-case. Computing & AI Connect*, 1 (Article ID: 0006). https://doi.org/10.69709/CAIC.2024.133109

Berihun, G., Abebe, M., Hassen, S., Gizeyatu, A., Berhanu, L., Teshome, D., Walle, Z., Desye, B., Sewunet, B., & Keleb, A. (2023). Drinking water contamination potential and associated factors among households with under-five children in rural areas of Dessie Zuria District, Northeast Ethiopia. *Frontiers in Public Health*, *11*, 1199314. https://doi.org/10.3389/FPUBH.2023.1199314/FULL

Bhadauria, A., Reddy, M. S. S., Asha, V., Nijhawan, G., Abdulhussein Hameed, A., & Pratap, B. (2024). Analytical survey on the sustainable advancements in water and hydrology resources with AI implications for a Resilient Future. *E3S Web of Conferences*. https://doi.org/10.1051/e3sconf/202455201074

Brouwer, S., Hofman-Caris, R., & van Aalderen, N. (2020). Trust in drinking water quality: understanding the role of risk perception and transparency. *Water*. https://doi.org/10.3390/W12092608

Chawishborwornworng, C., Luanwuthi, S., Umpuch, C., & Puchongkawarin, C. (2024). Bootstrap approach for quantifying the uncertainty in modeling of the water quality index using principal component analysis and artificial intelligence. *Journal of the Saudi Society of Agricultural Sciences*, 23(1), 17–33.

Czyczula Rudjord, Z., Reid, M. J., Schwermer, C. U., & Lin, Y. (2022). Laboratory development of an AI system for the real-time monitoring of water quality and detection of anomalies arising from chemical contamination. *Water*. https://doi.org/10.3390/w14162588

Dal Barco, M. K., Maraschini, M., Ferrario, D. M., Nguyen, N. D., Torresan, S., Vascon, S., & Critto, A. (2024). A machine learning approach to evaluate coastal risks related to extreme weather events in the Veneto region (Italy). *International Journal of Disaster Risk Reduction*. https://doi.org/10.1016/j.ijdrr.2024.104526

Daneshvarjejad, N., Khodaparast, P., Cho, Y., Paul, D., Farhadi, S., & Ershaghi, I. (2024). A Method for Detection of Methane Leaks from Idle and Orphaned Wells Using High-Precision Sensor and a Ventilation Canopy. In: *Society of Petroleum Engineers - SPE Canadian Energy Technology Conference and Exhibition, CET 2024*. https://doi.org/10.2118/218022-MS

Danilov, A., & Serdiukova, E. (2024). Review of methods for automatic plastic detection in water areas using satellite images and machine learning. *Sensors*. https://doi.org/10.3390/s24165089

Darmawan, I., Fatchan, M., & Firmansyah, A. (2024). Classification of drinking water potability with artificial neural network algorithm. *International Journal of Integrated Science and Technology*, 2(5), 506–515.

Dave, J., Patel, A., & Raval, H. (2024). Towards precise water quality assessment: a deep learning approach with feature selection in smart monitoring systems. *International Journal of Scientific Research in Computer Science, Engineering and Information Technology*, 10(4), 100–114.

De Jesus, K. L. M., Senoro, D. B., Dela Cruz, J. C., & Chan, E. B. (2022). Neuro-particle swarm optimization based in-situ prediction model for heavy metals concentration in groundwater and surface water. *Toxics*. https://doi.org/10.3390/toxics10020095

Dong, X., Wang, S., Liu, C., Dong, S., & Qing, H. (2023). Rapid detection of residual chlorpyrifos and pyrimethanil on fruit surface by surface-enhanced Raman spectroscopy integrated with deep learning approach. *Science Reports*, *13*(1), 19855.

Elvin, & Wibowo, A. (2024). Forecasting water quality through machine learning and hyperparameter optimization. *Indonesian Journal of Electrical Engineering and Computer Science*, 33(1), 496–506.

EPA Victoria State. (2021). *Point and Non-Point Sources of Water Pollution*.

Essahlaoui, F., Elhajrat, N., Halimi, M., Azrour, M., Kılıç, Z., & Abbassi, A. El. (2024). Intelligent real-time monitoring system for wastewater management using artificial neural network. *World Sustainability Series*. https://doi.org/10.1007/978-3-031-52419-6_2

Ezeorba, T. P. C., Okeke, E. S., Nwankwo, C. E., Emencheta, S. C., Enochoghene, A. E., Okeke, V. C., & Ozougwu, V. E. O. (2024). Emerging eco-friendly technologies for remediation of Per- and poly- fluoroalkyl substances (PFAS) in Water and Wastewater: A pathway to environmental sustainability. *Chemosphere*, *364*, 143168. https://doi.org/10.1016/j.chemosphere.2024.143168

Ferrario, D. M., Sano, M., Tiggeloven, T., Claasen, J., Petrovska, E., Maraschini, M., de Ruiter, M., Torresan, S., & Critto, A. (2024). Artificial Intelligence for climate change multi-risk assessment: a Myriad-EU case study in the Veneto Region. *EGU General Assembly, 2024*, 14–19.

Gatla, T. R. (2019). A cutting-edge research on AI combating climate change: innovations and its impacts. *International Journal of Innovations in Engineering Research and Technology*, 6(9), 1–8.

Geetha, T. S., Gandhimathi, G., Chellaswamy, C., & Thiruvalar Selvan, P. (2024). Comprehensive river water quality monitoring using convolutional neural networks and gated recurrent units: a case study along the Vaigai River. *Journal of Environmental Management*, 365, 121567.

Goncharova, I., Guichaoua, D., Taboukhat, S., Tarbi, A., Chtouki, T., Erguig, H., & Sahraoui, B. (2024). Laser-induced breakdown spectroscopy application for heavy metals detection in water: a review. *Spectrochimica Acta - Part B Atomic Spectroscopy*. https://doi.org/10.1016/j.sab.2024.106943

Guo, P., Wang, Y., Moghaddamfard, P., Meng, W., Wu, S., & Bao, Y. (2024). Artificial intelligence-empowered collection and characterization of microplastics: a review. *Journal of Hazardous Materials*. https://doi.org/10.1016/j.jhazmat.2024.134405

Hama Aziz, K. H., Mustafa, F. S., Omer, K. M., Hama, S., Hamarawf, R. F., & Rahman, K. O. (2023). Heavy metal pollution in the aquatic environment: efficient and low-cost removal approaches to eliminate their toxicity: a review. *RSC Advances*, *13*(26), 17595. https://doi.org/10.1039/D3RA00723E

Hernandez-Jaimes, M. L., Martinez-Cruz, A., Ramírez-Gutiérrez, K. A., & Feregrino-Uribe, C. (2023). Artificial intelligence for IoMT security: a review of intrusion detection systems, attacks, datasets and Cloud–Fog–Edge architectures. *Internet of Things*, 23, 100887.

Hosseinzadeh, A., Altaee, A., Li, X., & Zhou, J. L. (2023). Machine learning-based modeling and analysis of perfluoroalkyl and polyfluoroalkyl substances controlling systems in protecting water resources. *Current Opinion in Chemical Engineering*, *42*, 100983. https://doi.org/10.1016/j.coche.2023.100983

Illinois Department of Public Health. (2018). *Methane in Groundwater*. Illinois Department of Public Health. https://dph.illinois.gov/content/dam/soi/en/web/idph/files/methane-in-groundwater.pdf

Irakulis-Loitxate, I., Randles, C., Watine-Guiu, M., Mateo-Garcia, G., Vaughan, A., Demeter, M., Cifarelli, C., Guanter, L., Maasakkers, J. D., Aben, I., De Long, T. A., Sharma, S., Groshenry, A., Peyle, Q., Benoit, A., & Caltagirone, M. (2024). UNEP's Methane Alert and Response System (MARS): current status, new developments and case studies, EGU General Assembly 2024, Vienna, Austria, (pp. 14–19), EGU24-16048, https://doi.org/10.5194/egusphere-egu24-16048.

Iyer, S., Kaushik, S., & Nandal, P. (2023). Water quality prediction using machine learning. *International Journal of Engineering and Technology*, 10(1), 59–62.

Jada, I., & Mayayise, T. O. (2024). The impact of artificial intelligence on organisational cyber security: an outcome of a systematic literature review. *Data and Information Management*, 8(2), 100063.

Jayaraman, P., Nagarajan, K. K., Partheeban, P., & Krishnamurthy, V. (2024). Critical review on water quality analysis using IoT and machine learning models. *International Journal of Information Management Data Insights*. https://doi.org/10.1016/j.jjimei.2023.100210

Jin, C., Hu, Y., Lin, F., Zou, L., Li, R., & Hu, X. (2024). Research on water quality image classification method based on convolutional neural network. In: *IMCEC 2024- IEEE 6th Advanced Information Management, Communicates, Electronic and Automation Control Conference*. https://doi.org/10.1109/IMCEC59810.2024.10575513

Lakshmikantha, V., Hiriyannagowda, A., Manjunath, A., Patted, A., Basavaiah, J., & Anthony, A. A. (2021). IoT based smart water quality monitoring system. *Global Transitions Proceedings*, 2(2), 181–186. https://doi.org/10.1016/J.GLTP.2021.08.062

Kamyab, H., Khademi, T., Chelliapan, S., SaberiKamarposhti, M., Rezania, S., Yusuf, M., Farajnezhad, M., Abbas, M., Hun Jeon, B., & Ahn, Y. (2023). The latest innovative avenues for the utilization of artificial Intelligence and big data analytics in water resource management. *Results in Engineering*. https://doi.org/10.1016/j.rineng.2023.101566

Kokkinos, K., Lakioti, E., Samaras, P., & Karayannis, V. (2019). Evaluation of public perception on key sustainability indicators for drinking water quality by fuzzy logic methodologies. *Desalination and Water Treatment*. https://doi.org/10.5004/dwt.2019.24642

Kukralova, K., Miliutina, E., Guselnikova, O., Burtsev, V., Hrbek, T., Svorcik, V., & Lyutakov, O. (2024). Dual-mode electrochemical and SERS detection of PFAS using functional porous substrate. *Chemosphere*. https://doi.org/10.1016/j.chemosphere.2024.143149

Kumari, N., Kadave, K., Marandi, S., Pandey, S., & Kumar, G. (2023). Constructing the dynamics of water quality parameters using geospatial technology and in situ observations. In: *Geospatial Analytics for Environmental Pollution Modeling: Analysis, Control and Management*. https://doi.org/10.1007/978-3-031-45300-7_8

Kuzniewski, S. (2023, August 30). What is point source pollution? *Wastewater Digest*. https://www.wwdmag.com/what-is-articles/article/33010576/what-is-point-source-pollution

Ladjal, M., Bouamar, M., Brik, Y., & Djerioui, M. (2021). A decision fusion method based on classification models for water quality monitoring. *Environmental Science and Pollution Research*, 30, 22532–22549.

Leonila, T., Senthil, G. A., Geerthik, S., Sowmiya, R., & Nithish, J. (2024). Dynamic water quality monitoring via iot sensor networks and machine learning technique. *International Conference on Communication, Computing and Internet of Things, (IC3IoT)*. pp. 1–6, doi: 10.1109/IC3IoT60841.2024.10550224.

Leon-Medina, J. X., Tibaduiza, D. A., Burgos, J. C., Cuenca, M., & Vasquez, D. (2022). Classification of As, Pb and Cd heavy metal ions using square wave voltammetry, dimensionality reduction and machine learning. *IEEE Access*. https://doi.org/10.1109/ACCESS.2022.3143451

LG Sonic. (2024). *Water Contamination – Types, Sources, Effects, Solutions*.

Li, P., Huang, J., Zhang, S., & Qi, C. (2024). SecureEI: proactive intellectual property protection of AI models for edge intelligence. *Computer Networks*. https://doi.org/10.1016/j.comnet.2024.110825

Ligda, P., Mittas, N., Kyzas, G. Z., Claerebout, E., & Sotiraki, S. (2024). Machine learning and explainable artificial intelligence for the prevention of waterborne cryptosporidiosis and giardiosis. *Water Research*. https://doi.org/10.1016/j.watres.2024.122110

Lin, L., Yang, H., & Xu, X. (2022). Effects of Water Pollution on Human Health and Disease Heterogeneity: A Review. In A. El Nemr (Ed.), *Frontiers in Environmental Science* (Vol. 10). Frontiers Media S.A. https://doi.org/10.3389/fenvs.2022.880246

Liu, Y., Xu, J., Wu, Z., Cai, Y., Zhao, Z., & Qiu, J. (2024). Development of a fluorescent DNA sensor for dual detection of heavy metal ions utilising DAPI in distinct buffers. *Food Chemistry*. https://doi.org/10.1016/j.foodchem.2024.139390

Londhe, A., Apare, R., & Borhade, R. (2024). Aqua Status Prediction Using IoT and Optimization in Aquaculture: A Comprehensive Review. *MIT Art, Design and Technology School of Computing International Conference (MITADTSoCiCon)*, (pp. 1–4), Pune, India, doi: 10.1109/MITADTSoCiCon60330.2024.10575248.

Lowe, M., Qin, R., & Mao, X. (2022). A review on machine learning, artificial intelligence, and smart technology in water treatment and monitoring. *Water*. https://doi.org/10.3390/w14091384

Lu, H. Y., Cheng, C. Y., Cheng, S. C., Cheng, Y. H., Lo, W. C., Jiang, W. L., Nan, F. H., Chang, S. H., & Ubina, N. A. (2022). A low-cost AI buoy system for monitoring water quality at offshore aquaculture cages. *Sensors*. https://doi.org/10.3390/s22114078

Madhavireddy, V., & Koteswarrao, B. (2018). Smart water quality monitoring system using iot technology. *International Journal of Engineering & Technology*. https://doi.org/10.14419/ijet.v7i4.36.24214

Majnooni, S., Fooladi, M., Nikoo, M. R., Al-Rawas, G., Haghighi, A. T., Nazari, R., Al-Wardy, M., & Gandomi, A. H. (2024). Smarter water quality monitoring in reservoirs using interpretable deep learning models and feature importance analysis. *Journal of Water Process Engineering, 60*.

Makumbura, R. K., Mampitiya, L., Rathnayake, N., Meddage, D., Henna, S., Dang, T. L., Hoshino, Y., & Rathnayake, U. (2024). Advancing Water Quality Assessment and Prediction Using Machine Learning Models, Coupled with Explainable Artificial Intelligence (XAI) Techniques Like Shapley Additive Explanations (SHAP) For Interpreting the Black-Box Nature. *Results in Engineering*, *23*, 102831. https://doi.org/10.1016/j.rineng.2024.102831

Maliehe, K., Goulding, J., Alam, S., & Marsh, S. (2024). Earth observations, digital footprints and machine-learning: greenhouse gas stocktaking for climate change mitigation. *International Journal of Population Data Science*. https://doi.org/10.23889/ijpds.v9i4.2423

Manjakkal, L., Mitra, S., Petillot, Y. R., Shutler, J., Scott, E. M., Willander, M., & Dahiya, R. (2021). Connected sensors, innovative sensor deployment, and intelligent data analysis for online water quality monitoring. *IEEE Internet of Things Journal*. https://doi.org/10.1109/JIOT.2021.3081772

Marsden, P., Koelsman, B., Bourdon-Lacombe, J., Gouin, T., D'Anglada, L., Cunliffe, D., Jarvis, P., Fawell, J., & De France, J. (2019). *Microplastics in drinking-water*.

Mase, A. S., Babin, N. L., Prokopy, L. S., & Genskow, K. D. (2015). Trust in sources of soil and water quality information: implications for environmental outreach and education. *Journal of the American Water Resources Association*. https://doi.org/10.1111/1752-1688.12349

Miller, T., Cembrowska-Lech, D., Kisiel, A., Kołodziejczak, M., Krzemińska, A., Jawor, M., Lewita, K., Kozlovska, P., & Mosiundz, S. (2023). Advancing water quality monitoring through artificial neural networks: present insights and future opportunities in scientific exploration. *Scientific Collection «InterConf+»*, *32*(151), 399–409. https://doi.org/10.51582/interconf.19-20.04.2023.041

Mishra, R. K., Mentha, S. S., Misra, Y., & Dwivedi, N. (2023). Emerging pollutants of severe environmental concern in water and wastewater: A comprehensive review on current developments and future research. *Water-Energy Nexus*, *6*, 74–95. https://doi.org/10.1016/J.WEN.2023.08.002

Mohan, S., Kumar, B., & Nejadhashemi, A. P. (2025). Integration of Machine Learning and Remote Sensing for Water Quality Monitoring and Prediction: A Review. *Sustainability (Switzerland)*, *17*(3). https://doi.org/10.3390/SU17030998

Moley, M. A. A. (2023). Applications of artificial intelligence in the field of climate and environmental changes. *International Journal of Artificial Intelligence and Emerging Technology*, 2(1), 59–77.

Munagala, N. S., Kumar, V., Srividhya, S., Usha, G., & Jeyalakshmi, R. (2024). Micro-Objects Classification for Microplastic Pollution Detection using Holographic Images, *2024 International Conference on Advances in Computing, Communication and Applied Informatics (ACCAI)*, (pp. 1–7), Chennai, India, doi: 10.1109/ACCAI61061.2024.10602278.

Mondal, A., & Dubey, S. S. (2024). Machine learning-based water potability prediction: Model evaluation, and hyperparameter optimization. In A. K. Tripathi & V. Shrivastava (Eds.), *Advancements in communication and systems* (pp. 37–54). SCRS. https://doi.org/10.56155/978-81-955020-7-3-4

Mustafa, S. A., Al-Rudainy, A. J., & Salman, N. M. (2024). Effect of environmental pollutants on fish health: An overview. *Egyptian Journal of Aquatic Research*, *50*(2), 225–233. https://doi.org/10.1016/J.EJAR.2024.02.006

Nand Jha, N., Singh, R. K., Sharma, S., & Kumar, A. (2024). Computational machine learning analytics for prediction of water quality. *Communications on Applied Nonlinear Analysis*, *31*(4), 448–465.

National Institute of Environmental Health Sciences. (2024). *Perfluoroalkyl and Polyfluoroalkyl Substances (PFAS)*.

Nguyen, D. T., Luong, T. L., Nguyen, N. M., Nguyen, T. Q., Le, N. T. N., Truong, C. D., Lieu, N. T., Van Lam, H. V., Dinh, Q. K., Huynh, D. C., Vo, V., & Nguyen, T. V. H. (2024). Monitoring water quality parameters in aquaculture using edge computing. *Vietnam Journal of Catalysis and Adsorption*. https://doi.org/10.62239/jca.2024.047

Nutipalli, P., Sai, G. C. V., Abhinaya, G., Manozna, K. C., Sai, L. H., & Setamraju, R. A. (2024). An IoT- based measure guage for surveil of water quality by parametric quantity. *International Journal for Research in Applied Science and Engineering Technology*, 12(4), 3768–3773.

Oduro, S., Moss, E., & Metcalf, J. (2022). Obligations to assess: recent trends in AI accountability regulations. *Patterns*. https://doi.org/10.1016/j.patter.2022.100608

Oğuz, A., & Ertuğrul, Ö. F. (2023). A survey on applications of machine learning algorithms in water quality assessment and water supply and management. *Water Supply*, 23(2), 895–922.

Olawade, D. B., Wada, O. Z., Ige, A. O., Egbewole, B. I., Olojo, A., & Oladapo, B. I. (2024). Artificial intelligence in environmental monitoring: advancements, challenges, and future directions. *Hygiene and Environmental Health Advances*, 12, 100114.

Omeka, M. E. (2024). Exploring the recent trends, progresses, and challenges in the application of Artificial intelligence in water quality assessment and monitoring in Nigeria: A systematic review. *Research Square*. https://doi.org/10.21203/rs.3.rs-4429965/v1

Pokkuluri, K. S., Devi, U. N., & Khang, A. (2024). Integration of machine learning augmented with biosensors for enhanced water quality monitoring. In A. Khang (Ed.), *Agriculture and Aquaculture Applications of Biosensors and Bioelectronics* (pp. 184–195). IGI Global.

Pokkuluri, K. S., Usha Devi, N., & Khang, A. (2024). Integration of machine learning augmented with biosensors for enhanced water quality monitoring. In: *Agriculture and Aquaculture Applications of Biosensors and Bioelectronics* (pp. 184–195).

Prevedouros K, Cousins IT, Buck RC, & Korzeniowski SH (2006). Sources,Fate and Transport of Perfluorocarboxylates. Environ Sci Technology;40(1):32-44. https://pubs.acs.org/doi/10.1021/es0512475

Puneeth, G., Shashank, T., Vinit, P., Muzammil, S., & Karthikeya, N. (2024). Smart water pollution management: IoT for automatic detection and prevention. *International Journal of Advanced Research in Science, Communication and Technology (IJARSCT)*, 4(15), 532–548.

Rachid, E.-B., Salhi, A., Hafid, A., & Souad, R. (2025). Predicting water potability using a machine learning approach. *Environmental Challenges*, *19*, 101131. https://doi.org/10.1016/j.envc.2025.101131

Ramos, E., Lopes, A. G., & Mendonça, F. (2024). Application of machine learning in plastic waste detection and classification: a systematic review. *Processes*. https://doi.org/10.3390/pr12081632

Ramos-Cruz, B., Andreu-Perez, J., & Martínez, L. (2024). The cybersecurity mesh: a comprehensive survey of involved artificial intelligence methods, cryptographic protocols and challenges for future research. *Neurocomputing*, *581*, 127427.

Rane, N., Choudhary, S., & Rane, J. (2024). Acceptance of artificial intelligence: key factors, challenges, and implementation strategies. *SSRN Electronic Journal*. https://doi.org/10.2139/ssrn.4842167

Rao, G., Capone, D., Zhu, K., Knoble, A., Linden, Y., Clark, R., Lai, A., Kim, J., Huang, C.-H., Bivins, A., & Brown, J. (2024). Simultaneous detection and quantification of multiple pathogen targets in wastewater. *PLoS Water*. https://doi.org/10.1371/journal.pwat.0000224

Rodríguez, R., Pastorini, M., Etcheverry, L., Chreties, C., Fossati, M., Castro, A., & Gorgoglione, A. (2021). Water-quality data imputation with a high percentage of missing values: a machine learning approach. *Sustainability*. https://doi.org/10.3390/su13116318

Rothstein, J. C., Cui, J., Yang, Y., Chen, X., & Zhao, Y. (2024). Ultra-sensitive detection of PFASs using surface enhanced Raman scattering and machine learning: a promising approach for environmental analysis. *Sensors and Diagnostics*. https://doi.org/10.1039/d4sd00052h

Rouet-Leduc, B., & Hulbert, C. (2024). Automatic detection of methane emissions in multispectral satellite imagery using a vision transformer. *Nature Communications*. https://doi.org/10.1038/s41467-024-47754-y

Royer, S. J., Wolter, H., Delorme, A. E., Lebreton, L., & Poirion, O. B. (2024). Computer vision segmentation model – deep learning for categorizing microplastic debris. *Frontiers in Environmental Science*. https://doi.org/10.3389/fenvs.2024.1386292

Russo, G. S., Eftim, S. E., Goldstone, A. E., Dufour, A. P., Nappier, S. P., & Wade, T. J. (2020). Evaluating Health Risks Associated with Exposure to Ambient Surface Waters During Recreational Activities: A Systematic Review and Meta-Analysis. *Water Research, 176*, 115729. https://doi.org/10.1016/J.WATRES.2020.115729

Safitra, M. F., Lubis, M., Kusumasari, T. F., & Putri, D. P. (2024). Advancements in artificial intelligence and data science: models, applications, and challenges. *Procedia Computer Science*. https://doi.org/10.1016/j.procs.2024.03.018

Sahin, F., Celik, N., Camdal, A., Sakir, M., Ceylan, A., Ruzi, M., & Onses, M. S. (2022). Machine learning-assisted pesticide detection on a flexible surface-enhanced Raman scattering substrate prepared by silver nanoparticles. *ACS Applied Nano Materials*. https://doi.org/10.1021/acsanm.2c02897

Santos, M. R. C., & Carvalho, L. C. (2025). AI-driven participatory environmental management: Innovations, applications, and future prospects. *Journal of Environmental Management, 373*, 123864. https://doi.org/10.1016/j.jenvman.2024.123864

Sellamuthu, K., & Kaliappan, V. K. (2023). Q-learning-based pesticide contamination prediction in vegetables and fruits. *Computer Systems Science and Engineering*. https://doi.org/10.32604/csse.2023.029017

Shabani, F., Philamore, H., & Matsuno, F. (2021). An energy-autonomous chemical oxygen demand sensor using a microbial fuel cell and embedded machine learning. *IEEE Access*. https://doi.org/10.1109/ACCESS.2021.3101496

Sharma, S., & Bhattacharya, A. (2016). Drinking water contamination and treatment techniques. *Applied Water Science, 7*(3), 1043–1067. https://doi.org/10.1007/S13201-016-0455-7

Sharma, H., Sharma, G., & Kumar, N. (2024). AI-assisted secure data transmission techniques for next-generation HetNets: a review. *Computer Communications*. https://doi.org/10.1016/j.comcom.2023.12.015

Shashank, Y., Amith, S., Sahana, S., & Sastry, A. S. (2024). Monitoring of water quality using machine learning – a review. *International Journal for Research in Applied Science and Engineering Technology*. https://doi.org/10.22214/ijraset.2024.61578

Singh, R., Ryu, J., Hyoung Lee, W., Kang, J. H., Park, S., & Kim, K. (2024). Wastewater-born eviruses and bacteria, surveillance and biosensors at the interface of academia and field deployment. *Critical Reviews in Biotechnology*. https://doi.org/10.1080/07388551.2024.2354709

Singh, S., Halder, A., Sinha, O., Sarkar, P. K., Singh, P., Banerjee, A., Ahmed, S. A., Alharbi, A., Obaid, R. J., Ghosh, S. K., Mitra, A., & Pal, S. K. (2020). Nanoparticle-based "turn-on" scattering and post-sample fluorescence for ultrasensitive detection of water pollution in wider window. *PLoS One*. https://doi.org/10.1371/journal.pone.0227584

Singh, Y., & Walingo, T. (2024). Smart water quality monitoring with IoT wireless sensor networks. *Sensors, 24*(9). https://doi.org/10.3390/s24092871

Srivastava, S., & Sharma, V. (2021). Ultra-portable, smartphone-based spectrometer for heavy metal concentration measurement in drinking water samples. *Applied Water Science*. https://doi.org/10.1007/s13201-021-01519-w

Suma, S., Moon, R., Umer, M., Raju, K. S., Bhaskar, N., & Okali, R. (2023). A prediction of water quality analysis using machine learning. *International Conference on Distributed Computing and Electrical Circuits and Electronics (ICDCECE),* (pp. 1–6), Ballar, India, doi: 10.1109/ICDCECE57866.2023.10150940.

Tang, T., Liu, K., Li, C., Xiao, J., Liang, X., Luo, L., Li, J., Li, J., Tang, Y., & He, K. (2022). Multi-channel optical sensing system with a BP-ANN for heavy metal detection. *Optics Express*. https://doi.org/10.1364/oe.451817

Tharayil, S. M., Alomari, N. K., & Bubshait, D. K. (2024). Detecting anomalies in water quality monitoring using deep learning. In: *SPE Water Lifecycle Management Conference and Exhibition,* Abu Dhabi, UAE, March 2024.

Tushar, M.M.R., Pushan, Z.A., Aich, N. *et al.* Balancing sustainability goals and treatment efficacy for PFAS removal from water. *npj Clean Water* **7**, 130 (2024). https://doi.org/10.1038/s41545-024-00427-1

Ucun Ozel, H., Gemici, B. T., Gemici, E., Ozel, H. B., Cetin, M., & Sevik, H. (2020). Application of artificial neural networks to predict the heavy metal contamination in the Bartin River. *Environmental Science and Pollution Research*. https://doi.org/10.1007/s11356-020-10156-w

US Environmental Protection Agency. (2024). *Types of Drinking Water Contaminants*.

Vaivude, P. (2023). Artificial intelligence for water quality. *International Journal for Research in Applied Science and Engineering Technology*, 11(12), 1521–1533. JRASET Research and Development Corp., India. https://doi.org/10.22214/ijraset.2023.57670

Vald, G., Sermet, Y., Mount, J., Shrestha, S., Jackson Samuel, D., Cwiertny, D. M., & Demir, I. (2024). *Integrating Conversational AI Agents for Enhanced Water Quality Analytics: Development of a Novel Data Expert System*.

Wang, B., Chen, D., Weng, X., & Chang, Z. (2024). Development an electronic nose to recognize pesticides in groundwater. *Talanta*. https://doi.org/10.1016/j.talanta.2023.125506

WHO. (2023). *Drinking-Water*.

Wilkins, E. J., Miller, H. M., Tilak, E., & Schuster, R. M. (2018). Communicating information on nature-related topics: preferred information channels and trust in sources. *PLoS One*. https://doi.org/10.1371/journal.pone.0209013

William, P., Oyebode, O. J., Ramu, G., Gupta, M., Alzubaidi, L. H., & Srivastava, A. P. (2023). Architecture for AI-based validation of wastewater management using open data exchange technique. In: *International Conference on Circuit Power and Computing Technologies (ICCPCT)*, (pp. 1502–1507), Kollam, India. doi: 10.1109/ICCPCT58313.2023.10244857

Wilson, A. (2014). Identification of insecticide residues with a conducting-polymer electronic nose. *Chemical Sensors*, *4*(3), 5803.

Yang, X. E., Wu, X., Hao, H. L., & He, Z. L. (2008). Mechanisms and assessment of water eutrophication. *Journal of Zhejiang University. Science. B*, *9*(3), 197. https://doi.org/10.1631/JZUS.B0710626

Yang, L., Driscol, J., Sarigai, S., Wu, Q., Lippitt, C. D., & Morgan, M. (2022). Towards synoptic water monitoring systems: a review of AI methods for automating water body detection and water quality monitoring using remote sensing. *Sensors*, *22*(6), 2416.

Zainurin, S. N., Wan Ismail, W. Z., Mahamud, S. N. I., Ismail, I., Jamaludin, J., Ariffin, K. N. Z., & Wan Ahmad Kamil, W. M. (2022). Advancements in monitoring water quality based on various sensing methods: a systematic review. *International Journal of Environmental Research and Public Health,* 19(21). https://doi.org/10.3390/ijerph192114080

Zanfei, A., Menapace, A., & Righetti, M. (2023). An artificial intelligence approach for managing water demand in water supply systems. *IOP Conference Series: Earth and Environmental Science*, *1136*(1), 012004.

Zavala-Ariza, D., Hamburg, S. P., Calcan, A., Schwietzke, S., France, J. L., Randles, C., Baranski, M. R., Demeter, M., Kupers, R., Field, R., & Caltagirone, M. (2024). Transparent horizons: IMEO's methane data empowering global climate action. In: European Geosciences Union General Assembly 2024 (EGU24), held 14-19 April, 2024 in Vienna, Austria. Online at https://egu23.eu, id.20055.

Zhang, P., Yang, M., Lan, J., Huang, Y., Zhang, J., Huang, S., Yang, Y., & Ru, J. (2023). Water quality degradation due to heavy metal contamination: Health impacts and eco-friendly approaches for heavy metal remediation. *Toxics*, *11*(10), 828. https://doi.org/10.3390/TOXICS11100828

Zhang, Y., Li, J., Zhou, Y., Zhang, X., & Liu, X. (2024). Artificial intelligence-based microfluidic platform for detecting contaminants in water: a review. *Sensors*, *24*(13), 4350.

Zhao, W., & Whitfeld, C. (2024). An integrated framework for the detection and filtration of perfluoroalkyl substances from surface water in the Thames Basin. *ChemRxiv*.

Zhao, Y., Li, Z., & Lv, S. (2024). Enhancing AI system privacy: an automatic tool for achieving GDPR compliance in NoSQL databases. *Computers, Materials and Continua*. https://doi.org/10.32604/cmc.2024.052310

Zheng, Y., Wei, J., Zhang, W., Zhang, Y., Zhang, T., & Zhou, Y. (2024). An ensemble model for accurate prediction of key water quality parameters in river based on deep learning methods. *Journal of Environmental Management*, 366, 121932.

Ziani, K., Ioniță-Mîndrican, C. B., Mititelu, M., Neacşu, S. M., Negrei, C., Moroşan, E., Drăgănescu, D., & Preda, O. T. (2023). Microplastics: a real global threat for environment and food safety: a state of the art review. *Nutrients*. https://doi.org/10.3390/nu15030617

5 Artificial Intelligence in Water Resource Management

Durgam Venkatesh, Gaini Sairam, Gandi Rupa, Chidigae Sudha Rani, Chourat Sudhakar, and Talakanti Rajeshwar Reddy

5.1 INTRODUCTION

Artificial intelligence (AI) represents the fundamental and renowned sector of computer science dedicated to crafting intelligent systems and resolving challenges in a manner akin to human intellect. AI is swiftly revolutionizing numerous industries, and water resource management is no exception to this transformative wave. With global water requirements perpetually climbing amidst mounting issues like scarcity, pollution, and climate change, the demand for ground-breaking solutions becomes increasingly paramount (Hajirad, 2025). The primary objective of AI implementations is to amplify computing capabilities that resonate with human comprehension, including learning, problem-solving, reasoning, and perception (Danish, 2022). AI technologies furnish robust instruments to bolster the effectiveness and sustainability of water management methodologies, enabling stakeholders to make enlightened choices grounded in real-time data and predictive analytics (Krishnan et al., 2022). Water resource management entails methodically strategizing, developing, distributing, and overseeing water resources to fulfil human and ecological necessities. However, this domain confronts considerable obstacles as water scarcity impacts billions globally, worsened by population growth, urban sprawl, and inefficient consumption. Contamination from industrial and agricultural processes further muddles access to pristine water, while climate change ushers in erratic weather patterns that jeopardize both supply and quality. AI serves an essential function in addressing these difficulties by leveraging several technologies like machine learning, neural networks, and deep learning (Mortazavizadeh et al., 2025). These instruments can scrutinize enormous datasets to uncover trends, optimize resource distribution, and anticipate future water availability and demand. By weaving AI into water management frameworks, organizations can elevate operational efficiencies, refine decision-making procedures, and advocate for sustainable practices. The urgent nature of global water predicaments accentuates the necessity for embracing AI within this realm. The prospective advantages encompass improved surveillance of water quality, enhanced anticipation of demand fluctuations, and superior risk

DOI: 10.1201/9781003547099-6

management during extreme climatic events. As we navigate an increasingly complex landscape of water resource challenges, AI stands out as a critical ally in ensuring a sustainable future for this essential resource. Integrating artificial intelligence into water resource management is beneficial and essential for addressing the pressing challenges of scarcity, pollution, and climate change. By leveraging advanced AI technologies, stakeholders can enhance their capacity to manage this vital resource sustainably and efficiently. The necessity and effectiveness of AI in addressing global water crises and supporting sustainable resource management.

5.2 CURRENT CHALLENGES IN WATER RESOURCE MANAGEMENT

Water resource management faces significant global challenges, particularly in water scarcity, climate change, water quality, and efficiency in water use. Each of these factors has profound economic and social implications that affect communities worldwide.

5.2.1 Current Challenges in Water Resource Management

5.2.1.1 Global Water Scarcity Issues

Water scarcity is a pressing global concern, impacting over 2 billion people without safe drinking water access. The economic implications are staggering, with an estimated $260 billion lost annually due to inadequate water and sanitation facilities. This scarcity affects agricultural productivity, reducing crop yields and increasing food prices, disproportionately impacting vulnerable populations. In many regions, especially in developing countries, women and children spend significant time collecting water, which limits their educational and economic opportunities (Israilova et al., 2023). Furthermore, the competition for limited water resources can escalate tensions and conflicts among communities and sectors.

5.2.2 Climate Change Impact

Climate change exacerbates water scarcity by altering precipitation patterns and increasing the frequency of extreme weather events such as floods and droughts. These changes disrupt traditional water supply systems and threaten food security by affecting irrigation practices. For instance, diminished terrestrial water storage in soil and snow leads to unpredictable water availability. This unpredictability further complicates efforts to manage water resources effectively and sustainably. The impacts are particularly severe in regions already facing socio-economic challenges, where shifting climate patterns can lead to increased migration and social instability.

5.2.3 Water Quality

Water pollution remains a critical issue that complicates water resource management. Contaminants from industrial runoff, agricultural practices, and inadequate sanitation facilities lead to significant health risks for communities reliant on polluted water sources. The lack of effective monitoring systems hampers efforts to ensure safe drinking water. Contaminated water affects public health and has economic repercussions, as it can lead to increased healthcare costs and lost productivity due to illness. Ensuring clean water access requires robust monitoring frameworks to detect pollution levels proactively.

5.2.4 Efficiency in Water Use

Inefficient agricultural practices are a major contributor to water wastage. Agriculture accounts for approximately 72% of global freshwater withdrawals, yet many farming techniques remain outdated or unsustainable. Urban areas also face significant challenges related to water wastage due to aging infrastructure and poor management practices. For instance, leaks in distribution systems can result in substantial losses of treated water before it even reaches consumers (Dolan et al., 2021). Addressing these inefficiencies is crucial for conserving water resources and ensuring that available supplies meet growing demands.

The challenges facing water resource management are multifaceted and interconnected. Addressing global water scarcity requires comprehensive strategies considering the economic and social implications of inadequate access to clean water. Climate change further complicates these issues, necessitating adaptive management practices that can respond to shifting environmental conditions. Improved monitoring systems for water quality and more efficient use of resources in agriculture and urban settings are essential to a sustainable approach to managing this vital resource. Without coordinated action, the negative impacts of these challenges will only intensify, reinforcing cycles of poverty and inequality while threatening global stability. The comparison of traditional and AI-based water resource management is given in Table 5.1. Artificial Intelligence (AI) is increasingly recognized as a transformative force in water resource management, offering innovative solutions to complex challenges. This overview will delve into the fundamentals of AI, its application in environmental management, specifically, its role in water resource management.

5.3 AI FUNDAMENTALS IN WATER RESOURCE MANAGEMENT

AI encompasses a range of technologies designed to simulate human intelligence processes and the classification of AI techniques was given in Figure 5.1. The key concepts include:

Machine Learning (ML): A subset of AI that enables systems to learn from data and improve over time without being explicitly programmed.

Deep Learning: A specialized form of ML that utilizes neural networks with many layers to analyse various factors in large datasets.

Neural Networks: Computational models inspired by the human brain, used for pattern recognition and classification tasks.

Genetic Algorithms: Optimization techniques based on the principles of natural selection, useful for solving complex problems by evolving solutions over generations.

TABLE 5.1
Comparison of Traditional vs AI-Based Approaches in Water Resource Management

Aspect	Traditional Water Resource Management	AI-Based Water Resource Management
Data analysis	It relies on manual data collection and analysis, often leading to delays and inefficiencies	Utilizes real-time data analytics to monitor and manage water resources dynamically
Predictive capabilities	Limited predictive capabilities, often reactive rather than proactive	Employs predictive analytics to forecast demand, detect leaks, and optimize supply
Resource allocation	Based on historical data and fixed schedules, this can lead to overuse or shortages	Adapts resource allocation dynamically based on real-time consumption patterns and forecasts
Leak detection	Manual inspections and reports can miss small leaks, leading to significant water loss	Real-time monitoring with sensors and machine learning algorithms enables quick detection and response to leaks
Water quality monitoring	Involves periodic sampling and laboratory testing, which can delay responses to contamination	Continuous water quality monitoring through sensors that provide immediate alerts for contamination events
Infrastructure maintenance	Reactive maintenance is based on failures or scheduled checks, often leading to downtime	Predictive maintenance using AI analyses operational data to anticipate failures before they occur
Energy efficiency	Often, it does not optimize energy use, leading to higher operational costs	AI optimizes energy consumption by adjusting operations based on demand forecasts and energy costs
User engagement	Limited interaction with consumers regarding water usage; education is often one-way	Smart metres and apps provide consumers with real-time data usage, promoting conservation and informed choices
Environmental impact	May not effectively monitor or mitigate environmental impacts due to lack of real-time data	AI systems continuously monitor environmental conditions, helping protect ecosystems and reduce emissions
Cost efficiency	Often, it incurs higher operational costs due to inefficiencies and reactive measures	Reduces costs through optimized operations, predictive maintenance, and efficient resource allocation

These innovative methods can handle immense datasets, uncover trends, and forecast outcomes, which are vital for informed decision-making in water resource stewardship. The integration of AI in managing water resources not only tackles urgent issues but also fosters enduring sustainability. As AI technologies evolve, their role in optimizing water use and ensuring environmental protection will likely expand further.

5.4 AI TECHNIQUES RELEVANT TO WATER MANAGEMENT

AI is increasingly utilized in water management to address challenges such as water scarcity, inefficiencies, and environmental concerns. Comparison of AI Platforms and Tools for Water Resource Management is given in Table 5.2. Machine learning (ML) is a powerful tool in water management that can be categorized into supervised and unsupervised learning models:

Supervised Learning: This technique involves training algorithms on labelled datasets to predict outcomes based on input features. For instance, predictive models using historical data can forecast future water demand, enabling

TABLE 5.2
Comparison of AI Platforms and Tools for Water Resource Management

AI Platform/Tool	Key Features	Applications in Water Resource Management	Examples of Use
IBM water management solutions	Predictive analytics, real-time monitoring	Water quality monitoring, demand forecasting	Used by cities for efficient water resource management
Xylem smart water solutions	IoT integration, data analytics	Water and wastewater management	Implemented in various municipalities for operational efficiency
Microsoft AI for Earth	Support for environmental projects, machine learning	Water resource optimization, ecosystem monitoring	Grants and partnerships with organizations managing water resources
Google flood forecasting	Machine learning models for weather prediction	Flood prediction and management	Deployed in regions prone to flooding
Aqueduct	AI-driven optimization of distribution networks	Leak detection, flow monitoring	Used by utilities to enhance water distribution efficiency
Siemens digital water solutions	Integration of AI and IoT for real-time data analysis	Infrastructure maintenance, water quality monitoring	Applied in urban water systems to improve operational efficiency
Schneider electric eco structure	AI for energy management and resource optimization	Wastewater treatment, energy efficiency	Utilized in industrial settings to optimize water management
Ayyeka remote monitoring solutions	Real-time data collection and analysis	Water quality monitoring, flow measurement	Implemented by utilities for effective resource management
Veolia smart water management	AI-driven treatment processes, sustainability focus	Efficient water treatment, resource conservation	Used globally to enhance the sustainability of water systems

better resource allocation and planning. Research has shown that ML algorithms can significantly enhance the accuracy of demand predictions by analysing historical consumption patterns alongside variables like weather conditions and population growth (Ashoka et al., 2024).

Unsupervised Learning: This approach identifies patterns within data without predefined labels. It can cluster similar consumption behaviours or detect anomalies in water quality data. Such insights help utilities optimize operations and respond proactively to potential issues (Khatri et al., 2023).

Neural Networks: Neural networks are particularly adept at modelling complex non-linear relationships within large datasets.

Application in Water Management: These models can analyse multiple inputs simultaneously, such as flow rates and chemical concentrations, to provide accurate predictions about water quality and system performance. A study demonstrated that artificial neural networks (ANNs) effectively predict water quality indices by capturing the non-linear structures of complex time series data, leading to more reliable assessments of water quality (Ma et al., 2023).

5.4.1 Optimization Algorithms Are Essential for Improving Resource Allocation in Water Management Systems

Resource Allocation: Techniques such as genetic algorithms and particle swarm optimization help manage competing demands for water resources. These algorithms evaluate multiple objectives, minimizing costs while maximizing environmental benefits to derive optimal solutions for water distribution networks. For example, AI-driven optimization can significantly reduce unaccounted-for-water losses by strategically placing sensors and monitoring points within the infrastructure (Kikon et al., 2023).

Real-time Data Collection: IoT sensors continuously monitor flow rates, pressure levels, and water quality metrics. This real-time data collection allows utilities to detect leaks promptly and manage resources more effectively (Ashoka et al., 2024)

Big Data Analytics: The vast amounts of data generated by IoT devices necessitate advanced analytics techniques to extract actionable insights. AI algorithms analyse this data to identify trends, forecast demands, and optimize operations effectively. For instance, predictive analytics can inform utilities about potential future shortages or quality degradation based on historical patterns (Khatri et al., 2023). AI techniques such as machine learning, neural networks, optimization algorithms, and the integration of IoT with big data analytics are instrumental in enhancing the efficiency and sustainability of water management systems. These technologies improve operational efficiencies and significantly contribute to addressing global challenges related to water scarcity and environmental sustainability.

5.5 HOW AI IS TRANSFORMING WATER RESOURCE MANAGEMENT?

AI is significantly reshaping water resource management by enhancing efficiency, optimizing processes, and enabling real-time decision-making. This transformation is evident in various applications, including predictive analytics for water demand, leak detection, water quality monitoring, and flood management. Below is an overview of how AI is influencing this critical sector, supported by findings from recent research.

5.5.1 Predictive Analytics for Water Demand

AI algorithms analyse historical data, weather patterns, and usage trends to predict future water demand accurately. This capability allows water authorities to allocate resources more effectively and prevent shortages. For instance, predictive models can process vast datasets to forecast consumption based on population growth and seasonal variations (Signity Solutions, 2024). By optimizing distribution based on these predictions, utilities can minimize waste while ensuring a reliable supply to communities (Express Computer, 2024).

5.5.2 Leak Detection and Infrastructure Optimization

AI technologies are instrumental in identifying leaks within water distribution networks. ML algorithms analyse data from smart sensors to detect anomalies that indicate potential failures. This proactive approach allows us to address issues before they escalate into significant losses. Research shows that implementing AI-driven leak detection systems can substantially reduce water wastage and operational costs (Signity Solutions, 2024). For example, cities employing these technologies have reported improved efficiency in their water management systems (Express Computer, 2024).

5.5.3 Water Quality Monitoring

AI enhances water quality monitoring by enabling real-time analysis of contaminants through IoT-enabled sensors. Traditional methods often involve periodic sampling, which can delay responses to contamination events. In contrast, AI systems continuously process data from sensors monitoring parameters like pH levels and turbidity. When irregularities are detected, AI algorithms trigger immediate alerts for intervention (Express Computer, 2024). This capability is crucial for safeguarding public health and ensuring compliance with regulatory standards. A study highlighted the effectiveness of ML models in predicting algae blooms and detecting specific pollutants, such as heavy metals, in water bodies. These advancements allow for timely remediation actions and better management of water resources.

5.5.4 Flood Management and Disaster Preparedness

AI's role extends to flood management as well. Accurate forecasting is essential with climate change increasing the frequency of extreme weather events. AI algorithms can analyse vast datasets from weather forecasts and historical flood records to predict flood events accurately. This predictive capability aids governments and emergency response teams in preparing for disasters by allocating resources effectively and planning evacuations (Express Computer, 2024). AI fundamentally transforms water resource management by providing innovative solutions that enhance efficiency, improve monitoring capabilities, and optimize resource allocation. As the challenges related to water scarcity and environmental sustainability continue to grow, the integration of AI technologies will be crucial for creating resilient water management systems that ensure safe and sustainable access to this vital resource.

Infrastructure Monitoring: AI algorithms continuously analyse data from sensors installed in water supply systems to predict maintenance needs. By identifying patterns that precede equipment failures, utilities can schedule maintenance proactively, thus extending the lifespan of critical infrastructure while minimizing downtime (Signity Solutions, 2024). The integration of AI into water resource management is revolutionizing how utilities operate. Through predictive modelling, optimization of distribution networks, and enhanced monitoring capabilities, AI technologies enable more efficient use of water resources while addressing challenges related to scarcity and sustainability. As these technologies evolve, their impact on the sector will grow significantly.

5.6 AI APPLICATIONS IN WATER RESOURCE MANAGEMENT

AI transforms water resource management, particularly water quality monitoring. AI applications in different fields and their benefits, with examples, are briefly given in Table 5.3. This AI in water resource management involves real-time analysis, predicting contamination events, and suggesting treatment measures. Below is an overview of how AI contributes to these aspects, supported by findings from recent research, and different AI applications in water resource management are given in Figure 5.1.

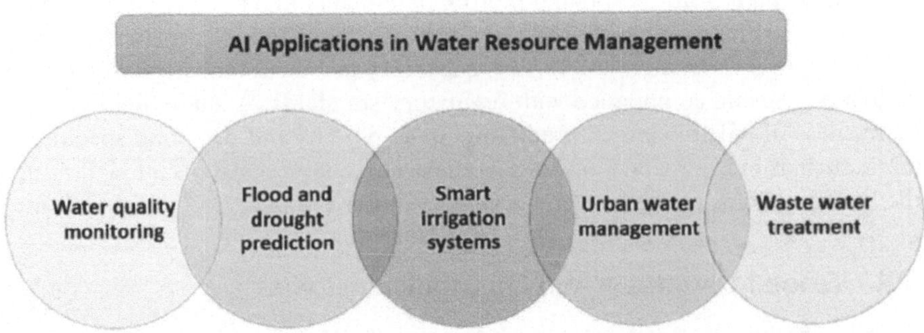

FIGURE 5.1 AI applications in water resource management.

TABLE 5.3
AI Applications in Different Areas of Water Resource Management

Application Area	AI Techniques Used	Key Benefits	Examples/Locations
Water quality monitoring	Machine learning, sensor data analysis	Continuous monitoring and immediate detection of contaminants	Urban water systems globally, e.g., Singapore
Water demand management	Predictive analytics, historical data analysis	Accurate forecasting, optimized supply distribution	Various urban areas, e.g., Los Angeles
Leak detection	Anomaly detection algorithms, real-time data analysis	Early identification of leaks, reduced water loss	Cities like Barcelona and Tucson
Groundwater management	Predictive modelling, data analytics	Informed extraction decisions, prevention of depletion	Arid regions in India and the U.S. Southwest
Irrigation optimization	IoT integration, AI algorithms	Precision irrigation, reduced water waste	Agricultural regions in California and Israel
Infrastructure maintenance	Predictive maintenance models	Reduced downtime, proactive maintenance	Water utilities in Europe and North America
Flood prediction and management	Weather data analysis, machine learning	Enhanced preparedness for flooding events	Coastal cities like New Orleans and Miami
Energy efficiency optimization	Real-time energy consumption analysis	Reduced operational costs and environmental footprint	Water treatment facilities worldwide

AI Applications in Water Quality Monitoring: AI technologies enhance water quality monitoring through real-time data analysis and predictive capability.

Real-Time Analysis: Traditional water quality monitoring methods rely on periodic sampling and laboratory analysis, which can delay responses to contamination events. AI systems utilize IoT-enabled sensors that continuously monitor water quality parameters such as pH, turbidity, and chemical concentrations. These sensors provide a constant stream of real-time data that AI algorithms analyse to detect anomalies or contaminant levels. For instance, a study highlighted the development of a digital system that processes data from sensors installed in a city's water distribution infrastructure, generating alerts when water quality metrics exceed safe thresholds (Wu et al., 2024).

Predicting Contamination Events: AI models can predict potential contamination events by analysing historical data alongside real-time sensor readings. ML algorithms have been shown to effectively forecast water quality indices (WQI) using parameters like dissolved oxygen and turbidity. A study demonstrated that advanced AI algorithms, including nonlinear autoregressive neural networks (NARNET) and long short-term memory (LSTM) models, achieved high accuracy in predicting WQI, thus enabling proactive measures to mitigate contamination risks (Rankovic et al., 2024).

Suggesting Treatment Measures: AI systems can recommend appropriate treatment measures based on the identified contaminants once a contamination event is detected or predicted. This capability allows for rapid response actions to safeguard public health. For example, AI-driven systems can analyse data from multiple sources to determine the most effective treatment protocols for various pollutants encountered in water bodies (Gazzaz et al., 2023). Integrating AI in water quality monitoring enhances the efficiency and effectiveness of detecting and responding to contamination events and supports informed decision-making regarding treatment measures. As these technologies continue to evolve, they hold the potential to improve public health outcomes and environmental sustainability significantly.

5.6.1 AI Applications in Flood and Drought Prediction

AI models have shown significant promise in improving the forecasting of extreme weather conditions, particularly floods and droughts. Traditional forecasting methods often struggle, especially in ungauged areas where data is sparse. A study published in *Nature* demonstrated that an AI model trained on 5,680 existing stream gauges could predict daily streamflow in ungauged watersheds with a lead time of up to five days. This model outperformed existing systems like the Global Flood Awareness System (GloFAS) in both short-term and long-term flood predictions, providing critical early warnings for often overlooked regions (Nearing et al., 2024). In another study on drought prediction, researchers developed a new meteorological drought index using various AI models, including decision trees and artificial neural networks (ANNs). The findings indicated that these AI-based indices significantly outperformed conventional drought indices, achieving higher correlations with key drought indicators such as soil moisture levels. This advancement underscores the potential of AI to enhance drought monitoring and forecasting capabilities (Zhang 2024).

5.6.1.1 AI-Driven Early Warning Systems for Flood and Drought Preparedness

AI-driven early warning systems are essential for effective flood and drought preparedness. These systems utilize real-time data from diverse sources, weather forecasts, satellite imagery, and historical records to alert at-risk communities promptly. For instance, the United Nations World Food Programme is collaborating with researchers from the University of Oxford to develop an AI weather forecasting system specifically designed for regions like East Africa. This hybrid model combines ML with physical weather models to improve predictions of extreme weather events. The system can produce forecasts that allow communities to receive warnings up to 48 hours in advance, enabling proactive measures such as evacuations or resource allocation.

Additionally, AI applications in flood management include advanced modelling techniques that simulate various flood scenarios based on real-time hydrological data.

These models help authorities assess potential impacts and develop strategic responses. For example, systems like aqua FLOOD leverage real-time data to track flood progression and inform decision-making processes regarding emergency responses (Vassar Labs, 2024). AI technologies are revolutionizing flood and drought prediction by enhancing forecasting accuracy and enabling effective early warning systems. These advancements not only improve community preparedness but also contribute to more resilient water resource management practices.

5.6.2 AI Applications in Smart Irrigation Systems

AI significantly enhances agricultural water efficiency through smart irrigation systems. These systems leverage AI technologies to optimize irrigation practices, predict crop water requirements, and improve weather forecasting, ultimately leading to more sustainable farming practices. Below is an overview of these applications, supported by findings from recent research.

5.6.2.1 Role of AI in Improving Agricultural Water Efficiency

Precision Irrigation: AI-fuelled intelligent irrigation systems harness real-time insights from diverse sources, such as soil moisture detectors, meteorological stations, and satellite imagery, to refine water application techniques. For instance, research has shown that embedding AI into irrigation frameworks can trim water consumption by as much as 25% while simultaneously boosting crop production. These systems automate irrigation schedules based on precise crop needs and environmental conditions, ensuring water is applied only when necessary (Ashoka et al., 2024).

Crop Water Requirement Predictions: AI algorithms analyse historical weather data and current environmental conditions to accurately predict the water requirements of different crops. A systematic approach combining data collection and AI-enabled models has been employed to enhance water usage efficiency in legume agriculture. This method addresses water shortages and promotes sustainable farming practices by ensuring that crops receive the optimal amount of water for growth (Vassar Labs, 2024).

5.6.2.2 Weather Forecasting

AI technologies also play a critical role in improving weather forecasting for agricultural purposes. ML models can provide more accurate rainfall and temperature fluctuation predictions by analysing vast datasets from meteorological stations and satellite observations. This capability allows farmers to adjust their irrigation schedules proactively, thereby conserving water and reducing the risk of crop stress due to drought or excessive moisture (Signity Solutions, 2024). For instance, a recent study demonstrated that ML models could effectively predict irrigation water requirements based on limited meteorological variables. The findings revealed that these models could significantly enhance water use efficiency (WUE) in arid regions by providing timely insights into optimal irrigation practices (Alharbi et al., 2023). Integrating AI into smart irrigation systems represents a transformative approach to managing agricultural water resources. By enabling precision irrigation, predicting crop water requirements accurately, and improving weather forecasting, AI technologies significantly enhance agricultural productivity while promoting sustainability. As the demand for efficient water use grows amid climate change challenges, adopting AI-driven solutions will be crucial for the future of agriculture.

5.6.3 AI Applications in Urban Water Management

AI is increasingly utilized in urban water management to optimize water distribution networks and predict water demand. These applications enhance resource allocation, reduce wastage, and improve overall efficiency in water supply systems. Below is an overview of these applications, supported by findings from recent research.

5.6.3.1 Smart Water Distribution Networks

AI technologies are crucial for optimizing water supply systems and minimizing wastage. Using ML algorithms, utilities can analyse data from sensors embedded in the water distribution network to monitor flow rates, pressure levels, and usage patterns. This real-time analysis enables the detection of leaks and inefficiencies within the system.

Leak Detection and Prevention: AI algorithms can identify anomalies in water flow that may indicate leaks. For instance, a study highlighted that ML models could predict potential failures in water distribution systems by analysing historical failure data alongside environmental conditions. This proactive approach allows utilities to address issues before they escalate, significantly reducing water loss and operational costs (Environmental Finance Center Network, 2024).

Optimizing Operations: AI can optimize the operation of pumps and valves within the distribution network to ensure efficient water delivery while minimizing energy consumption. By analysing real-time data, AI systems can adjust flow rates dynamically based on demand fluctuations and system conditions (ADB, 2024).

5.6.3.2 Predicting Water Demand in Cities

Accurate predictions of urban water demand are essential for effective resource allocation and management. AI models utilize historical data, weather patterns, population growth trends, and other variables to forecast future water needs.

Water Demand Forecasting: AI-Forecast has been developed to provide short-term water demand predictions using state-of-the-art ML models. This tool allows water utilities to manage resources more effectively by providing reliable forecasts that help prevent shortages or over-supply situations (Bakker et al., 2022). The system incorporates data pre-processing, outlier detection, and model tuning to enhance prediction accuracy.

Hybrid Models for Improved Accuracy: Research indicates that hybrid models combining multiple AI techniques such as ANN, support vector machines (SVM), and fuzzy logic can yield more accurate demand forecasts compared to traditional methods. A recent review found that dynamic neural networks consistently outperformed conventional models in predicting urban water demand across different time scales (Muhammad et al., 2024). Integrating AI into urban water management systems significantly enhances the efficiency of water distribution networks and improves the accuracy of water demand predictions. By optimizing operations and minimizing wastage through smart technologies, cities can better manage their water resources amidst increasing demands and environmental challenges.

5.6.4 AI Applications in Wastewater Treatment

AI transforms wastewater processing by fine-tuning treatment methodologies, anticipating maintenance requirements, and curbing energy use. These innovations not only bolster operational efficacy but also advance environmental sustainability.

5.6.4.1 Optimizing Wastewater Treatment Processes

AI technologies are being implemented to optimize various aspects of wastewater treatment processes. ML algorithms analyse real-time data from sensors throughout treatment plants to adjust operational parameters dynamically, leading to improved treatment efficiency and effluent quality.

Dynamic Process Optimization: For example, the Marselisborg Wastewater Treatment Plant in Aarhus, Denmark, employs an AI-driven dynamic process optimization system that continuously analyses data from multiple sensors. This system adjusts treatment processes based on changes in influent characteristics and environmental conditions, resulting in enhanced nutrient removal and reduced energy consumption (Diva Envitec, 2024).

Energy Efficiency: AI can significantly reduce energy use in wastewater treatment plants. Research indicates that AI-driven optimizations can lead to a 30%–40% reduction in aeration energy consumption. By dynamically adjusting the operation of blowers and other energy-intensive equipment based on real-time data, facilities can achieve substantial cost savings while maintaining compliance with water quality standards (Environmental Finance Centre Network, 2024).

5.6.4.2 Predicting Maintenance Needs

AI also plays a crucial role in predictive maintenance for wastewater treatment facilities. By analysing historical data and real-time operational metrics, AI algorithms can identify potential equipment failures before they occur.

Predictive Analytics: A study demonstrated that predictive analytics could detect anomalies and predict maintenance needs in wastewater networks. For instance, the city of Tucson, Arizona, implemented ML technology to evaluate data on soil, weather, and land use to develop targeted predictions for pipe failures. This proactive approach allows utilities to focus resources on critical assets and reduce downtime.

Resource Recovery Optimization: AI can optimize resource recovery processes such as anaerobic digestion and nutrient removal. By analysing data on organic matter content and nutrient levels, AI systems can enhance the timing and dosing of chemicals used in treatment processes, thereby improving overall efficiency (Chawishborwornworng et al., 2023).

5.6.4.3 Reducing Energy Use

The integration of AI into wastewater treatment processes not only optimizes operations but also contributes to significant reductions in energy consumption.

Energy Consumption Reduction: Implementing AI solutions has been shown to result in energy savings ranging from 10% to 30% across various facilities. For example, the HARVI system used in Calgary optimized pump schedules to minimize operational costs while ensuring compliance with regulatory requirements (Global Infrastructure Hub, 2024).

Chemical Usage Optimization: AI applications have also led to reductions in chemical usage by optimizing dosing profiles based on live data. Facilities have reported savings of up to 30% in chemical usage through improved modelling techniques that maintain compliance while minimizing excess chemical application (Jacobs, 2024). AI applications in wastewater treatment are transforming how facilities operate by optimizing processes, predicting maintenance needs, and reducing energy consumption. These innovations not only enhance efficiency but also support sustainable practices within the water sector.

5.6.4.4 Ethical, Economic, and Social Considerations

AI is making significant strides in water resource management, bringing forth a range of ethical, economic, and social considerations that must be addressed to ensure its responsible and effective implementation.

5.6.4.5 Ethical Considerations

AI systems must operate transparently to build trust among stakeholders. Clear guidelines on collecting, processing, and using data are essential to ensure accountability in decision-making processes. This transparency is particularly important in public utilities, where decisions can significantly impact communities. AI often involves collecting vast amounts of data, including personal information. Implementing robust data protection measures to safeguard privacy while ensuring compliance with regulations is crucial. AI models can inadvertently perpetuate existing biases if not carefully designed. Ensuring these systems are fair and equitable is vital, especially when addressing issues like water distribution and access in marginalized communities.

5.6.4.6 Economic Considerations

By optimizing water distribution networks and predicting maintenance needs, AI can significantly reduce operational costs for utilities. For instance, predictive maintenance can prevent costly failures and extend the lifespan of infrastructure. While the initial investment in AI technologies may be substantial, the long-term savings and operational efficiencies often justify the costs. Governments and organizations should consider funding research and development to foster innovation in this field.

The integration of AI will likely transform job roles within the water sector. While some positions may become obsolete, new opportunities will arise that require a blend of technical skills and domain knowledge.

5.6.4.7 Social Considerations

Engaging communities in the decision-making process is critical for successfully implementing AI technologies. Stakeholder involvement ensures that the systems developed meet the needs of those affected by water management policies. AI can potentially improve access to clean water by optimizing distribution networks. However, ensuring these benefits are equally distributed across different socio-economic groups is essential to avoid exacerbating existing inequalities. Real-time monitoring capabilities enabled by AI can significantly enhance public health by ensuring safe drinking water. Timely alerts regarding contamination can prevent health crises, underscoring the importance of integrating AI into public health strategies. While AI presents transformative opportunities for enhancing water resource management, it also necessitates careful consideration of ethical, economic, and social factors. Stakeholders must collaborate to develop frameworks that promote responsible AI use while addressing potential risks. By prioritizing transparency, fairness, community engagement, and equitable access, we can harness the power of AI to create sustainable water management systems that benefit all members of society.

5.7 CASE STUDIES IN AI FOR WATER RESOURCE MANAGEMENT

5.7.1 Hybrid Models for Hydrological Forecasting

A study by Chen et al. (2022) explored hybrid models combining the Sparrow Search Algorithm (SSA) and Moore–Penrose generalized inverse methods for hydrological forecasting. This approach was applied in the Zhanghe Irrigation District, where it effectively predicted hydrological time series, aiding water managers in making informed decisions regarding resource allocation under uncertain conditions. The results demonstrated improved accuracy in forecasting, which is crucial for sustainable water management strategies.

5.7.2 ML for Wastewater Flow Estimation

In Louisville, Kentucky, Li et al. (2022) developed an ML model to estimate past wastewater influent flow rates based on various factors such as weather data and demographics. The model achieved an accuracy of 91.7%, significantly improving predictions over traditional methods. This research highlights the potential of AI in public health applications, particularly in wastewater-based epidemiology, by providing timely insights into community health trends through wastewater analysis.

5.7.3 Smart Irrigation Systems

A review by Ashoka et al. (2024) emphasizes the role of AI in optimizing irrigation practices for sustainable agriculture. AI-driven systems utilize real-time data from soil moisture sensors and weather forecasts to automate irrigation schedules. These systems enhance water efficiency and improve crop yields by ensuring that plants receive the right amount of water at critical growth stages. The study underscores the importance of addressing challenges like data privacy and adaptability to different agricultural environments.

5.7.4 Real-Time Water Quality Monitoring

AI technologies are revolutionizing water quality monitoring through real-time data analysis. Systems equipped with sensors measure parameters like pH levels and turbidity, while ML algorithms process this data to identify irregularities indicative of contamination or operational issues. For instance, predictive models can forecast the spread of contaminants, allowing for preventive measures to safeguard drinking water quality and protect public health.

5.7.5 Groundwater Resource Management

AI models are also being applied to improve groundwater management by predicting availability based on historical data and various environmental factors. These models help prevent over-extraction and facilitate strategic planning for sustainable resource management, especially in regions heavily reliant on groundwater

5.7.6 Predictive Maintenance in Water Infrastructure

AI is also being applied to enhance the maintenance of water infrastructure through predictive analytics. By continuously monitoring equipment conditions with IoT sensors, AI algorithms can predict potential failures before they occur. This proactive maintenance approach minimizes downtime and reduces maintenance costs while ensuring uninterrupted water supply services. For example, predictive models can assess the lifespan of pumps and valves based on operational loads and environmental conditions. These case studies illustrate the transformative impact of AI on water resource management across various applications, improving irrigation efficiency, enhancing wastewater treatment processes, and ensuring water quality monitoring. As these technologies continue to develop, they hold great promise for addressing global water scarcity and sustainability challenges.

5.7.7 Smart Water Management- Case Study

In order to put in place an intelligent water management system, Korea Water Resources Corporation became a member of the International Water Resources Association (IWRA). In order to intelligently address water concerns worldwide, this system employs Information and Communication Technology (ICT) to provide

IWRA with real-time water data. Using AI processes, the smart water management system also offers solutions for leak detection, effective irrigation, water quality monitoring, and intelligent water management during floods. The system includes real-time satellite data, GIS monitoring engines, and IoT sensors. In order to facilitate effective decision-making, AI also teaches the system to offer automated services throughout various occurrences (Rocher, 2018).

5.7.8 Smart Water Management: The Way to (Artificially) Intelligent Water Management

In a summarised report, Nickum et al. (2020) highlight various SWMs in different countries. Mexico, Korea, and France have created a smart flood monitoring system that analyzes predictions using AI and the IoT network. Most nations have created smart water management (SWM) systems, which include a variety of technologies and policies for modern, dynamic water management. Data monitors for enhancing water quality and lowering overall water use based on an intelligent water resource network were part of a Mexican initiative called "PUMAGUA." South Korean researchers created an intelligent SWM known as the Hydro Intelligent Toolkit, which uses variables like groundwater levels, flood analysis, rainfall predictions, and hydrological data. These metrics are determined by analysing data from the intelligent IoT network.

5.7.9 Using AI for SWM Systems

ICT implementation has surfaced across all domains. In order to reduce costs and improve water distribution, intelligent data analysis can provide an effective water management system. Techniques from AI can be used to make efficient decisions about how much water to use for different purposes. The Sustainable Development Goals for water management and sanitization would be easier to accomplish with the help of ICT and AI. AI in water management would also assist in developing policies to reduce water leakage and address the issue of water shortage while considering population density. Jenny et al., (2020).

5.8 CHALLENGES AND LIMITATIONS OF AI IN WATER RESOURCE MANAGEMENT

AI has the potential to significantly enhance water resource management, but its integration also presents various challenges and limitations. Here are some key points derived from recent research.

5.8.1 Challenges in AI for Water Resource Management

5.8.1.1 Data Quality and Quantity

One of the foremost challenges in applying AI to water resource management is the quality and quantity of data. Effective AI models require large volumes of diverse and accurate datasets to produce meaningful insights. In hydrogeological contexts, obtaining comprehensive datasets can be complex and sometimes unfeasible. The lack of sufficient data can hinder the performance of AI models, leading to unreliable predictions and decisions.

5.8.2 Interpretability Issues

AI models often operate as "black boxes," making it difficult to understand how they arrive at specific conclusions. This lack of interpretability can be problematic in water management, where decisions can have significant implications for public health and ecosystems. Stakeholders need to grasp the predictions made by AI and the rationale behind them to assess their reliability effectively.

5.8.3 Technical Expertise Requirements

Implementing AI in water resource management necessitates specialized technical expertise, which can hinder its adoption. Cross-disciplinary collaboration is often essential for successful integration, requiring stakeholder training and capacity-building. This need for expertise can slow down the deployment of AI solutions in various water management scenarios.

5.8.4 Limitations of AI in Water Resource Management

AI systems are susceptible to security vulnerabilities, posing risks in sensitive areas like water management. Ensuring data integrity and protecting against cyber threats are critical issues that need addressing before widespread adoption. AI systems often struggle with context-aware data analysis, crucial for effective decision-making in dynamic environments like water resources. The ability to adapt to changing conditions is essential for real-time monitoring and predictive modelling, yet many existing systems lack this capability. While AI offers transformative potential for water resource management, addressing the challenges related to data quality, interpretability, technical expertise, security concerns, and context-aware analysis is crucial for its successful implementation.

5.9 FUTURE PROSPECTS AND RESEARCH DIRECTIONS

The future prospects and research directions of AI in water resource management are promising as AI technologies continue to evolve and integrate into various aspects of water management.

5.9.1 Enhanced Predictive Modelling

AI's ability to utilize historical data for predictive modelling significantly advances water resource management. By forecasting groundwater levels and quality, AI can facilitate proactive decision-making, leading to optimized resource allocation and improved sustainability practices. This capability is crucial for addressing water scarcity and contamination risks, making it an essential tool for future water management strategies.

5.9.2 Real-Time Monitoring and IoT Integration

Integrating AI with IoT devices allows continuous real-time monitoring of water resources. This synergy enhances the ability to dynamically assess water quality and quantity, enabling rapid responses to potential contamination or resource depletion. Real-time data analytics can significantly improve operational efficiency by allowing for timely interventions and adjustments in management practices.

5.9.3 Collaborative Decision Support Systems

AI-driven decision support systems can empower stakeholders by providing actionable insights from complex datasets. These systems can facilitate collaboration among scientists, policymakers, and local communities, promoting a more integrated approach to water resource management. Future research should focus on developing these collaborative platforms to enhance stakeholder engagement and ensure sustainable practices.

5.9.4 Research Directions

5.9.4.1 Addressing Data Quality Issues

One of the critical areas for future research is improving data quality and availability. Many AI applications in water management face challenges related to the reliability of input data. Research should aim at developing methods for better data collection, integration, and validation to enhance the performance of AI models.

5.9.5 Ethical Considerations and Policy Development

As AI technologies become more prevalent in water management, ethical considerations regarding data privacy, security, and human oversight must be addressed. Future studies should explore frameworks that ensure responsible AI deployment while protecting stakeholder interests. This includes developing standards and best practices for data management and usage.

5.9.6 Adaptation to Climate Variability

Research should also delve into how AI can adjust to fluctuating climatic conditions impacting water resources. Grasping how AI models can amalgamate socioeconomic data with environmental factors will be essential for developing resilient water management strategies capable of enduring climate shifts. The incorporation of AI into water resource management presents substantial potential for boosting efficiency, sustainability, and cooperation among diverse stakeholders. Nonetheless, overcoming challenges related to data integrity, ethical dimensions, and adaptability to climatic variability will be crucial to fully harnessing AI's capabilities in this sector.

5.10 CONCLUSION: AI IN WATER RESOURCE MANAGEMENT

The integration of AI into water resource management represents a transformative shift towards more efficient, sustainable, and resilient systems. As discussed throughout this chapter, AI's prowess in real-time data analysis, predictive modelling, and the optimization of water distribution is crucial in tackling the increasing intricacies faced by water utilities today. AI amplifies operational efficiency by perpetually observing systems, spotting anomalies in peak performance, and enabling proactive upkeep. This culminates in diminished waste and enhanced resource management. The synergy of AI and IoT technologies permits ongoing surveillance of water quality and volume, allowing swift action in response to potential problems like leaks or pollution. AI-driven maintenance strategies can anticipate infrastructure demands, empowering utilities to prolong component lifespans and mitigate unforeseen breakdowns. AI systems act as decision support instruments that bolster human acumen, delivering actionable insights that encourage collaborative initiatives among water management stakeholders. By refining water usage and advancing infrastructure oversight, AI plays a pivotal role in sustainable practices while ensuring resilience against climate fluctuations and population increases. AI's remarkable advantages in managing water resources necessitate that governments, utilities, and organizations embrace these cutting-edge technologies. Adopting AI not only enhances operational prowess but also supports the primary aim of sustainable water management amid escalating challenges such as climate change and urban expansion. Furthermore, continued exploration and innovation are vital for unlocking the complete potential of AI within this arena. Efforts should concentrate on enhancing data quality, improving model clarity, and addressing ethical issues related to AI implementation. By nurturing an atmosphere that encourages innovation, we can secure AI's position as a fundamental element of effective global water resource management strategies. As we advance towards an AI-integrated future in water management, it is essential to recognize the collaborative nature of this technology, one that complements human decision-making rather than replacing it. By leveraging AI's capabilities responsibly and ethically, we can build a more sustainable and resilient future for our water resources.

LIST OF ABBREVIATIONS

AI	Artificial intelligence
ANNs	Artificial neural networks
GloFAS	Global Flood Awareness System
IoT	Internet of things
LSTM	Long short-term memory
ML	Machine learning
NARNET	Nonlinear autoregressive neural networks
SSA	Sparrow search algorithm

SVM Support vector machines
WQI Water quality indices
WUE Water use efficiency

REFERENCES

ADB. 2024. Using artificial intelligence for smart water management systems. https://www.adb.org/sites/default/files/publication/614891/artificial-intelligence-smart-water-management-systems.pdf

Alharbi, O., Gazzaz, A. and Alzahrani, M. 2023. Toward sustainable farming: implementing artificial intelligence to predict optimum water and energy requirements for sensor-based micro irrigation systems powered by solar PV. *Agronomy*, 13(4): 1081.

Ashoka, P., Devi, B. R., Sharma, N., Behera, M., Gautam, A., Jha, A. and Sinha, G. 2024. Artificial intelligence in water management for sustainable farming. *Journal of Scientific Research and Reports*, 30(6): 511–525.

Bakker, M. 2022. AI-forecast: an innovative tool for short-term water demand forecasting. *Water Supply,* 24(4): 1352–1360.

Chawishborwornworng, A. 2023. Artificial intelligence and machine learning for the optimization of pharmaceutical wastewater treatment systems. *Journal of Environmental Management*, 319(1): 115703.

Chen, Y. 2022. Theoretical analysis and applications of artificial intelligence in hydrology. *Water Supply*, 23(4): iii–xxx.

Danish, M. 2022. Artificial intelligence and machine learning in water resources engineering. *Current Directions in Water Scarcity Research*, 7: 3–14.

Diva Envitec. 2024. How AI is being used in wastewater treatment plants? https://www.linkedin.com/pulse/how-ai-being-used-wastewater-treatment-plants-diva-envitec-pvt-ltd

Dolan, F., Lamontagne, J., Link, R., Hejazi, M., Reed, P. and Edmonds, J. 2021. Evaluating the economic impact of water scarcity in a changing world. *Nature Communications,* 12(1): 1915.

Express Computer. 2024. Leveraging AI for water management. https://www.expresscomputer.in/guest-blogs/leveraging-ai-for-water-management/116145/

Gazzaz, A., Alharbi, O. and Alzahrani, M. 2023. Machine learning approaches for water quality assessment: a review. *Journal of Cleaner Production*, 385(1): 135–150.

Global Infrastructure Hub. 2024. AI for process optimisation for water treatment. https://www.gihub.org/infrastructure-technology-use-cases/case-studies/ai-for-process-optimisation-for-water-treatment/

Hajirad, I. 2025. Role of artificial intelligence in improving water resource management: from demand forecasting to waste reduction and water crisis mitigation. *Journal of Nutrition and Food Processing*, 8(5). https://doi.org/10.31579/2637–8914/305

Israilova, E., Voronina, A. and Shatila, K. 2023. Impact of water scarcity on socio-economic development. *E3S Web of Conferences*, 458: 08027.

Jacobs. 2024. AI and machine learning for wastewater treatment optimization. https://www.jacobs.com/newsroom/onewater-reflections/ai-and-machine-learning-wastewater-treatment-optimization

Jenny, H., Alonso, E. G., Wang, Y. and Minguez, R. 2020. *Using Artificial Intelligence for Smart Water Management Systems*. Asian Development Bank: Mandaluyong, Philippines.

Khatri, R., Kumar, A. and Kikon, N. 2023. Exploring the rise of AI-based smart water management systems. *AQUA – Water Infrastructure, Ecosystems and Society*, 72(11): iii–iv.

Krishnan, S. R., Nallakaruppan, M. K., Chengoden, R., Koppu, S., Iyapparaja, M., Sadhasivam, J. and Sethuraman, S. 2022. Smart water resource management using artificial intelligence-a review. *Sustainability*, 14: 13384.

Li, X. 2022. Machine learning for estimating wastewater influent flow rates. *Water Research*, 205: 117–125.

Ma, Y., Zhang, Y. and Liu, J. 2023. Water quality assessment with artificial neural network models. *Environmental Science and Pollution Research*, 50: 369–387.

Mortazavizadeh, F., Bolonio, D., Mirzaei, M., Ng, J. L., Mortazavizadeh, S. V., Dehghani, A., Mortezavi, S. and Ghadirzadeh, H. 2025. Advances in machine learning for agricultural water management: a review of techniques and applications. *Journal of Hydroinformatics*, 27(3): 474–492.

Muhammad, A. U., Li, X. and Feng, J. 2024. Artificial intelligence approaches for urban water demand forecasting: a review. In: *Advances in Artificial Intelligence* (pp. 51–67). Springer. 10.1007/978-3-030-32388-2_51

Nearing, G., Cohen, D., Dube, V., Gauch, M., Gilon, O., Harrigan, S., Hassidim, A., Klotz, D., Kratzert, F., Metzger, A. and Nevo, S. 2024. Global prediction of extreme floods in ungauged watersheds. *Nature*, 627(8004): 559–563.

Nickum, J. E., Kuisma, S., Bjornlund, H. and Stephan, R. M. 2020. Smart water management: the way to (artificially) intelligent water management, or just another pretty name? *Water International*, 45: 515–519.

Rankovic, A., Stojanovic, J. and Milinkovic, M. 2024. Advanced AI algorithms for predicting water quality index: a case study. *Water Research*, 220: 118–130.

Rocher, V. 2018. *Smart Water Management—Case Study Report*. K-Water: Deajeon, Korea.

Signity Solutions. 2024. AI for water management: smart solutions for conservation. https://www.signitysolutions.com/tech-insights/ai-for-water-management

Vassar Labs. 2024. Flood management with AI and advanced water resource analytics. https://vassarlabs.com/flood-management-with-ai/

Wu, Y., Zhang, L. and Chen, X. 2024. Real-time water quality monitoring using IoT and AI technologies. *Environmental Science and Technology*, 58(3): 1234–1245.

Zhang, Y. 2024. Drought prediction using artificial intelligence models based on meteorological indices. *Scientific Reports*, 14: 19700.

6 Water Management in the Context of Net Zero Emissions

Sherwin P. De Jesus, Claudette Joyce B. Mendina, and Joan Cecilia C. Casila

6.1 INTRODUCTION

Water is one of the most essential resources for sustaining life, ecosystems, industries, and agriculture. However, challenges have emerged regarding managing this resource during the 21st century, some of which are most pressing as they affect many people globally. With the continuous global population rise and catastrophic impacts of climate variability, water resources face significant threats related to both availability and quality. Furthermore, achieving net zero emissions complicates these challenges, as it necessitates the integration of sustainable water practices to reduce emissions from sectors that heavily depend on water.

6.1.1 Water as a Global Resource

Water is fundamental to life on Earth, supporting biological systems and ensuring food security, industrial production, and public health. Agriculture alone utilizes approximately 70% of global freshwater, meaning water is a critical resource for food production (FAO, 2018). Furthermore, water is used for hydropower generation and industrial processes such as cooling and sanitation, which are all significant for the economic sector (Daneshgar & Zahedi, 2022). The World Health Organization (WHO) underscores that access to safe and clean water is essential for sustaining human health, as it is a critical factor for sustainable development (WHO, 2022). As the population continues to rise, water demand comes with it, making the need for efficient and equitable management necessary.

6.1.2 Water Management and Net Zero Emissions

Pressing global issues in the 21st century have emerged, directly impacting water management. Population growth, urbanization, and climate change—the increasing intensity of extreme weather conditions such as droughts and floods—have degraded water quality and heightened scarcity worldwide. The United Nations (UN, 2018) has projected that by 2025, half of the world's population will live in areas dealing with water stress.

DOI: 10.1201/9781003547099-7

Achieving net zero emissions is one of the global concerns, as outlined in the Paris Agreement, and reducing greenhouse gas (GHG) emissions is required to limit global warming to 1.5°C. However, water management is directly connected to this goal. Many energy and agriculture sectors are water-intensive, significantly contributing to GHG emissions (Intergovernmental Panel on Climate Change [IPCC], 2023). Simultaneously, water treatment processes that require large amounts of energy, such as desalination and wastewater treatment, contribute to emissions, making it necessary to adapt water management practices that support emissions reduction efforts.

Water management and net zero emissions are two distinct yet interconnected concepts. Achieving net zero emissions will demand a shift from fossil fuels, which could affect water-intensive industries, especially energy generation. Traditional power plants, especially those depending on coal or natural gas, require massive amounts of water for cooling, while renewable sources such as wind and solar use much less (Jin et al., 2019).

Addressing the combined challenges for water management and net zero emissions will require innovative and integrated solutions. Sustainable water management practices, such as promoting water efficiency, recycling, and renewable energy integration into water treatment, will be vital in minimizing emissions while safeguarding water availability. Integrating AI demonstrated potential in optimizing water resource management and reducing energy use, aiding in achieving sustainable development goals (SDGs), specifically SDG 6 (Clean Water and Sanitation), SDG 13 (Climate Action), SDG 14 (Life Below Water), and SDG 17 (Partnerships for the Goals) simultaneously.

6.1.3 Renewable Energy

Sectors such as industries, municipalities, localities, and agricultural systems must consider utilizing water conservation techniques that match net zero emission goals. High energy-intensive desalination plants can reduce carbon emissions by adapting renewable energy sources like solar and wind power (Goosen et al., 2023). Additionally, low-energy water treatment technologies can reduce water pollution and GHG emissions (Li et al., 2023) by effectively removing hazardous chemical and microbial components and promoting better sanitation practices while utilizing small amounts of energy.

This chapter will explore the critical water management challenges that emerged in the 21st century in the context of net zero emissions. Specifically, it will focus on the connections between climate change, water scarcity, pollution, and distribution inequalities while emphasizing the potential solutions, underscoring the role of innovative technologies—such as AI, renewable energy, and energy-efficient systems—in decreasing the environmental impact of water-intensive sectors. This chapter aims to provide readers with a holistic understanding of how sustainable water management can contribute to achieving net zero emissions while establishing long-term water security and resilience.

6.2 WATER SCARCITY

The imbalance between the supply and demand of freshwater, where demand exceeds supply, is called water scarcity. An estimated four billion people are challenged by water stress at least one month per year, and this trend will continue to grow due to many factors, such as population growth and climate variability. This will inflict more challenges to social integrity and economic development. Water scarcity directly affects land, energy, and water sectors (Birnbaum et al., 2022).

6.2.1 Physical vs. Economic Water Scarcity

When water demand in an area—including agriculture, domestic, industries, and environment—is not met due to water shortage or stress, that is what we call physical water scarcity. Evidence of physical water scarcity includes intense environmental degradation, decreased groundwater tables, and disputes in water allocation. On the other hand, economic water scarcity refers to the lack of means of an area in terms of materials for the management and governance of water resources (Vallino et al., 2020), with a focus on the funding. Evidence of economic water scarcity includes inadequate infrastructure, inequitable water distribution systems, and high vulnerability to seasonal water fluctuations. Regardless of the category, water scarcity affects many people.

6.2.2 Water Security and Global Distribution of Water Scarcity

According to UN-Water (2019), water security is the capability of an area to safeguard sustainable access to sufficient and acceptable quality of water for sustaining economic activities, human well-being, and socioeconomic development, for establishing protection against pollution, water-related disasters, and preserving ecosystems in a peaceful and politically stable manner. According to Omolere (2024), over 2 billion people are still experiencing water scarcity, or they are categorized as water insecure.

Water scarcity mainly affects the poor and vulnerable areas (UN-Water, 2019). It is more pronounced in regions where climate conditions are semi-arid, arid, tropical, and agricultural (Morante-Carballo et al., 2022). They are also challenged by poverty (Pacheco-Treviño & Guadalupe, 2024). Changing accessibility of water resources due to many factors contributes to challenges in this water issue (Leveque et al., 2024).

6.2.3 Climate Change and Water Scarcity

Climate change and human activities, population growth, economic development, and urbanization have increased energy, water, and food demands that have heightened the pressures on land and water resources (Aghsaei et al., 2020). Increased water demand for food production due to rising population, mismanagement and overuse of water resources (Bozorg-Haddad et al., 2024), and lack of water-related infrastructures (Vallino et al., 2020) are interconnected factors that advance water scarcity. According to Estrela (2024), as the population grows, there is also an increase in food demand. Thus, food production also increases, which is mainly water-dependent. As a response, developing water-related infrastructures such as dams is crucial but often not met due to lack of funding (Vallino et al., 2020). Mismanagement and overuse of water resources such as surface and groundwater are caused mainly by population growth, which results in reduced runoff and groundwater recharge (Bozorg-Haddad et al., 2024). It explains the relationship between the factors mentioned above.

6.2.4 Strategies to Address Water Scarcity

Climate change significantly affects global water resources by changing significant hydrological events. According to IPCC (2023), it is one of the reasons why it is challenging to meet SDGs, specifically SDG 6 (Clean Water and Sanitation).

The pivotal role of various approaches in addressing water-related problems sets the goal and emergence of policy changes. The 2030 Agenda of the UN sets the vision for global environmental change and human development. SDGs are incorporated into this agenda, some aligned and focused on water resources and climate action (Breuer & Spring, 2020).

In recent decades, it is evident that water resources have continuously deteriorated due to human activities and climate variabilities, which affects sustainable development (Voitko et al., 2024). To address this issue, firm policy changes focusing on restoration, infrastructure investments, and international cooperation should be highlighted, aligning with the SDGs (Tsani et al., 2020). The European Union formalized its comprehensive water management policy that addresses the stressors of water bodies of the EU in 2000, titled the Water Framework Directive (Tsani et al., 2020). Investment in water-related infrastructure is crucial for securing water resources and promoting economic development and resilience (Yusupova et al., 2024). Additionally, international cooperation is essential to enhance water governance and foster economic growth (Voitko et al., 2024), aligning with SDG 17 (Partnerships for the Goals).

6.3 WATER POLLUTION: CAUSES AND CONSEQUENCES

Water pollution arises from various sources, significantly impacting ecosystems and human health. It encompasses the contamination and degradation of water bodies due to human activities and industrial processes, leading to environmental and health issues (Singh, 2024).

6.3.1 Sources of Water Pollution

Critical contributors to water pollution include untreated agricultural runoff, industrial waste, and domestic pollutants, each introducing harmful substances into water bodies. Understanding these sources is crucial for effective management and mitigation strategies.

Untreated agricultural runoff could stem from soil contamination through pesticides and fertilizers. Using agricultural chemicals leads to runoff that carries nitrates and phosphates into water systems, causing eutrophication and disrupting ecosystems (Singh, 2024). Another key source is the industrial waste from factories and industrial plants. The pollutants discharged into waterways, such as heavy metals, chemicals, and toxins, can severely harm aquatic life and contaminate drinking water supplies. Lastly, domestic pollutants like everyday household waste (e.g., detergents, oils) often end up in the water system through improper disposal or drainage, contributing to pollution. These sources combine to create complex water pollution challenges that require coordinated efforts to manage and reduce.

6.3.2 Water Pollution and Net Zero Emissions

Water is essential for energy production, and energy is crucial for water treatment and distribution. Water-energy nexus is critical in understanding the energy demands for water treatment and its implications for GHG emissions. Wastewater treatment plants (WWTPs) are significant energy consumers and GHG emitters, necessitating innovative strategies to achieve net zero emissions. The key aspects of this nexus include:

a. Energy Consumption in Wastewater Treatment (Del Río-Gamero et al., 2020)
    - WWTPs consume substantial energy, with one case study reporting 2,956 MWh/year, highlighting the need for energy-efficient technologies.
    - Implementing renewable energy sources, such as anaerobic digestion and solar panels, can meet or exceed energy needs, potentially generating 3,396 MWh/year.

b. GHG Emissions and Mitigation Strategies
    - Yapıcıoğlu and Demir (2021) categorize GHG emissions from WWTPs into direct emissions from treatment processes and indirect emissions from energy consumption.
    - Strategies like operational efficiency improvements and energy recovery from wastewater can reduce GHG emissions by 25%–100% (Guo et al., 2024).

c. Water Conservation and Sustainable Practices (Novotny, 2011)
    - Sustainable water management practices, including conservation and reuse, are essential for minimizing energy use and GHG emissions.
    - A marginal water/energy chart can assist in planning for future sustainable cities, emphasizing the importance of integrated water-energy management.

While the focus on reducing GHG emissions through energy efficiency and renewable sources is crucial, it is also essential to consider the potential trade-offs, such as increasing water consumption associated with renewable energy technologies, which may complicate the path to achieving net zero emissions (Chlela & Selosse, 2024).

6.3.3 Mitigation Approaches

Tackling water pollution and emissions involves a complex, multifaceted approach involving improved waste management, stricter regulations, and clean technologies. Improving waste management includes proper disposal of waste to prevent contaminants from entering water systems, upgraded treatment plans that could handle more complex pollutants, and public awareness that can foster better waste management practices at the community level. Stricter regulations include enforcing standards that ensure industrial and domestic activities comply with environmental laws, pollution control in which regulatory bodies can impose fines and penalties on those that fail to meet standards, encouraging industries to adopt sustainable practices, and providing incentives. Finally, cleaner technologies involve advanced treatment processes like membrane filtration, UV treatment, biological treatment, renewable energy use to reduce carbon footprint, and smart systems like Internet of Things (IoT) and AI to monitor and manage water resources in real time.

6.4 CLIMATE CHANGE AND WATER MANAGEMENT

Climate variabilities are challenging water management. Climate-induced water is disrupting the quality and availability of water stresses, such as floods and droughts, by altering water cycles. Global warming causes extreme weather conditions, threatening food security, socioeconomic development, and ecosystems. As climate change advances, increased frequency and severity of droughts are expected, highlighting the different stressors to water resources (Pacheco-Treviño & Guadalupe, 2024).

6.4.1 Climate Change-Induced Water Stress

Many factors can be included as a source of water stress, but two climate-driven factors contribute the most. Droughts can cause many strains on water management, affecting water-dependent sectors and society (Estrela, 2024). Additionally, floods can alter the operation of many industries and increase the contamination rate, resulting in economic losses and risks to human health (Pradhan et al., 2022). Aside from these adverse effects, water distribution is the most devastating. Estrela (2024) emphasized that the latter is essential for allocating water resources and attaining sustainable water use.

6.4.2 Impact on Agriculture and Urban Areas

The agriculture sector is one of the most vulnerable to climate change. Climate-induced stresses can negatively affect a crop's physiological growth, reducing yield (Zollinger, 2024). Urban areas are populated, making them energy-dependent, consuming 67%–76% of global energy and an essential driver of GHG emissions. This trend is more likely to continue for decades, making it difficult for countries to meet the target set by the 2015 Paris Climate Agreement and the SDGs. These contributions of urban areas to climate change will also affect them regarding water-related infrastructures. Climate-related disasters destroy water resources and appurtenant infrastructures

(Ferdowsi et al., 2024), affecting water quality and availability in agriculture and urban areas.

6.4.3 Climate Adaptation Strategies

Climate adaptation strategies, including water conservation and resilient infrastructure, are crucial in water management. Adaptation strategies should potentially reduce or directly influence the impacts of climate variabilities. To create a sound adaptation strategy in mitigating the adverse effects of climate variabilities, it is essential to understand uncertainty, determine the current and possible future impacts, and facilitate policy development. To adequately address water stressors in water management, sacrifices such as progressive water demand reduction and appropriate reallocation of water resources are necessary. The development of climate-resilient hydraulic infrastructures is pivotal to making water available to essential sectors in the face of uncertainties caused by climate change (Sordo-Ward et al., 2019).

Without climate mitigation and adaptation measures, alteration of hydrological cycles (Sordo-Ward et al., 2019), extreme temperatures that cause drought, and excessive rainfall that causes floods globally will become more severe and frequent (Zollinger, 2024).

6.5 THE ROLE OF ARTIFICIAL INTELLIGENCE IN WATER MANAGEMENT

Integrating AI into water systems represents a significant advancement in addressing global challenges related to water scarcity, quality, and infrastructure resilience. As technology continues to evolve, the potential for AI to revolutionize water management practices will only grow, paving the way for a more sustainable future in resource management. By embracing these innovations responsibly, stakeholders can ensure that water resources are managed wisely for future generations.

6.5.1 Optimizing Irrigation and Water Distribution through AI

The integration of AI into smart irrigation systems has been increasing, making it the bedrock of automation and revolutionizing water management in agriculture (How AI can solve agriculture's water management issues, 2024). Several critical components in AI-driven smart irrigation are real-time data collection, automated decision-making, user interfaces and notifications, predictive analytics, and integration with other technologies, such as drone imagery and satellite data. Utilizing AI alongside the IoT, these irrigation systems could optimize various irrigation processes, enhance crop yield, and conserve water resources.

While integrating AI in smart irrigation systems presents numerous advantages, challenges remain. These include the need for reliable sensor technology, data security concerns, and the initial costs associated with implementing advanced systems. Ongoing research aims to enhance machine learning models for better predictive capabilities and to develop more cost-effective solutions tailored to various agricultural contexts (Alvim et al., 2022).

In conclusion, AI's role in smart irrigation systems and water distribution networks is pivotal for modern agriculture, promoting sustainable practices and enhancing efficiency and operations. As technology continues to evolve, its impact on irrigation and water management practices is expected to grow even further, addressing critical global challenges related to water scarcity, food security, and resource allocation.

6.5.2 Water Recycling and Conservation Using AI

Another application of AI in water management is the optimization of water recycling and conservation efforts. AI-powered systems can analyze water quality data, identify potential contaminants, and develop customized treatment solutions to ensure the safe reuse of water.

Some key ways AI is being utilized nowadays include predictive analytics, where AI analyzes historical data to predict future water demand, helping in efficient water distribution and reducing waste; leak detection, where AI algorithms monitor water distribution networks to detect leaks early, preventing water loss and saving costs (Waterwise, 2023); optimized irrigation, where AI helps in smart agricultural irrigation by analyzing soil moisture and weather data to apply the right amount of water at the right time (Kalotra, 2024); water quality monitoring, where AI systems monitor water quality in real time, detecting pollutants and ensuring safe drinking water; and smart water grids, where AI integrates data from various sources to manage water resources more effectively, balancing supply and demand (Sand Technologies, 2024).

6.5.3 Predictive Models for Climate-Driven Water Challenges

AI-powered predictive models can analyze historical data on precipitation, temperature, and other climate variables to forecast future water availability and extreme events, such as droughts and floods. One important use of these predictive models is for hydro-climatic forecasting. These models blend climate predictions with AI to forecast variables like rainfall, temperature, streamflow, and droughts (Slater et al., 2023). AI models also play a role in drought prediction since they can predict future drought conditions with high accuracy, which is crucial for regions prone to water scarcity. This then calls for water quality monitoring, in which machine learning algorithms analyze data to predict water quality, detect pollutants, and ensure safe drinking water (Karthick et al., 2023). Lastly, AI is beneficial for hybrid forecasting. This approach combines data-driven methods with traditional climate models to improve prediction accuracy and reliability (Slater et al., 2023). These models are essential for sustainable water management and mitigating the impacts of climate change on water resources.

6.6 EXISTING SUSTAINABLE WATER MANAGEMENT PRACTICES

Sustainable water management practices such as the construction of appurtenant infrastructures, employment of water-saving technologies, creation of solid policies, and water-energy nexus have been countries' responses to water problems. However, some of these practices would serve a better purpose if they were designed within the context of climate change and net zero emissions. In this regard, Estrela (2024) emphasized that sustainable water management is challenged by climate change, water security, pollution, and the destruction of water bodies. Aging infrastructure and lack of maintenance and skilled experts are also affecting water management practices through infrastructure failures (Pamidimukkala et al., 2021). Thus, the consideration of these factors in water management is necessary.

6.6.1 Water-Saving Technologies

Traditional water-saving practices have changed due to climate variability, urbanization, and population increase (Lakhiar et al., 2024). Efficient water-saving innovations that revolutionized water management (Ray & Majumder, 2024) were developed. These technologies aim to conserve water and maximize use, minimize wastage, save energy, and optimize crop yield in agriculture while reducing the environmental footprint.

Ray and Majumder (2024) listed some water-saving technologies and techniques that consider climate variabilities in agriculture and urban areas. The technologies include drip irrigation (e.g., subsurface drip irrigation), precision irrigation, weather-based irrigation scheduling, automation and control systems (e.g., valves and pumps), smart irrigation controllers, and IoT-based irrigation and distribution systems. Techniques include rainwater harvesting, on-farm water storage, recycling and reuse, and soil moisture conservation practices.

6.6.2 Policies for Sustainable Water Use

Research on sustainability is increasingly prioritized, particularly in the context of sustainable water use, as countries develop relevant policies. For instance, the Asian Water Development Outlook (AWDO) 2020 offers recommendations for managing water in light of climate variability (Panella et al., 2020), while EU member states have enacted legislation addressing water supply, reuse, and wastewater treatment (Cipolletta et al., 2021). However, significant gaps remain in effectively promoting sustainable water use globally. Initiatives like SDG 6 aim to create comprehensive frameworks for sustainable water management.

Transboundary disputes, such as those surrounding the Mekong River—which flows through multiple Southeast Asian countries—highlight the urgency of addressing these challenges. Upstream dam projects in China and Laos disrupt water flow, negatively impacting downstream agriculture and fisheries in Vietnam and Cambodia. While the Mekong River Commission (MRC) facilitates some cooperation, the absence of binding agreements limits effectiveness (Ngoc & Khanh, 2020). Therefore, fostering mutually beneficial solutions, such as benefit-sharing and adaptive governance, is essential for resolving conflicts and ensuring sustainable management of shared water resources (Hossen et al., 2023).

6.6.3 Water-Energy Nexus

Water and energy are two significant resources contributing to sustainable development. They are essential and interconnected for sustaining significant industries that maintain and secure a future for humanity. The connection between these resources falls under two categories: water for energy generation and energy for water production. It is generally called a water-energy nexus. It is being disrupted due to an exponentially growing global population and energy demand since water is a limited resource (Politano et al., 2024). The global water demand will continue to increase as the global population increases. With this, strategies to solve water issues become essential.

Renewable energy sources and recent technologies, such as low-energy desalination systems, support the water-energy nexus. The introduction of renewable energy systems is focused on reducing climate change impacts on water resources and GHG emissions. The integration of renewable energy systems—wind, solar, and geothermal—in the desalination technology paved the way toward sustainability, matching economic efficiency while considering the environment.

Desalination technology separates dissolved salts from saline water, thus creating freshwater (Panagopoulos, 2021). There are three types of desalination technology: Multi-Stage Flash (MSF), Low-temperature multi-effect distillation (MED), and reverse osmosis (RO). RO is the most utilized among these technologies, accounting for 63.60% of the technology. These desalination technologies are being developed to become more energy-efficient and have less carbon footprint by modifying specific processes. For MSF desalination plants, the application of the forward osmosis (FO) process for pretreatment of feed solution and vacuum membrane distillation (VMD) enhances the efficiency, thus energy consumption (Thabit et al., 2019). MED can be more environmentally friendly by incorporating direct contact spray-assisted evaporation and condensation (DCSEC) (Alrowais et al., 2020). For the RO or Seawater Reverse Osmosis (SWRO), it is necessary to design the optimal configuration to attain the lowest energy consumption. These energy-efficient and low carbon footprint technologies are significant for water-energy nexus and the reduction of carbon emissions.

The adaptation of innovations is necessary to attain net zero emissions. New technologies and strategies concerning water management are incorporated into this goal. The importance of using renewable energy and technologies in the processes involved in sustainable water management highlights the significance of reducing carbon emissions to mitigate the impacts of climate change.

6.7 ALIGNING WATER MANAGEMENT STRATEGIES WITH NET ZERO EMISSIONS

Water management is a critical component of the global effort to achieve net zero emissions. Sustainable water use is essential for mitigating climate change, as water-intensive activities contribute significantly to GHG emissions. To align water management strategies with net zero emissions, a multifaceted approach is required that considers factors such as water scarcity, energy consumption, and the socioeconomic implications of water-related policies. This involves integrating sustainable practices across various aspects of water supply and treatment. This alignment is crucial for reducing GHG emissions while ensuring efficient water use and management.

6.7.1 Challenges in Reducing Water Sector Emissions

Reducing emissions in energy-intensive water treatment processes is indeed a complex challenge. One key issue is the high energy demand. Water treatment processes, such as desalination and wastewater treatment, require significant amounts of energy, leading to high GHG emissions. Another issue is the outdated infrastructure, as many water treatment facilities use outdated technologies that are less energy-efficient, making it difficult to reduce emissions without major upgrades or replacements. The reliance on carbon-intensive energy sources in water treatment processes also contributes to high carbon emissions. Ultimately, the financial constraints in upgrading to more energy-efficient technologies and infrastructures pose a financial challenge for many water utilities and management systems.

Addressing these challenges requires a combination of technological innovation, policy support, and financial investment. Transitioning to renewable energy sources, adopting energy-efficient technologies, and improving operational practices are essential steps toward reducing emissions in the water sector.

6.7.2 Pathways to Achieving Net Zero in Water Management

Achieving net zero emissions in water management requires a multifaceted approach integrating technological advancements, operational optimization, and policy changes. Various strategies focusing on energy efficiency, resource recovery, and innovative treatment methods can realize the significant potential for decarbonization in the water sector.

Upgrading to energy-efficient technologies and optimizing existing processes can significantly reduce energy consumption and emissions. There is also the matter of maximizing on-site renewable capacities. Integrating renewable energy sources like solar and wind directly into water treatment facilities can help offset energy use. Harvesting energy from treated effluent can also be a critical pathway in reducing harmful emissions. Technologies that capture and utilize energy from wastewater treatment processes, like biogas production, can contribute to energy self-sufficiency (Rani et al., 2022). Utilizing natural systems or nature-based solutions, in general, to manage water and sequester carbon can provide multiple environmental benefits. All these must be under proper policy implementation and regulation. It is crucial to continuously develop policies that incentivize low-emission practices and investments in sustainable water management. Lastly, raising awareness and engaging the public in water conservation efforts can drive collective action toward net zero goals.

6.7.3 Global Case Studies

Countries and regions have successfully implemented various water and emission reduction policies, showcasing innovative approaches to environmental management. Notably, China has adopted stringent emission controls in the coal-fired power sector, significantly reducing pollutants while managing water use (Wang et al., 2024). In this chapter, the authors investigated the water, energy, and emissions nexus

(WEEN), focusing on end-of-pipe treatment technologies for the coal-fired power industry in the Beijing-Tianjin-Hebei region. Similarly, Henan Province has developed a total emission pollutant control model that aligns water quality improvement with pollutant reduction, achieving notable reductions in chemical oxygen demand and ammonia-nitrogen levels. A water total emission pollutant allocation model was built based on water environmental carrying capacity (WECC) in order to make TEPC truly serve the improvement of water quality, and the model was used to construct a water pollutant emission control allocation scheme for 28 cities in Henan Province, China. In Europe, the focus on micro-pollutant emissions has led to the implementation of comprehensive measures to maintain water quality, as mandated by the Water Framework Directive (Fuchs & Toshovski, 2016). Japan's prefectures have also explored renewable energy policies, demonstrating that increased renewable electricity consumption can substantially lower carbon emissions (Sato & Narita, 2023).

6.8 POLICY FRAMEWORKS FOR SUSTAINABLE WATER AND EMISSION MANAGEMENT (REGIONAL AND GLOBAL POLICIES)

The integration of sustainable water and emission management policies is crucial for addressing the interdependencies between water and energy resources. Various frameworks have emerged at both regional and global levels, emphasizing the need for coordinated approaches to achieve sustainability goals.

6.8.1 International Agreements and Water Governance

One key example of a global policy aimed at promoting sustainable water and emission management is the United Nations SDGs. As mentioned, SDG 6 emphasizes integrated water resources management, water quality, and efficiency. The UN-Water Policy Brief on Water and Climate highlights the link between water and climate change, advocating for integrated approaches to manage water resources and mitigate climate impacts. Another important example is the Paris Agreement, which encourages countries to adopt measures that reduce GHG emissions, which can include sustainable water management practices.

For example, regional policies include the European Union Water Framework Directive (WFD), the African Ministers' Council on Water (AMCOW), and the Asia-Pacific Water Forum (APWF). The WFD aims to achieve good qualitative and quantitative status for all European water bodies, promoting sustainable water use and reducing pollution. AMCOW supports sustainable water management and development in Africa, focusing on integrated water resources management and climate resilience. APWF promotes regional cooperation and policy dialogue on water security, aiming to achieve sustainable water management and reduce emissions in the Asia-Pacific region. These frameworks provide a foundation for countries to develop and implement policies that promote sustainable water management and reduce emissions.

6.8.2 Collaborative Water Management

Partnerships between governments, industries, and NGOs are crucial for effective water management. One advantage is the shared resources and expertise. Governments provide regulatory frameworks, industries bring innovation and technology, and NGOs offer community engagement and environmental advocacy. These combined resources and expertise lead to more comprehensive solutions. Moreover, the combined resources and expertise lead to more comprehensive solutions, enhancing efficiency and increasing funding. Partnerships streamline efforts, reducing duplication and ensuring effective use of resources. Collaborative projects can access diverse funding sources and address technological and financial barriers in water management. From governmental budgets and personnel to private sector investments and manpower and NGO bodies and grants, sustainable financing and collaborative efforts ensure that water management strategies are inclusive, considering economic, social, and environmental perspectives.

6.9 AGRONOMIC AND TECHNOLOGICAL INNOVATIONS FOR WATER SECURITY IN THE NET ZERO ERA

The demand for agronomic and technological innovations for water management to attain water security is increasing, specifically the more efficient use of water resources, which enhances water quality for humans and the environment, improves water resources planning, and enhances adaptability to extreme weather events in the context of net zero emissions. The importance of integrating climate change stabilization techniques to aim for net zero GHG emissions in agriculture is essential (Brears, 2023).

6.9.1 Emerging Water Technologies

As water demands continue to increase globally due to exponential population growth, expansion of economies, and increased per capita water consumption, technologies to improve and conserve limited resources have emerged. Emerging water technologies such as sustainable water use, desalination, wastewater treatment, and smart water management systems have gained focus in the last decades.

Sustainable water use (Estrela, 2024) and desalination technologies have contributed to decreasing water scarcity (Wang et al., 2024), and low-energy desalination plants also lessen the carbon footprint of the technology. Population growth and increased demand have pushed limited resources—specifically water resources—to their limits. Thus, eutrophication and contamination have risen (Janczukowicz & Rodziewicz, 2024). These issues pose significant threats to human health and the environment. Additionally, inadequate management and wastewater treatment only contribute to the problem. However, the effects of these problems can be mitigated through modern crop production technologies, sustainable wastewater management, and the integration of innovative approaches regarding wastewater treatment, reuse, and nutrient recovery (Al-Hazmi et al., 2023). Wastewater treatment is paramount in water safety and security, and this should be included as a top priority. Additionally, wastewater management and sanitation are aligned in the interconnected SDGs, especially SDG 6 (Larsen et al., 2021).

The conservation of water has become a necessity. Thus, smart water management systems are essential to achieve this. Water management systems have been revolutionized through sustainable techniques and the integration of emerging technologies such as machine learning, AI, and IoT. Machine learning and AI aid in understanding the behavior of water cycles (Gucek et al., 2019). Furthermore, AI and IoT contribute to efficient water management (Kotwal et al., 2024).

6.9.2 The Role of Renewable Energy in Water Management

The energy source of traditional water systems is fossil fuels, which contribute to GHG emissions, thus causing global warming and climate change. The latter are the main drivers of integrating renewable energy sources in water management systems—including water pumping, supply, and distribution systems—and are significant (Zakariazadeh et al., 2024). The concept of water-energy nexus is again emphasized. In this case, it is known as energy for water production or management. Aside from water (including ocean and hydropower), renewable energy sources like solar, wind, biomass, and geothermal are utilized in water management systems, specifically wastewater treatment. Del Rio-Gamero et al. (2020) state that integrating different renewable energy in wastewater treatment plants can reduce and eliminate even GHG emissions.

6.9.3 AI and Big Data in Water Security and Future Trends in Water Technologies

AI and big data analytics revolutionize water security by enabling more efficient water resources management. As pollution and water scarcity continuously threaten water resources, AI and big data can provide tools to optimize water usage, ensure quality control, and enhance distribution. Furthermore, the German Water Partnership (GWP) has defined the term "Water 4.0" or water revolution 4.0 (W4.0), which underscores the integration of sensors, computer models, and real-time control, together with heavy use of intelligent networks and internet (Gaskin et al., 2023). These technologies will advance smart decision-making for sustainable water use, thus attaining water security.

6.9.4 Future Trends in Water Technologies

Sustainable solutions that reduce stress in water while minimizing emissions and environmental impacts are the focus of future trends in water technologies. It aims to address the global challenge of water scarcity while maintaining energy-efficient and eco-friendly practices. Key technologies such as advanced desalination systems that utilize innovative energy-saving processes to extract freshwater from saltwater (Wang et al., 2024), smart water systems which employ real-time monitoring and data analytics to optimize water use and reduce wastewater (Kotwal et al., 2024), and wastewater treatment and water recycling innovations are contributing to close the loop in water use, making unusable water resources accessible for industrial, agricultural, and even potable purposes. These technologies are essential for securing adequate water supply for future generations, especially in water-stressed regions, while aiming to maintain low carbon emissions and energy efficiency.

6.10 CONCLUSION AND FUTURE DIRECTIONS

Water scarcity, pollution, and inequality are all profoundly interconnected, and addressing these issues is crucial for achieving net zero emissions. Climate change exacerbates water scarcity by altering rainfall patterns and increasing temperature. This leads to more frequent droughts and reduced water availability, affecting agriculture, energy production, and overall water supply (Valero et al., 2023). On the other hand, pollution from agricultural runoff, industrial waste, and domestic pollutants contaminates water sources, making them unsafe for consumption and harming ecosystems. This pollution can also contribute to GHG emissions, as contaminated water requires more energy-intensive treatment processes. Water scarcity and pollution disproportionately affect marginalized communities, exacerbating existing inequalities. Poorer regions and vulnerable populations often lack access to clean water and sanitation, which impacts their health, livelihoods, and ability to adapt to climate change. Therefore, sustainable water management practices, such as using renewable energy for water treatment, reducing agricultural runoff, and improving water infrastructure, can help reduce emissions and promote equitable access to water.

Addressing water management from a holistic perspective should consider the entire water cycle, including supply, demand, quality, and ecological health. The role of innovation in overcoming water challenges includes utilizing renewable energy sources to reduce carbon emissions, water recycling, and reuse, development and implementation of advanced technologies such as AI and IoT, and data-driven decision-making. Leveraging big data and IoT for real-time monitoring and management of water resources improves efficiency and response to water-related challenges.

Policies could include regulatory frameworks that ensure sustainable water use, water quality protection, and promotion of equitable access. Providing financial incentives for industries and communities to adopt water-saving technologies and practices encourages widespread adoption. Finally, collaborative policies and agreements between countries sharing water resources ensure fair and sustainable water management.

The future of water security and emission reduction hinges on continued research and development. Continuous R&D fuels the development of cutting-edge technologies like advanced filtration systems, AI for water management, and renewable energy integration in water treatment. These innovations are essential for making water systems more efficient and less carbon-intensive.

With climate change impacting water availability and quality, R&D helps develop strategies and technologies to enhance the resilience of water systems. This includes everything from drought-resistant crops to improved flood management techniques. Moreover, research fosters collaboration across borders. Shared knowledge and joint efforts can lead to more effective water management solutions and emission reduction strategies on a global scale. Ultimately, understanding the socioeconomic aspects of water management is crucial. Research helps craft policies and technologies that are effective and equitable, ensuring that vulnerable communities are not left behind.

REFERENCES

Aghsaei, H., Mobarghaee Dinan, N., Moridi, A., Asadolahi, Z., Delavar, M., Fohrer, N., & Wagner, P. D. (2020). Effects of dynamic land use/land cover change on water resources and sediment yield in the Anzali wetland catchment, Gilan, Iran. *Science of the Total Environment,* 712, 136449. https://doi.org/10.1016/j.scitotenv.2019.136449

Al-Hazmi, H. E., Mohammadi, A., Hejna, A., Majtacz, J., Esmaeili, A., Habibzadeh, S., Saeb, M. R., Badawi, M., Lima, É. C., & Mąkinia, J. (2023). Wastewater reuse in agriculture: prospects and challenges. *Environmental Research*, 236, 116711–116711. https://doi.org/10.1016/j.envres.2023.116711

Alrowais, R., Qian, C., Burhan, M., Ybyraiymkul, D., Wakil Shahzad, M., & Choon NG, K. (2020). A greener seawater desalination method by direct-contact spray evaporation and condensation (DCSEC): experiments. *Applied Thermal Engineering*, 115629. https://doi.org/10.1016/j.applthermaleng.2020.115629

Alvim, S. J. T., Guimarães, C. M., De Sousa, E. F., Garcia, R. F., & Marciano, C. R. (2022). Application of artificial intelligence for irrigation management: a systematic review. *EngenhariaAgrícola*, 42(spe). https://doi.org/10.1590/1809-4430-eng.agric.v42nepe20210159/2022

Birnbaum, A., Lamontagne, J., Wild, T., Dolan, F., & Yarlagadda, B. (2022). Drivers of future physical water scarcity and its economic impacts in Latin America and the Caribbean. *Earth's Future*, 10(8). https://doi.org/10.1029/2022ef002764

Bozorg-Haddad, O., Bahrami, M., Gholami, A., Chu, X., & Loáiciga, H. A. (2024). Investigation and classification of water resources management strategies: possible threats and solutions. *Natural Hazards*. https://doi.org/10.1007/s11069-024-06589-y

Brears, R. (2023). *Financing Water Security and Green Growth*. Oxford University Press. https://books.google.com.ph/books?hl=en&lr=&id=8Y64EAAAQBAJ&oi=fnd&pg=PP1&dq=Agronomic+Practices+and+Technological+Innovations+for+%22Water+Security%22+in+the+Net+Zero+Era.+references+from+2019-2024&ots=M-UfKgEj_M&sig=XN8B7baTtWc7gbjWyfSPN7k--Io&redir_esc=y#v=onepage&q&f=false

Breuer & Spring. (2020). The 2030 Agenda as Agenda setting event for water governance? Evidence from the Cuautla River Basin in Morelos and Mexico. *Water*, 12(2), 314. https://doi.org/10.3390/w12020314

Chlela, S., & Selosse, S. (2024). Water use in a sustainable net zero energy system: what are the implications of employing bioenergy with carbon capture and storage? *International Journal of Sustainable Energy Planning and Management*, 40, 146–162. https://doi.org/10.54337/ijsepm.8159

Cipolletta, G., Ozbayram, E. G., Eusebi, A. L., Akyol, Ç., Malamis, S., Mino, E., & Fatone, F. (2021). Policy and legislative barriers to close water-related loops in innovative small water and wastewater systems in Europe: a critical analysis. *Journal of Cleaner Production*, 288, 125604. https://doi.org/10.1016/j.jclepro.2020.125604

Daneshgar, S., & Zahedi, R. (2022) Investigating the hydropower plants production and profitability WHO using system dynamics approach. *Journal of Energy Storage*. https://doi.org/10.1016/j.est.2021.103919

Del Río-Gamero, B., Ramos-Martín, A., Melián-Martel, N., & Pérez-Báez, S. (2020). Water-energy nexus: a pathway of reaching the zero net carbon in wastewater treatment plants. *Sustainability*, 12(22), 9377. https://doi.org/10.3390/su12229377

Estrela, T. (2024). Introduction to section 3. *Water International*, 49(3–4), 466–475. https://doi.org/10.1080/02508060.2024.2343177

Ferdowsi, A., Piadeh, F., Behzadian, K., Mousavi, S.-F., & Ehteram, M. (2024). Urban water infrastructure: a critical review on climate change impacts and adaptation strategies. *Urban Climate*, 58, 102132. https://doi.org/10.1016/j.uclim.2024.102132

Food and Agriculture Organization. (2018). *Water for Sustainable Food and Agriculture: A Report for the G20 Presidency of Germany*. chrome-extension://efaidnbmnnnibpcajpcglclefindmkaj/https://openknowledge.fao.org/server/api/core/bitstreams/b48cb758-48bc-4dc5-a508-e5a0d61fb365/content

Fuchs, S., & Toshovski, S. (2016). *Modelling of Micro Pollutant Emissions to Water and Evaluation of Reduction Measures*. SciSpace – Paper. https://typeset.io/papers/modelling-of-micro-pollutant-emissions-to-water-and-4tfaea28gs

Gaskin, C., Stockil, K., Track, T., Horan, W., Lucht, A., & Conheady, P. (2023). *Industrial Water 4.0-A Framework for Catchment-based Digitally Integrated Industrial Water Stewardship*. https://www.epa.ie/publications/research/circular-economy/Research_Report_435.pdf

Goosen, M., Mahmoudi, H., Alyousef, Y., & Ghaffour, N. (2023). Solar desalination: a review of recent developments in environmental, regulatory and economic issues. *Solar Compass*, 5, 100034. https://doi.org/10.1016/j.solcom.2023.100034

Gucek, A., Costa, J. P., Massri, M., Costa, J. S., Rossi, M., del Busto, I., & Mocanu, I. (2019). *Addressing Climate Change Preparedness from a Smart Water Perspective*. https://www.academia.edu/download/109772173/Guceketal_SiKDD_2022.pdf

Guo, D., Li, B., Yu, W., Baroutian, S., & Young, B. R. (2024). A system engineering perspective for net zero carbon emission in wastewater and sludge treatment industry: a review. *Sustainable Production and Consumption*, 46, 369–381. https://doi.org/10.1016/j.spc.2024.02.033

Hossen, M. A., Connor, J., & Ahammed, F. (2023). How to resolve transboundary river water sharing disputes. *Water*, 15(14), 2630. https://doi.org/10.3390/w15142630

Intergovernmental Panel on Climate Change. (2023). Summary for policymakers. In H. Lee & J. Romero (Eds.), *Climate Change 2023: Synthesis Report. Contribution of Working Groups I, II and III to the Sixth Assessment Report of the Intergovernmental Panel on Climate Change* (pp. 1–34). IPCC. https://doi.org/10.59327/IPCC/AR6-9789291691647.001

Janczukowicz, W., & Rodziewicz, J. (2024). Water and wastewater management in agriculture. *Applied Sciences*, 14(6), 2588. https://doi.org/10.3390/app14062488

Jin, Y., Behrens, P., Tukker, A., & Scherer, L. (2019). Water use of electricity technologies: a global meta-analysis. *Renewable and Sustainable Energy Reviews*, 115, 109391. https://doi.org/10.1016/j.rser.2019.109391

Kalotra, S. (2024). *AI for Water Management: Smart Solutions for Conservation*. OpenAI. https://www.signitysolutions.com/tech-insights/ai-for-water-management?form=MG0AV3

Karthick, K., Krishnan, S., & Manikandan, R. (2023). Water quality prediction: a data-driven approach exploiting advanced machine learning algorithms with data augmentation. *Journal of Water and Climate Change*. https://doi.org/10.2166/wcc.2023.403

Kotwal, S., Pati, S., & Patil, Y. (2024). Review on AI and IoT based integrated smart water management and distribution system. *Educational Administration: Theory and Practice*, 30(4), 594–605. https://doi.org/10.53555/kuey.v30i4.1507

Lakhiar, I., Yan, H., Zhang, C., Wang, G., He, B., Hao, B., Han, Y., Wang, B., Bao, R., Syed, T. N., Chauhdary, J. N., & Md. Rakibuzzaman. (2024). A review of precision irrigation water-saving technology under changing climate for enhancing water use efficiency, crop yield, and environmental footprints. *Agriculture*, 14(7), 1141–1141. https://doi.org/10.3390/agriculture14071141

Larsen, T. A., Gruendl, H., & Binz, C. (2021). The potential contribution of urine source separation to the SDG agenda – a review of the progress so far and future development options. *Environmental Science: Water Research & Technology*. https://doi.org/10.1039/d0ew01064b

Leveque, B., IrakizaShyaka, A., Ndong, M., Jalbert, J., Burnet, J.-B., Kammoun, R., Dorner, S., & Bichai, F. (2024). Assessing the vulnerability of urban drinking water intakes to water scarcity under global change: a bottom-up approach. *Environmental Challenges*, 15, 100885. https://doi.org/10.1016/j.envc.2024.100885

Li, D., Wang, Z., Yang, Y., Liu, H., Fang, S., & Liu, S. (2023). Research status and development trend of wastewater treatment technology and its low carbonization. *Applied Sciences*, 13(3), 1400. https://doi.org/10.3390/app13031400

Ngoc, T. T. H., & Khanh, T. P. (2020). Transboundary issues of water Governance in Mekong River Basin. *International Journal of Advanced Science and Technology*, 29(8s), 4290–4305. https://sersc.org/journals/index.php/IJAST/article/view/25460

Novotny, V. (2010). Water and energy link in the cities of the future – achieving net zero carbon and pollution emissions footprint. *Water Science & Technology*, 63(1), 184–190. https://doi.org/10.2166/wst.2011.031

Morante-Carbollo, F., Montalván-Burbano, N., Quiñonez-Barzola, X., Jaya-Montalvo, M., Carrión-Mero, P. (2022). What do we know about water scarcity in semi-arid zones? a global analysis and research trends. *Water,* 14(17), 2685. https://doi.org/10.3390/w14172685

Omolere, M. P. (2024). *Why Global Water Security Matters in 2024*. Earth.org. https://earth.org/global-water-crisis-why-the-world-urgently-needs-water-wise-solutions/

Pacheco-Treviño, S., & Guadalupe, M. (2024). Review of water scarcity assessments: highlights of Mexico's water situation. *WIREs Water*. https://doi.org/10.1002/wat2.1721

Pamidimukkala, A., Kermanshachi, S., Adepu, N., & Safapour, E. (2021). Resilience in water infrastructures: a review of challenges and adoption strategies. *Sustainability*, 13, 12986. https://doi.org/10.3390/ su132312986

Panagopoulos, A. (2021). Water-energy nexus: desalination technologies and renewable energy sources. *Environmental Science and Pollution Research*, 28(17), 21009–21022. https://doi.org/10.1007/s11356-021-13332-8

Panella, T., Illescas, F., Cardascia, S., Beek, van, Guthrie, L., Hearne, D., Wiberg, D., Pandey, P., Subsasinghe, C., Velpuri, M., Naga, K., S. J., Allan, J., Mogg, W., Jawad, M., Sochacka, B., Strazzabosco, A., Stewart-Koster, B., Harte, J., Ndehedehe, C., & O'Mara, K. (2020). *Asian Water Development Outlook 2020: Advancing Water Security across Asia and the Pacific | QUT ePrints*. Qut.edu.au. https://eprints.qut.edu.au/207557/

Politano, A., Al-Juboori, R., Alnajdi, S., Alsaati, A., Athanassiou, A., Bar-Sadan, M., Beni, A., Campi, D., Cupolillo, A., & D'Olimio, G. (2024). 2024 roadmap on membrane desalination technology at the water-energy nexus. *Journal of Physics*, 6(2). https://doi.org/10.1088/2515-7655/ad2cf2

Pradhan, P., Pham, T.T.H., Shrestha, S., Loc, H. H., & Park, E. (2022). Projecting the impact of human activities and climate change on water resources in the transboundary Sre Pok River Basin. *Climatic Change*, 172(26). https://doi.org/10.1007/s10584-022-03367-2

Rani, A., Snyder, S. W., Kim, H., Lei, Z., & Pan, S. (2022). Pathways to a net-zero-carbon water sector through energy-extracting wastewater technologies. *Npj Clean Water*, 5(1). https://doi.org/10.1038/s41545-022-00197-8

Ray, S., & Majumder, S. (2024). Water management in agriculture: innovations for efficient irrigation. In: Sil, P., Chhetri, P., Majumder, S., & Santosh D.T., (Eds.) *ResearchGate* (pp. 169–185), Satish Serial Publishing House, Azadpur, Delhi. https://www.researchgate.net/publication/381867727

Sand Technologies. (2024). Water AI: 8 ways AI in water management creates a better future. https://www.sandtech.com/insight/water-ai-8-ways-ai-in-water-management-creates-a-better-future/?form=MG0AV3

Sato, I., & Narita, D. (2023). Use of nested multi-regional input–output analysis for the evaluation of subnational emission reduction policies: a case of Japanese prefectures. *Sustainable Production and Consumption*, 37, 294–305. https://doi.org/10.1016/j.spc.2023.03.003

Singh, V. (2024). Water pollution. In: *Textbook of Environment and Ecology* (pp. 253–266). https://doi.org/10.1007/978-981-99-8846-4_17

Slater, L. J., Arnal, L., Boucher, M., Chang, A. Y., Moulds, S., Murphy, C., Nearing, G., Shalev, G., Shen, C., Speight, L., Villarini, G., Wilby, R. L., Wood, A., & Zappa, M. (2023). Hybrid forecasting: blending climate predictions with AI models. *Hydrology and Earth System Sciences*, 27(9), 1865–1889. https://doi.org/10.5194/hess-27-1865-2023

Sordo-Ward, A., Granados, A., Iglesias, A., Garrote, L., & Bejarano, M. D. (2019). Adaptation effort and performance of water management strategies to face climate change impacts in six representative Basins of Southern Europe. *Water*, 11(5), 1078. https://doi.org/10.3390/w11051078

Thabit, M. S., Hawari, A. H., Ammar, M. H., Zaidi, S., Zaragoza, G., & Altaee, A. (2019). Evaluation of forward osmosis as a pretreatment process for multi stage flash seawater desalination. *Desalination*, 461, 22–29. https://doi.org/10.1016/j.desal.2019.03.015

Tsani, S., Koundouri, P., & Akinsete, E. (2020). Resource management and sustainable development: a review of the European water policies in accordance with the United Nations' sustainable development goals. *Environmental Science & Policy*, 114, 570–579. https://doi.org/10.1016/j.envsci.2020.09.008

United Nations. (2018). *World Water Development Report 2018: Nature-Based Solutions for Water*. UNESCO Publishing. https://unesdoc.unesco.org/ark:/48223/pf0000261424

UN-Water. (2019). *Climate Change and Water UN—Water Policy Brief* (p. 28). The UN-Water Expert Group on Water and Climate Change UN-Water Policy Brief. https://www.unwater.org/publications/un-water-policy-brief-climate-change-and-water

Valero, D., Cook, J., Lee, A., Browne, A. L., Ellis, R., Pancholi, V. S., & Hoolohan, C. (2023). Addressing water poverty under climate crisis: implications for social policy. *Social Policy and Society*, 22(4), 747–762. https://doi.org/10.1017/s1474746423000258

Vallino, E., Ridolfi, L., & Laio, F. (2020). Measuring economic water scarcity in agriculture: a cross-country empirical investigation. *Environmental Science & Policy*, 114, 73–85. https://doi.org/10.1016/j.envsci.2020.07.017

Voitko, S., Kukharuk, A., & Korohodova, O. (2024). An international cooperation-based approach to assessing the recreational potential of the danube region territories. *Věda a Perspektivy,* 5(36). https://doi.org/10.52058/2695-1592-2024-5

Wang, Y., Morosuk, T., & Cao, W. (2024). Carbon footprint of seawater desalination technologies: a review. *ASME Journal of Energy Resources Technology*, 146, 080801. https://asmedigitalcollection.asme.org/energyresources

Waterwise. (2023). AI and water management: a new era of water conservation. *Waterwise Innovations*. https://waterwiseinnovations.com/ai-and-water-management-a-new-era-of-water-conservation/?form=MG0AV3

World Health Organization. (2022). *Guidelines for Drinking-Water Quality: Fourth Edition Incorporating the First and Second Addenda*. https://www.who.int/publications/i/item/9789240045064

Yapıcıoğlu, P., & Demir, Ö. (2021). Minimizing greenhouse gas emissions of an industrial wastewater treatment plant in terms of water–energy nexus. *Applied Water Science*, 11(12). https://doi.org/10.1007/s13201-021-01484-4

Yusupova, G. N., Yugai, N. A., Choguldurov, M. D., & Khubieva, S. A. (2024). Development of water resources infrastructure in the Kyrgyz Republic: conflict potential. *BIO Web of Conferences*, 83, 04001. https://doi.org/10.1051/bioconf/20248304001

Zakariazadeh, A., Ahshan, R., Al Abri, R., & Al-Abri, M. (2024). Renewable energy integration in sustainable water systems: a review. *Cleaner Engineering and Technology*, 18, 100722. https://doi.org/10.1016/j.clet.2024.100722

Zollinger, G. (2024). *Estimating the Impact of Climate Change on Maize Cultivation in Switzerland Based on Agricultural Climate Indicators*. Zora.uzh.ch. https://www.zora.uzh.ch/id/eprint/262357/1/Thesis_Zollinger.pdfLiu

7 Artificial Intelligence for Global Water Security and Management

Saikat Biswas, Pritha Kundu, and Rupa Das

7.1 INTRODUCTION

Climate change is definitely one of the biggest challenges at hand, since it affects almost all spheres of the environment, the economy, and society as a whole. One of the most critical effects occurs on water resources, which threatens global water security. Water, which is essential to all organisms, has been made scarcer due to climate change that modifies its amount, spatial and temporal distribution, and its quality. Global warming, changes in precipitation, and an increase in evaporation, mainly in dry and semi-arid regions, are rapidly depleting the freshwater resources of the planet (Dao et al., 2024). These alterations are already aggravating the situation of geographical water scarcity and droughts, and may contribute to the increase in frequency and intensity of natural disasters, which negatively impact water sources and supply systems. As water sources shrink, the demand for this resource escalates, starting from agriculture to industry and eventually, human health and food safety. Agriculture, a branch that thrives on sufficient and steady water quantity, is at high risk, as low water volumes can lower agricultural productivity and compromise food quality (Bouteska et al., 2024). This scarcity further burdens rural economies and the food supply chain, creating a ripple effect that affects both urban and rural inhabitants equally. Meanwhile, public health is compromised due to a scarcity of clean drinking water, increased exposure to contaminants, and inadequate sanitation (Balasooriya et al., 2023). In light of these issues, creative and sustainable approaches to water management have become crucial for ensuring equal distribution and efficient use of water resources.

Amidst the growing needs, artificial intelligence (AI) is a game-changing technology that can enable new and effective water management solutions (Goralski and Tan, 2020). It enhances the precision and accuracy of forecasts for water demand, adapts resource management models, and provides predictive insights into climate-related water scarcity. Advanced AI algorithms can process massive amounts of climate, hydrological, and environmental data to predict water availability patterns, enhance agricultural irrigation methods, and identify leaks or inefficiencies in water distribution networks (Javaid et al., 2023). AI-powered models can also predict drought conditions, assess water quality, and provide early warning systems, enabling communities to prepare better for water-related difficulties (Olawade et al.,

DOI: 10.1201/9781003547099-8

2024). This chapter explores the relationship between water security, AI, and climate change and explores how these components may interact to create novel, sustainable water management strategies. It attempts to comprehensively review AI's capacity for enhancing water security through its ability to increase resilience to climate-driven water scarcity and facilitate effective resource distribution. This analysis focuses on the crucial role of technology in developing resilient water systems to withstand the difficulties associated with the rapidly changing global environment, using AI capability for data-driven insight, adaptability, and predictive accuracy.

7.2 THE IMPACT OF CLIMATE CHANGE ON WATER RESOURCES

Climate change is a gradual shift in climate variabilities over a period of time, mainly due to anthropogenic activities that alter the natural balance of the Earth. It negatively affects human activities, including agriculture, industrial manufacturing, transportation, and health infrastructure. It is predicted that the regions that have the least ability to adapt to climate change will be most severely affected by it (Sasmita and Mohanty, 2009). In the scenario of an increasing global population, there's an exponential increase in the global demand for food, climate change coupled with the intensification of extreme weather, is a great challenge in ensuring quality food and water security for all. Water supplies are impacted by climate change in a number of ways. Climate change modifies precipitation patterns, leading to erratic rainfall and prolonged dry spells, making water resource management challenging. This results in lower crop yields and reduced freshwater supply (Bharambe et al., 2023). Rising temperatures also increase evaporation, further reducing the freshwater supply. Additionally, glaciers melt, affects millions of people. Glacier retreat may temporarily increase water supply, but long-term effects are uncertain. Extreme weather events like hurricanes and floods pose risks to water security, contaminating sources and affecting communities, necessitating sustainable management (Bolan et al., 2023). It is estimated that by the end of 21st century, as many as 650 million people globally would have to migrate as a result of increased sea level and/or recurrent flooding due to climate change which causes oceans to warm and expand, triggering vast influxes of water from melting of glaciers and ice sheets (Akerlof et al., 2017). Climate change affects the temporal and spatial availability of water resources by altering the frequency, intensity, and timing of precipitation, which ultimately affects the hydrological cycle (Melese, 2016). Moreover, changes in hydrological systems will also affect water quality, water temperature, evapotranspiration by crops and natural vegetation, soil moisture content, and groundwater recharge. Climate change is projected to reduce the renewable surface water as well as groundwater significantly in many regions of the world, thereby intensifying competition for water use and increasing water scarcity worldwide (Dao et al., 2024). The existing climatic impact on water regime is further worsening other environmental and social water management systems, including reduced river runoff that concentrates pollutant content, exacerbating the spread of water-borne diseases, agricultural production processes, and industrial and urban development. The effect of climate change on different water resources and its effect on human health and the environment is presented in Figure 7.1. In the

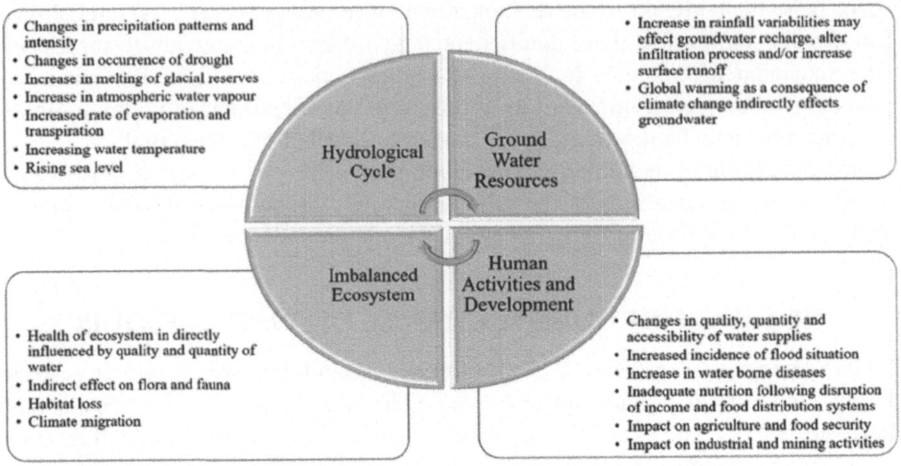

FIGURE 7.1 Effect of climate change on water resources and its impact on the environment and human life.

scenario of an increasing global population, there's an exponential increase in the global demand for food, climate change coupled with the intensification of extreme weather, is a great challenge in ensuring quality food and water security for all (Tchonkouang et al., 2024). Mitigation strategies to adapt to climate change need to carefully consider competing water uses and their various implications for food and water security, as adverse effects on one could exacerbate another.

7.3 VULNERABILITIES IN WATER SECURITY

Water is essential for human lives and socio-economic development as it is required to fulfil personal and household needs, for energy generation, industrial production, and to maintain water-dependent ecosystems and ecosystem services. Additionally, water is an integral input in the global food system, from production in fields all the way up in the value chain. With increasing demand, there's a rising challenge for water supply chain as the planet's water resources are under increasing stress due to poor harvesting and management techniques, increased pollution and contamination, and climate change as well (Cosgrove and Loucks, 2015). Vulnerable people are disproportionately affected by climate change, especially those in poor nations who do not have the means to adapt. Changing climate, in addition to rising global temperatures, is causing an intensification of the hydrological cycle, resulting in wetter rainy season as well as drier dry season, which significantly impacts the availability and accessibility of water. Melting of glaciers and rising sea level increase the risk of more frequent and extreme floods, while prolonged and extreme dry periods result in drought occurrences and reduce the dry-season water supplies to an estimated one-sixth of the global population (Melese, 2016). Prediction suggests that as an outcome of climate change, the availability of freshwater will increase in temperate regions of the world, whereas it will decrease significantly in regions close to the equator, including prominent agricultural and heavily irrigated areas in China, India,

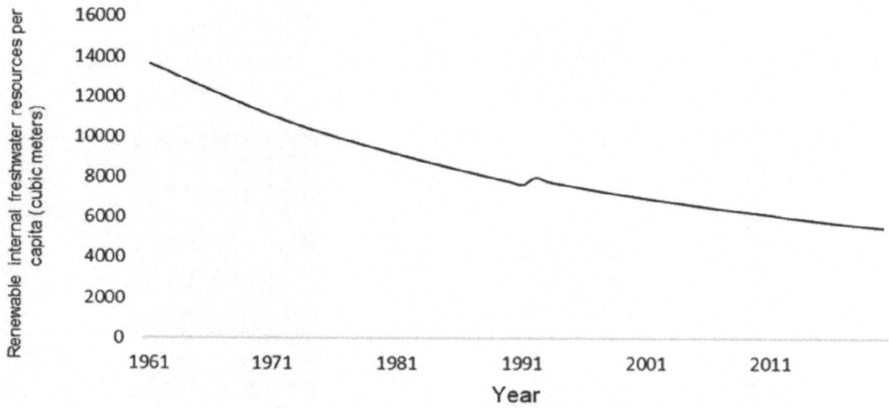

FIGURE 7.2 Global per capita renewable internal freshwater resource over the years (World Bank Open data, 2024).

and Egypt (Dao et al., 2024). By 2050, the world needs to produce an estimated 60% more food to ensure global food security while conserving and improving the natural resource base (FAO, 2024). The severely damaged communities are frequently those that rely on agriculture for a living since unpredictable weather patterns throw off the timing of planting and harvesting. Changes in water regime directly affect agricultural production by reducing crop yields and allied sectors like fisheries and livestock management, which will result in insufficient food supplies to hundreds of millions of people from around the globe, driving families to relocate in pursuit of better prospects. This can intensify social unrest and disputes over diminishing resources (Koren et al., 2021).

Global per capita renewable internal freshwater resource over the years indicates a declining trend of freshwater availability (Figure 7.2). Constraints in freshwater availability as a result of climate change impacts and increased weather-induced variability may lead to a huge reduction in the overall share of irrigated agricultural production in various parts of the world (Bharambe et al., 2023). To ensure the availability of sufficient water for crop production, transparency in the irrigation management systems should be maintained in addition to achieving efficiency and equity in water accessibility. This demands changes in the behaviour and attitude of the farmers, coupled with institutional changes by means of institutional restructuring, infrastructure modernization, capacity building, and upgrading of the technical capabilities of farmers and water managers (FAO, 2024). The effects of water scarcity in different sectors are presented in Figure 7.3. Presently, economic development and demographic growth are putting unprecedented pressure on the renewable but finite water resources, particularly in arid regions (Mishra et al., 2021). It is estimated that by 2025, 1,800 million people will be living in 'absolute' water scare regions (<500 m^3 per year per capita) while two-thirds of the global population will be living under 'stress' condition (between 500 and 1,000 m^3 per year per capita) (FAO, 2024). The situation is likely to worsen as rapid urbanization puts pressure on neighbouring water resources. Further, climate change and bio-energy demands are expected

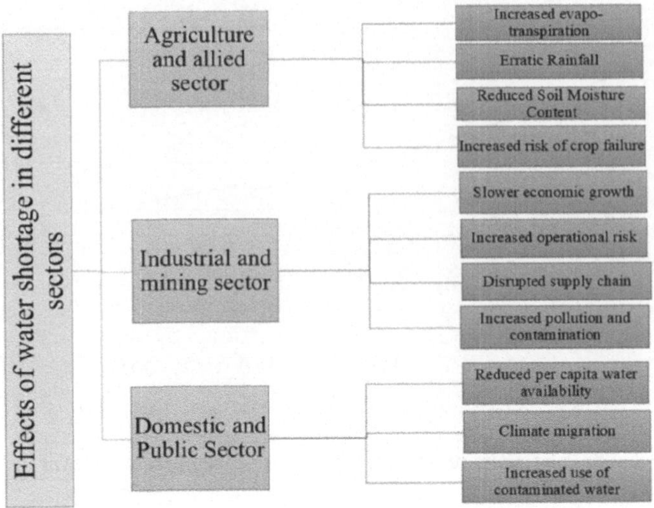

FIGURE 7.3 Effect of reducing water resources on different sectors.

to amplify the existing complex relationship between water demand and global development. Water scarcity is fundamentally dynamic where an imbalance between 'demand' and 'supply' varies according to the local situations. The several dimensions of water scarcity include non-availability of fresh water of acceptable quality, in-access to water services either due to lack of adequate infrastructure, financial constraint, or inadequate supply (Giordano et al., 2019). Besides, deterioration in the quality of water for millions of global citizens will expand the range of water-borne diseases as well as some dangerous vector-borne diseases like malaria, dengue, etc.

The infrastructure that is in place in many places is inadequate to meet the demands of climate change. Addressing the concern of efficient water management practices would need a well-articulated roadmap to move forward. It is very crucial to develop alternative and feasible ways to improve water security since conflict is more likely to arise as competition for scarce water resources grows. The adaptation options should indicate planned interventions to cope with water scarcity situation under climate change, considering both demand and supply side management strategies along with local needs to formulate an effective water management adaption plan (Iglesias and Garrot, 2015).

7.4 ARTIFICIAL INTELLIGENCE

The agricultural sector of the 21st century should adopt modern ways of working, information and communication technology (ICT), and AI, if it is to be able to efficiently and sustainably meet the global demand for food (Jha et al., 2019). The goal of AI is to devise institutions and systems of equations that solve problems like a person and thus relieve a person from routine control and adjustment operations (Shankar et al., 2020; Vyas et al., 2022). Usually, AI is deployed to make water supply decisions; however, it is also useful in water resource management approaches

TABLE 7.1
Use, Benefits and Limitations of Various Artificial Intelligence (AI) and Integrated IoT Systems/Models in Water Security

AI and Integrated IoT Systems/Models	Use	Benefits	Limitations	References
Convolutional Neural network (CNN), Long Short-Term Memory (LSTM)	Can be used to simulate water levels and water quality parameters like organic carbon, nitrogen and phosphorus content	CNN for prediction of water level LSTM for monitoring water quality	Use of limited data set Other parameters like dissolved oxygen, faecal bacteria and algae needs to be studied further	Baek et al. (2020)
Support Vector Machine (SVM), Support Vector Regression (SVR), Radial Basis Function Kernel, Random Forest Regression	Proposed IoT smart system for automation of agriculture by classification and quantitative predictions of soil type, crop type and amount of irrigation required by crops with an accuracy of 81.6%	Smart system for automating the agriculture sector	It does not support dynamic systems Limited data set Low accuracy	Vij et al. (2020)
U-Net, Tensor Flow Libraries, CNN	This proposed deep learning automatic detection models is able to locate, map, and quantify the centre pivot irrigation systems within the study area, while also detecting pivots of different sizes with variable land cover and spatial texture using a modified U-Net image segmentation algorithm	Helps to determine the centre pivot irrigation systems efficiently	The proposed model was deployed on short area and needs to be studied further at different location Time consuming	Saraiva et al. (2020)

(Continued)

TABLE 7.1 (*Continued*)
Use, Benefits and Limitations of Various Artificial Intelligence (AI) and Integrated IoT Systems/Models in Water Security

AI and Integrated IoT Systems/Models	Use	Benefits	Limitations	References
Deep Neural Networks (DNNs) Feed-Forward Deep Neural Networks (FF-DNNs), RMSprop(Root Mean square Propagation) optimization algorithm	The proposed model provide information in real time and comprise a powerful decision support system (DSS) for monitoring of the water quality parameters contributes in management by (a) controlling water quality for irrigation, (b) monitoring atmospheric conditions, (c) determining microclimate indicators, (d) issuing a warning in case of crises caused by extreme events, (e) providing continuous knowledge of the state of the water bodies, and (f) maintaining the ecological balance of the ecosystems and water resources of the region	Real-time water quality and monitoring model	Use of limited data set Need to improve the accuracy	Karamoutsou and Psilovikos (2021)
SVM, Long Short-Term Memory (LSTM)	The proposed IoT system can be used for monitoring real-time water quality in aquaculture and fisheries, as well as for forecasting quality indicators such as salinity, pH, temperature, and DO that will help in getting early warning and can help the farmers to manage the quality of the water	Efficient water quality monitor for aquaculture and fisheries	Implemented on limited data set Simple LSTM deployed, dynamic systems not considered	Thai-Nghe et al. (2020)
AlexNet, GoogleNet, ResNet, VGG16 (Visual geometry group of university of Oxford), and SqueezeNet (deep learning architectures)	This study proposed a proximal sensing system by using a colour camera towards smart irrigation based on computer vision and deep learning to identify water requirements of three soil texture classes under different illumination conditions	Offer more efficient water use in irrigated agriculture. By determining irrigation requirements of various soil texture classes in production fields using sophisticated sensing technologies	Limited number of studies have been conducted with respect to detecting irrigation requirements of soil based on deep learning and machine vision	Kurtulmus et al. (2022)

(*Continued*)

TABLE 7.1 (*Continued*)

Use, Benefits and Limitations of Various Artificial Intelligence (AI) and Integrated IoT Systems/Models in Water Security

AI and Integrated IoT Systems/Models	Use	Benefits	Limitations	References
U-Net architecture with a Resnet-34 backbone	Deep learning model to classify irrigation systems at a regional scale using remote sensing imagery	Effective and accurate remote sensing-based machine learning model to distinguish various irrigation systems	Further research needs to be done on the proposed model especially in wild land and urban areas intermingled with agricultural fields	Raei et al. (2022)
Principal Component Analysis (PCA), Random forest	This is a deep neural network-based complex system to formulate a water consumption prediction model with the aid of PCA and the integrated deep neural network, UWM-Id, for urban water management	Useful efficient Urban Water Management	Used limited data set Quality Parameters not considered	Gao et al. (2020)

(*Continued*)

TABLE 7.1 (*Continued*)
Use, Benefits and Limitations of Various Artificial Intelligence (AI) and Integrated IoT Systems/Models in Water Security

AI and Integrated IoT Systems/Models	Use	Benefits	Limitations	References
Multi-agent deep enforcement earning (MADRL) with LCA (Life Cycle Analysis) optimization	Low-cost camera-based automated system for the observation of FOG (fat, oil and grease) layer dynamics in wastewater pumping stations at high-frequency (minutes) over extended time windows (months) using optical imagery is processed through a deep learning computer vision routine that helps describing FOG layer dynamics (e.g. accumulation rate and changes in shape) and various hydraulic processes in the pump sump (e.g. the water level, surface flow velocity fields, vorticity, or circulation)	This monitoring system can be a source of information for the management of (waste) water pumping stations (e.g. detection of free-surface vortices and scheduling of sump cleaning operations) that facilitates the collection of standardized high-frequency FOG layer dynamics data. The system can perform in-camera image processing, thus allowing the transfer of compressed-processed datasets when deploy in remote locations (Edge AI computing), which can be utilized for the hydro-ecological monitoring community	Implemented on limited data set	Moreno-Rodenas et al. (2021)
K-Nearest Neighbour (KNN), SVM	This proposed system can help replace the classification and monitoring of the traditional method for determining water quality and can save human beings from various infections and diseases caused by the unsafe use of water	Automation water quality monitoring system	Uses different sensors such as pH, temperature, turbidity, and conductivity	AlZubi (2022)

(*Continued*)

TABLE 7.1 (*Continued*)
Use, Benefits and Limitations of Various Artificial Intelligence (AI) and Integrated IoT Systems/Models in Water Security

AI and Integrated IoT Systems/Models	Use	Benefits	Limitations	References
Multilayer Perceptron (MLP), Decision Tree (DT), Support Vector Machine (SVM), Logistic Regression (LR),Random Forest (RF), Extreme Gradient Boosting (XGBoost), Categorical Boosting (CATBoost)	This study utilized various AI algorithms and machine learning classifiers and their stacking ensemble models for classifying the Water Quality (WQ) data via the Water Quality Index (WQI) over an extended period for developing a dependable approach for forecasting water quality as accurately as possible	Using stacking models can help extract every bit of performance and can combine the capabilities of several high-performing models to make predictions that outperform any single model in the ensemble on a classification or regression task	Further research using more classifiers, neural networks, and other AI techniques needs to be used on the same or different datasets to enhance water monitoring on a global scale	Nasir et al. (2022)
Refined stochastic gradient descent (SGD)	Proposed IoT-based framework named SmartWater combines cutting-edge technologies in the field of sensor clouds, deep learning, knowledge reasoning, and data processing and analytics to support smart water monitoring and management	Provide efficient monitoring and management facilities for organizations in both the public and private sectors to ensure high-quality water resources and services	Spatial time-series forecasting techniques need to be explored to tackle the scalability problem caused by processing complex and large amounts of data collected by IoT networks	Mezni et al. (2022)

TABLE 7.2
Global Application of AI and Other Related Models/Technology in Water Security and Management in Various Sectors

Location/Country	Learning and Modelling Technique (Intelligent Models, Methods, and Technology Utilized)	Application(s)	Sector Influenced	References
Philippines	MATLAB, Neural Network Toolbox	Smart Farm Automated Irrigation system to ensure optimum water resource by ensuring water usage optimization	Agricultural Sector	Dela Cruz et al. (2017)
Africa	Radial Basis Function Network (RBFN)	Solar-powered smart irrigation system is designed using Internet of Everything (IoE) environment that predicts the expected water level values, weather forecasts, humidity, temperature, and irrigation data	Agricultural Sector	Adenugba et al. (2019)
Spain	Arduino microcontroller	A low-cost automated irrigation system for green walls has been designed that reduces energy consumption, increases irrigation efficiency and saves time	Agricultural Sector	Rivas-Sánchez et al. (2019)

(Continued)

TABLE 7.2 (*Continued*)
Global Application of AI and Other Related Models/Technology in Water Security and Management in Various Sectors

Location/Country	Learning and Modelling Technique (Intelligent Models, Methods, and Technology Utilized)	Application(s)	Sector Influenced	References
India	Fuzzy Logic-based AI	The projected work relies on independent IoT enabled wireless sensor network (WSN) framework consisting of soil moisture (MC) probe, soil temperature measuring device, environmental temperature sensor, environmental humidity sensing device, CO_2 sensor, daylight intensity device (light dependent resistor) to acquire real-time farm information through multi-point measurement for generating necessary action throughout the whole course of farming	Agricultural Sector	Keswani et al. (2019)
Greece	TelosB and the IRIS motes	Reference network architecture, full-scale smart irrigation system developed for growing strawberry under greenhouse environment primarily aimed towards developing smart irrigation	Agricultural Sector	Angelopoulos et al. (2020)

(*Continued*)

TABLE 7.2 (*Continued*)
Global Application of AI and Other Related Models/Technology in Water Security and Management in Various Sectors

Location/Country	Learning and Modelling Technique (Intelligent Models, Methods, and Technology Utilized)	Application(s)	Sector Influenced	References
Manila, Philippines	k-NN (k-nearest neighbour algorithm), SVM	Using images of crops for generating information on classification of growth stage (vegetative, head development, and harvestable) in lettuce grown using aquaponics	Aquaponics, Urban Indoor Farming	Lauguico et al. (2020)
--	Smart Sensing and Control	Decision support design for a novel farming methodology in which many different production units can be synchronized together to respond to consumer demand in a way that minimizes waste based digital twin and machine learning approach for generating predictive decision analytics to predict production from urban farming	Aquaponics, Urban Indoor Farming	Ghandar et al. (2021)

(*Continued*)

TABLE 7.2 (*Continued*)
Global Application of AI and Other Related Models/Technology in Water Security and Management in Various Sectors

Location/Country	Learning and Modelling Technique (Intelligent Models, Methods, and Technology Utilized)	Application(s)	Sector Influenced	References
India	Novel AI monitoring system, Smart Sensing and Control	To integrate intelligent agents in the management of hydroponic systems as well as collecting live data by means of development and execution of a hydroponic structure equipped with intelligent agents for internet-enabled monitoring, data collection and storage. These technologies allow for increased automation and refined control of hydroponic systems	Hydroponics, Urban Indoor Farming	Vanipriya et al. (2021)
--	Self-Learning AI	To be able to cope with increasing interference complexity of fertilizer solutions in greenhouses, a NPK (nitrogen, phosphorus, potassium) spectroscopy-based, AI-supported by a robust self-learning artificial intelligence was developed for obtaining reliable, robust and accurate output	Hydroponics, Urban Indoor Farming	Silva et al. (2021)

(*Continued*)

TABLE 7.2 (*Continued*)
Global Application of AI and Other Related Models/Technology in Water Security and Management in Various Sectors

Location/Country	Learning and Modelling Technique (Intelligent Models, Methods, and Technology Utilized)	Application(s)	Sector Influenced	References
Bangladesh	Random Forest and Random Subspace models	Used for prediction of groundwater potential	Agricultural Sector, Environmental Sector	Sarkar et al. (2022)
India	ANN, ANFIS	Predicting dissolved oxygen (DO) concentration in upstream, midstream and downstream using pH, biological oxygen demand (BOD), water temperatures, DO as input variables	Water Quality Assessment, Environmental Sector	Elkiran et al. (2018)
USA	Extreme learning machine (ELM), ANN, AFIS, RF	Predicting DO concentration using hourly temperature, pH, specific conductivity as input variables	Water Quality Assessment, Environmental Sector	Kisi et al. (2020)
Malaysia	ANN, Recurrent neural network (RNN)(LSTM), ELM, HW, ANN-RF, RNN(LSTM)-RF, ELM-RF, HW-RF	Using BOD, chemical oxygen demand (COD), pH, NH_3, temperature, chlorine, calcium, sodium, and total solids concentrations for predicting DO concentration	Water Quality Assessment, Environmental Sector	Abba et al. (2020)
Tabriz, Iran	ANN, ANFIS, SVM	Daily influent BOD/COD, TSS, pH, previous BOD/COD effluent as input variables for determining BOD, COD effluent	Waste Water Treatment, Environmental Sector	Nourani et al. (2021)
Malaysia	ANN, ANFIS	For predicting the future water level	Water Quality Assessment, Environmental Sector	Zakaria et al. (2021)

(*Continued*)

TABLE 7.2 (*Continued*)
Global Application of AI and Other Related Models/Technology in Water Security and Management in Various Sectors

Location/Country	Learning and Modelling Technique (Intelligent Models, Methods, and Technology Utilized)	Application(s)	Sector Influenced	References
Palla, India	ANN, ANFIS	Using DO, pH, BOD, NH_4, water temperature as input variables for determining water quality index	Water Quality Assessment, Environmental Sector	Gaya et al. (2020)
India	ANN, ANFIS, SVM	Determining DO concentration using pH, BOD, COD, flow rate, NH_3 concentration, water temperature as input variables	Water Quality Assessment, Environmental Sector	Elkiran et al. (2019)
Thailand	SVM with varying kernel functions	Setting up water quality prediction model for better water resource management using BOD, DO, faecal coliform bacteria, total coliform bacteria, ammonia concentration, salinity as input variables	Water Quality Assessment, Environmental Sector	Ahmed et al. (2019)
Malaysia	ANN	To predict time-series data of water quality parameters (WQP), i.e., DO, BOD, COD, pH, ammonia nitrogen (NH_3-NL), suspended solids (SS)	Water Quality Assessment, Environmental Sector	Hayder et al. (2020)

(Continued)

TABLE 7.2 (*Continued*)
Global Application of AI and Other Related Models/Technology in Water Security and Management in Various Sectors

Location/Country	Learning and Modelling Technique (Intelligent Models, Methods, and Technology Utilized)	Application(s)	Sector Influenced	References
Mesa, Arizona	RNN(LSTM)	To predict Blue-green algae conc., conductivity, chlorophyll conc., DO, pH, turbidity in algae cultivation pond using microbial potentiometric sensor measured open-circuit potentials	Water Quality Assessment, Environmental Sector	Saboe et al. (2021)
China	CNN (Convolutional Neural Network), LSTM (Long Short-Term Memory), hybrid CNN-LSTM model	Helps to determine water level	--	Zhang et al. (2023)
Hong Kong, China	ANN, Fuzzy Network	Useful for predicting soil moisture content	Agricultural Sector	Tsang and Jim (2016)
Johor state, Malaysia	ANFIS	Using temperature, conductivity, salinity, nitrate, turbidity, phosphate, chloride, potassium, sodium, magnesium, iron, and *E. coli* concentrations to predict water quality	Water Quality Assessment, Environmental Sector	Sillberg et al. (2021)
Hesse, Germany	Regression trees, Random Forest, and Boosted Regression trees	Used for prediction of nitrogen concentration in ground water	Agricultural Sector, Environmental Sector	Knoll et al. (2019)
Wailuku River, Hilo, Hawaii	ANN, ELM, SVR	The performance of machine learning models applied for forecasting of water quality parameters in the river-flow-induced state	Water Quality Assessment, Environmental Sector	Alizadeh et al. (2018)

employing data analytics, regression models, and algorithm (Table 7.1). It is relieving most of the activities from a human being to AI due to the fact that AI helps in resolving aspects such as soil, crops establishment, production and monitoring, weed management, irrigation, and water control, and so forth (Kim et al., 2008). These agricultural challenges can also be managed through AI by assessing historical weather data and satellite images, creating an outlook for the future, and monitoring indicators such as farm health, rainfall, and sunshine. In the present climate change scenario, it provides a modern component that is meant to promote accessibility of water (Pigola et al., 2021). One of the central elements of AI is machine learning (ML), which incorporates a number of mathematical and statistical models that help make more accurate and timely predictions in enhancing productivity (Patel et al., 2021). It mainly focuses on enhancing the agricultural systems and resource usage efficiency by providing precision, prediction, and performance evaluation algorithms for pest recognition and irrigation management (Table 7.2).

Devices that fall under the Internet of Things (IoT) are intelligent devices containing embedded systems like microcontrollers, sensors, and communication hardware with the ability to connect (Oliveira et al., 2024). In addition, IoT uses the information obtained from the environment through gadgets to collect, distribute, and take actions. These devices communicate with each other and take actions based on the information they exchange. Biswas et al. (2021) have shown how IoT data was utilized for predicting the water needs. It was realized that AI types of technologies and models available in India are lessening the age-old farm labour dependency and ensuring stable income and crop yields (Talaviya et al., 2020). Considerable advances have been made in the development of such AI models in the context of agriculture namely those relating to the prediction of water requirements. AI uses both the historical data of water consumption as well as the current situation in order to make predictions about the development of the water supply systems and the demand for water. This enhances the effectiveness and the reliability of the water management systems, especially for the regions that are facing a lot of scarcity of water resources. AI also involves predictive maintenance that anticipates outages or service requirements of the parts of the water system. This allows for effective scheduling of equipment changes to reduce costs and minimize losses from unscheduled outages (Rojek et al., 2023).

7.5 THE ROLE OF AI IN WATER MANAGEMENT

Water resource management is key for biodiversity maintenance in nature and for treating complex issues that affect the stability and resilience of ecosystems (Bratsberg and Rogeberg, 2018). These require new paradigms such as protection, efficiency, and monitoring. In this regard, AI plays an important role, particularly in treating large amounts of data (Figure 7.4). However, an elaborative plan is needed to investigate and develop integrated strategies for the whole hydrological cycle. Further, viable AI-based numerical tools and approaches responding to global water management issues evoked by climate change are used to improve techniques of water resource management (Krishnan et al., 2022). It helps in tariff management, protects older water resources from ageing out, and identifies growth-restraint

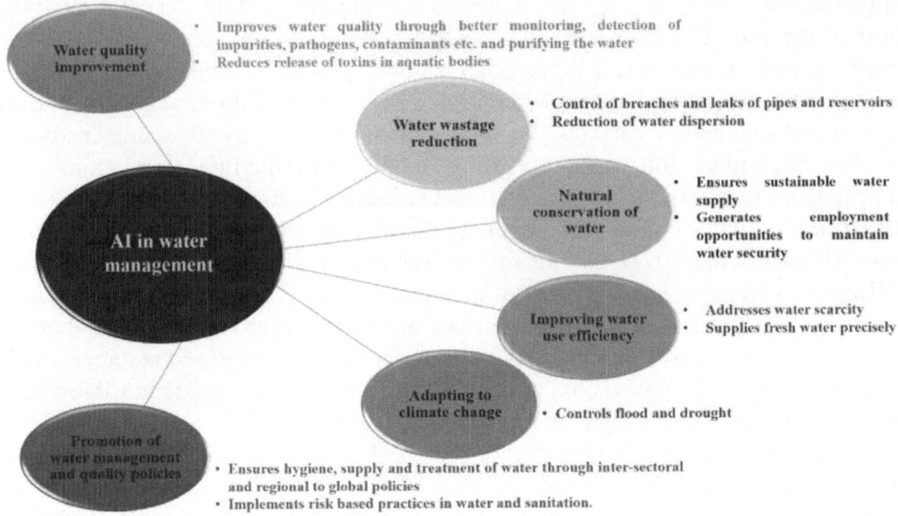

FIGURE 7.4 Role of artificial intelligence (AI) in water management.

factors in water resources as well. It also develops soil-water-plant relationships in agriculture, incorporates the impact of urbanization and climate change, etc. It can generate some accurate estimations of how much water is available for use and how much might be in demand through analytics of gigantic data coming through sensor networks, satellite pictures, and climate models (Nourani et al., 2014). This ability is important in the efficient management of water resources; it enables resource managers and policymakers to make decisions. AI algorithms can process large data sets to explore patterns and trends in usage, precipitation, and evaporation.

In addition, AI helps to proactively manage resources where ML techniques can forecast future available water based on historical data. AI models can predict drought conditions months ahead to prepare communities by providing enough time to plan and adjust the consumption rate of water usage (Saleh et al., 2024). Through real-time analysis of usage patterns as well as environmental factors, it can optimize the distribution of water supplies. AI-integrated decision support systems can provide decision-makers with useful information for sustainable water management (Sulistyaningsih et al., 2021). The system permits stakeholders to analyse the probable consequences of their choices through simulation of different scenarios using varying climate conditions and approaches to water management. This cooperative method encourages adaptable methods in the face of uncertainty and well-informed decision-making. In the following, the role of AI in water management, supply, quality, treatment, etc. is explored.

7.5.1 AI in Agricultural Water Management

Proper use of water in agriculture serves as the key to magnifying crop yield besides lowering costs, and this will significantly contribute to environmental sustenance. The quality and quantity of the crop can be much enhanced with this irrigation system. Irrigation management employs AI to schedule, identify problems, optimize water usage, and increase application accuracy. AI reduces the need for freshwater resources in the agriculture supply chain, thereby significantly improving efficiency. AI-based precision irrigation replicates the characteristics of natural ecosystems through ML, data analytics, and real-time monitoring (Akhter and Sofi, 2022; Upreti, 2022). With the use of AI, smart irrigation systems can calculate the right amount of water that should be supplied to crops based on the needs of the plants, the amount of moisture present in the soil, and the weather conditions. This means that crops can receive the right amount of moisture, increasing agricultural productivity and saving water at the same time.

AI-based irrigation methods can also monitor soil moisture level for precise water application. This will possibly decrease water usage and even monitor the wastage of water, which may become vital for checking the global water scarcity (Jha et al., 2019). So as to determine farmland management decisions, soil characteristics, precipitation, and evapotranspiration, AI techniques, such as artificial neural networks (ANNs) and regression modelling, are applied. Regression models fetch and analyse data collected by sensors that measure various soil properties and moisture levels, and so on (Morellos et al., 2016). ANNs are very much used in agriculture for parallel reasoning-based prediction and forecasting rather than the traditional programming method where a person has to learn input to do some work (Jain et al., 1996). AI-based irrigation analytics, such as ANNs, can predict the soil moisture availability and its use for irrigation planning and the generation of moisture availability maps. It can also enhance the quality of water (Dursun and Özden, 2014). Arif et al. (2013) predicted the levels of soil moisture using ANN-based models and made provisions for irrigation. ML algorithms enhance the quality of water as described by Jayaraman et al. (2022) and the temporal availability of soil moisture for irrigation planning is predicted by Kumar et al. (2023). Parra et al. (2020) used AI to predict the productive efficiency of an orchard using proper irrigation. Dela Cruz et al. (2017) used the feed-forward and back-propagation technology of ANN for optimizing the water resources of a smart farm. Agricultural water management requires highly intensive usage of sensors and IoT to monitor the resources that can be controlled and improved. Oberascher et al. (2021) developed an intelligent approach for the de-centralized water infrastructure via low-cost sensors. Gonçalves et al. (2020) put forward the concept of an architecture known as REFlex Water architecture that helps bring self-sufficiency within water management.

Decision support systems (DSS) represent the computer-based technologies used for smart and intelligent irrigation schedule decisions in relation to data and models of prediction (Saggi and Jain, 2022). The data collected from weather stations, soil moisture sensors, satellite pictures, and crop models are analysed to maximize irrigation techniques. Thus, applied in decision-making tasks through AI learning techniques, their applications provide water-smart solutions, improve harvests using less water, and manage agricultural water sustainably. DSS can lower water intake by 20% but retain crop yield (Gu et al., 2021). Some of the breakthroughs in AI include aerial photogrammetry, unmanned aerial vehicles, and 3-D maps for soil testing before planting. Scarcity of water uses thermal infrared images, meteorological data, and information on the soil to understand drought stress, soil water potential, and plant water content. Models include convolutional neural networks, random forests, and gradient-boosted decision trees (Saggi and Jain, 2019; Zhou et al., 2021). Further, the use of agricultural drones can water, apply fertilizer, herbicides, and pesticides, take pictures, and create real-time maps to help farmers make management decisions (Rejeb et al., 2022).

AI breakthroughs in weather forecasting have helped farmers to adjust their irrigation systems more effectively for natural rain or compensate for droughts. Remote sensing technology with AI and satellite imaging offers many possibilities in monitoring agricultural areas and water supply control. Moreover, AI aids in the assessment of past data to predict the amount of irrigation needed for the future and update the recommendations with changing conditions. The use of AI in irrigation management has revolutionized irrigation management because it can provide the development of automated and accurate irrigation systems. The cycles of watering are controlled by automatic systems that make use of sensors and data for optimizing water usage and waste (Obaideen et al., 2022). ML algorithms for drought prediction work well when used with sensors, ZigBee, and Arduino microcontrollers (Arvind et al., 2017; Goap et al., 2018).

7.5.2 AI in Detection of Water Contamination and Quality Water Assessment

Clean water is vital for human health. AI ensures quality water through object recognition using ML, followed with deep learning. AI-powered technology uses sensors that can detect pollutants and anomalies (Ramadan et al., 2024). Real-time alerts allow immediate response to issues, but predictive models forecast future pollution events or changes in water quality that allow preventive measures to maintain water quality. Two of the AI model examples that may be implemented for monitoring surface water quality to improve water resource management and prevent pollution are: long short-term memory (LSTM) algorithms and ANNs. Furthermore, convolutional neural networks (CNNs), with IoT, allow performing real-time analysis and detecting contaminants even without internet facilities (Said et al., 2021). The IoT is increasingly applied to the monitoring of water quality because diverse parameters, such as pH and total dissolved solids (TDS), reflect different outcomes (Aldhyani et al., 2020). CNN-based models can identify foam and algae in water, cleanse them of impurities, and convert them to waste. Sensors monitor different environmental factors like pH, nitrate, alkali metal, halide, surfactant, pesticide, and fertilizer concentration. Potentiometric sensors may detect organic pollutants, heavy metals, and organic contamination. SGM (semiconductor/graphene/metal) thin-film devices may detect organic pollutants. Laser interferometry sensors detect chemical contaminants by measuring water refractive index changes. MEMS (Micro- Electro - Mechanical Systems) acceleration sensors are low-power devices that measure water flow fluctuations. MEMS sensors with enzymatic amplification detect bacterial agents (Massarelli et al., 2021). AI, when integrated with novel urban sewer efficiency management systems, can reduce contamination hazards (Bogataj et al., 2017). AI models can quickly and accurately process large amounts of data, identifying trends in water quality data that traditional methods may struggle to detect. This technology has the potential to significantly improve surface water quality assessment and management.

Early simulation models like the Streeter-Phelps model and Bi-S-SRU are also being explored for accurate water quality forecasting (Ahmad et al., 2016).

7.5.3 AI in Wastewater Management

AI has improved the effectiveness and efficiency of wastewater management. Analysis of chemical and physical monitoring data assists AI in making a complete contribution to understanding the composition of wastewater. It has also been used in data collection for wastewater management, ensuring greater safety in the operation of dams, and minimizing the risk of flooding. Zhao et al. (2020) studied AI applications in wastewater treatment, where it has been proven to be efficient and cost-effective. They found that ANN and federated learning (FL) are the most efficient techniques. Other AI techniques in this regard are adaptive neuro-fuzzy inference systems (ANFIS), support vector regression (SVR), and feed-forward neural networks (FFNN) (Krishnan et al., 2022).

7.5.4 AI in Domestic and Commercial Water Use

AI algorithms can predict the amount of water to be consumed by using past and current measurements. Several models have been developed for the real-time monitoring of water usage and estimation of the demand for water by houses and businesses, including SCEM-UA ANN (Shuffled Complex Evolution Metropolis Algorithm), FFBP-ANN (Feed-Forward Back-Propagation), SMA-ANN (Slime Mould Algorithm), BSA-ANN (Backtracking Search Algorithm), and CSA-ANN (Crow Search Algorithm) (Morain et al., 2024). The use of deep learning neural networks with CNN, or LSTM has enhanced the accuracy of forecasts of water usage. These include fuzzy models in combination with AI models. Hybrid models are commonly used, such as ANFIS, combining the features of neural networks and fuzzy inference systems, which are used in forecasting urban water consumption. Drogkoula et al. (2023) have looked into the application of ML to water management and have found that AI in its various forms has significantly evaluated water use in domestic and commercial sectors. They studied AI-enabled technologies as enablers or barriers to achieving the sustainable development goals (SDGs), including smart water management and AI applications in household, industry, agriculture, and sanitation services.

7.5.5 AI in Adopting Global Warming and Improving Water Storage Structures

Precision irrigation by AI helps different sectors to lower carbon footprint and drive global sustainability goals, reducing the destructive impacts of climate change (Kim and Alzubi, 2024). In addition, AI can monitor water infrastructure in a real-time cycle, noting potential problems ahead of their occurrence. Predictive AI-based maintenance can identify breaks in pipelines and review the integrity of dam structure for proactive repairs (Hallaji et al., 2022). Communities can help avoid the failure of infrastructure, hence retain access to clean water and minimize the costs incurred on repairs.

7.6 INTEGRATED APPROACHES TO WATER RESOURCE MANAGEMENT

Though AI facilitates advancement in water security greatly, it should be integrated as an enhancement to existing practices rather than being viewed as a replacement. The incorporation of AI technologies with conventional frameworks, such as the Integrated Water Resource Management (IWRM) framework, enhances performance in accomplishing effective water security plans (Bilalova et al., 2023). IWRM focuses on the relationships among water, the environment, and human activities. It can enhance the ability of water resource managers and policymakers to adopt holistic and flexible strategies to the complex challenges posed by climate change and growing demand for water resources by incorporating AI (Bibri et al., 2024). Such integration ensures a more contemporary pattern of monitoring, forecasting, and decision-making within water resource management, such as through AI's integration into IWRM. AI, specifically, can offer prediction models to guide water allocations on the basis of both real-time conditions and projected future water demands so that shortages do not occur and resources are not overexploited. Additionally, AI can assist in the performance of ecosystem monitoring whereby stakeholders can use AI in order to monitor pollution, the status of biodiversity, and the welfare of water ecosystems (Ditria et al., 2022).

In addition to these challenges, the implementation of AI and IWRM principles necessitates the urgent development of modern, resilient infrastructure that is designed to cope with the effects of climate change (Krishnan et al., 2022). Water systems in many places across the globe are old and ineffective resulting in acceptable levels of water wastage due to leakage, evaporation, and poor system design. These should be upgraded to reduce the wear and tear of the systems and to reduce water wastage and promote water efficiency by the government and private sector agencies. To help save water and also provide water for use when it is required, smart infrastructure changes like smart metres, leak detection systems, and efficient water technologies for irrigation should be introduced (Gupta et al., 2020). Water management at the community level is equally very important. Since water management involves people, taking them through the processes assures that solutions are fitting into their particular needs and the context of those most affected by water scarcity. Allowing local communities to be a part of the discussion regarding water and its resources reclamation by stakeholders will help bring about efficient and sustainable water management regimes (Ahmadov, 2020). This, in turn, fosters accountability, as well as healthy relationships between the organizations and common inhabitants leading to greater and sustainable results.

Not only the traditional methods, but also the integration of AI with emerging smart technologies can aid in effective water resource management. Converging AI and big data analytics are transforming agricultural water management by giving deeper insights and better decision-making skills. AI algorithms can evaluate massive amounts of data from sources such as satellite photos, weather stations, soil sensors, and IoT devices to forecast irrigation requirements and optimize water usage (Wei et al., 2024). This information can assist in alleviating the effects of droughts and other water-related issues. Integrating AI with big data analytics enables precision agriculture, which reduces resource waste and increases agricultural yields. Blockchain technology improves transparency, data security, and traceability in water management, changing systems into safer and more efficient operations (Soori et al., 2023). Blockchain and AI are transforming water management and supply chain management, reducing discrepancies in water access and billing (Ashoka et al., 2024). This integration can also optimize the use of renewable energy in irrigation systems, reducing carbon footprint and dependency on non-renewable sources. AI algorithms can predict optimal times for pump operation, minimizing reliance on stored power or grid electricity. Real-time data can also be used to adjust irrigation system operations, ensuring efficiency under varying environmental conditions. In India, integrating AI with solar-powered irrigation systems resulted in a 50% reduction in water and energy use compared to traditional methods (Thakur et al., 2022).

7.7 GLOBAL COLLABORATION FOR SUSTAINABLE SOLUTIONS

The issues regarding water security and climate change are global and require a global response. As climate change increases the extent of aridity, alters rainfall, and brings about droughts, it does not place itself squarely on the shoulders of any one country (Adamo et al., 2022). Innovative concepts and benchmark exchanges

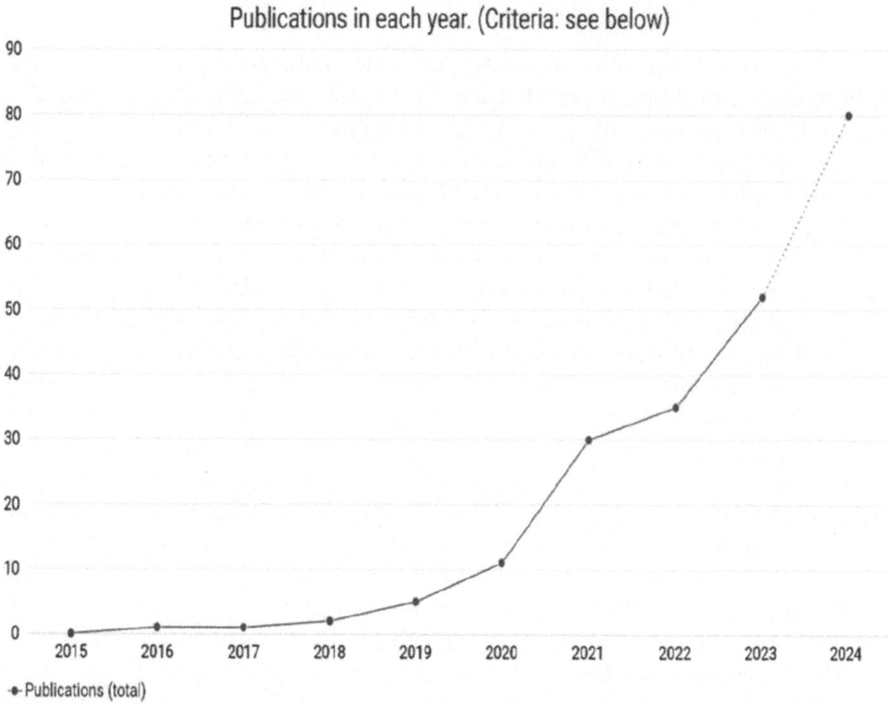

FIGURE 7.5 Publication trend on artificial intelligence AND Water Security AND water management from 2015 to 2024.

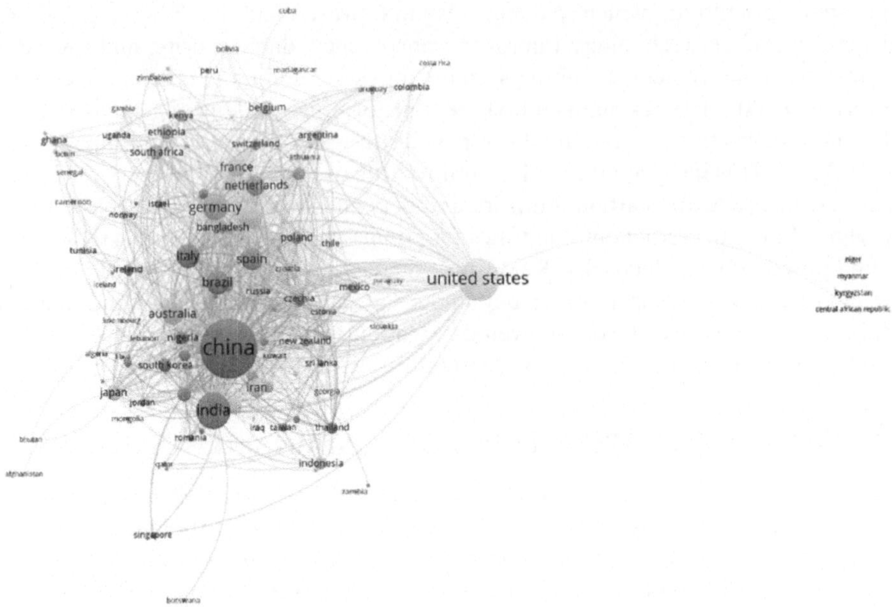

FIGURE 7.6 Countries collaboration in research and publications on artificial intelligence AND water Security AND water management.

must come from all segments of society: the state, non-state actors, academia, and the business sector. Gaps in knowledge and resources that cannot be rectified at the national level underscore the necessity of cooperation, particularly with respect to those groups who face significant water issues but do not have the capacity or the resources to address them (Agrawal and Gibson, 1999). Cooperation allows wealthy countries to render their assistance through technology and knowledge transfer, as well as through implementing regional water management projects in poorer countries. For instance, international agencies can provide sophisticated monitoring systems and AI weather forecasting, and modern economic efficient agriculture to areas that face acute water shortages (Pandey and Mishra, 2024). Such collaborations help to alleviate the communities' problems while at the same time ensuring that the solutions geared towards water security are not only imaginative but available to all at a reasonable cost. In this regard, a bibliometric analysis has been made to observe the trend of publication from 2015 to 2024 on 'Artificial intelligence AND water management AND water security' and countries' collaboration in research publications. The data has been collected from 'Dimensions' database (Dimensions, 2024), and network analysis has been made in VOSviewer (ver. 1.6.20, Van Eck and Waltman, 2014). A gradual increase in publications (a total of 217 publications consisting of 120 articles, 75 book chapters, 19 proceedings, and 14 edited book) over the years has been noted, peaking in 2024 (>50 publications), which indicates the growing prospects and applications of AI on water security and management (Figure 7.5). Further, the network analysis on the collaboration of countries shows that there are 122 countries that have collaborated at least once. The highest collaboration has been found by China, followed by the USA and India, indicating the top three countries executing collaborative research on AI and water security and management (Figure 7.6).

Furthermore, the implementation of campaigns, strategies, and initiatives that intend to educate the public on the need to conserve water or adopt water-friendly practices contributes immensely to increasing water security (Lebu et al., 2024). Such awareness campaigns promote partnerships with local authorities and their agencies, non-government organisations (NGOs), and community-based organizations in order to reach out to as many people as possible. In this way, water use is viewed as a precious resource that should be managed wisely. Besides, schools and other learning institutions contribute a lot to helping students understand the need to preserve water for the coming generations. This can be accomplished by integrating these elements into the teaching syllabus so that, at an early age, students' understanding of global warming, care for water, and living sustainably is entrenched. Moreover, involving youth in such activities as tree planting, waste clean-up within watersheds, and water conservation efforts enhances environmental consciousness and nurtures sustainable practices in them, even in the years to come.

7.8 CONCLUSIONS

The nexus of AI, water security, and climate change offers very significant opportunities and challenges. A study on the impact of climate change on water supplies, besides the utilization of AI tools, can lead to increased resilience and sustainable development. The chapter has tried to provide an exhaustive discussion on these issues while underlining the necessity of integrated approaches, collaboration, and involvement of community members. Innovative technologies can render a resilient and sustainable future in an ever-changing environment and can protect water resources for both the present and the future. Low-cost sensors that are accessible to a wide number of people ensure that citizens as well as decision-makers can easily access critical data, bringing the gap between local problems and global solutions much closer. Such a system will be able to offer precise water management insights aligned with specific users and areas, thus enhancing efforts toward conservation through the use of AI technologies that can evaluate huge volumes of data. However, it is still required to explore and utilize AI in water security, which can help the communities adapt to climate change concerns.

REFERENCES

Abba, S. I., Linh, N. T., Abdullahi, J., Ali, S. I., Pham, Q. B., Abdulkadir, R. A., Costache, R., Nam, V. T., & Anh, D. T. Hybrid machine learning ensemble techniques for modeling dissolved oxygen concentration. *IEEE Access*, 2020, 8, 157218–157237.

Adamo, N., Al-Ansari, N., Sissakian, V., Fahmi, K. J., & Abed, S. A. Climate change: droughts and increasing desertification in the Middle East, with special reference to Iraq. *Engineering*, 2022, 14(7), 235–273.

Adenugba, F., Misra, S., Maskeliūnas, R., Damaševičius, R., & Kazanavičius, E. Smart irrigation system for environmental sustainability in Africa: an Internet of Everything (IoE) approach. *Mathematical Biosciences and Engineering*, 2019, 16(5), 5490–5503.

Agrawal, A., & Gibson, C. C. Enchantment and disenchantment: the role of community in natural resource conservation. *World Development*, 1999, 27(4), 629–649.

Ahmad, Z., Rahim, N. A., Bahadori, A., & Zhang, J. Improving water quality index prediction in Perak River basin Malaysia through a combination of multiple neural networks. *International Journal of River Basin Management*, 2016, 15, 79–87.

Ahmadov, E. Water resources management to achieve sustainable development in Azerbaijan. *Sustainable Futures*, 2020, 2, 100030.

Ahmed, A. N., Othman, F. B., Afan, H. A., Ibrahim, R. K., Fai, C. M., Hossain, M. S., Ehteram, M., & Elshafie, A. Machine learning methods for better water quality prediction. *Journal of Hydrology*, 2019, 578, 124084.

Akerlof, K., Covi, M., & Rohring, E. Communicating sea level rise. In von Storch, H. (ed.) *Oxford Research Encyclopedia of Climate Science*, Oxford, UK: Oxford University Press, 2017.

Akhter, R., & Sofi, S. A. Precision agriculture using IoT data analytics and machine learning. *Journal of King Saud University – Computer and Information Sciences*, 2022, 8, 5602–5618.

Aldhyani, T. H. H., Alrasheedi, M., Alqarni, A. A., Alzahrani, M. Y., & Bamhdi, A. M. Intelligent hybrid model to enhance time series models for predicting network traffic. *IEEE Access*, 2020, 8, 130431–130451.

Alizadeh, M. J., Kavianpour, M. R., Danesh, M., Adolf, J., Shamshirband, S., & Chau, K. W. Effect of river flow on the quality of estuarine and coastal waters using machine learning models. *Engineering Applications of Computational Fluid Mechanics*, 2018, 12, 810–823.

AlZubi, A. A. IoT based automated water pollution treatment using machine learning classifiers. *Environmental Technology*, 2022, 45(12), 1–18.

Angelopoulos, C. M., Filios, G., Nikoletseas, S., & Raptis, T. P. Keeping data at the edge of smart irrigation networks: a case study in strawberry greenhouses. *Computer Networks*, 2020, 107039.

Arif, C., Mizoguchi, M., & Setiawan, B. I. Estimation of soil moisture in paddy field using artificial neural networks. arXiv preprint, arXiv:1303.1868, 2013.

Arvind, G., Athira, V. G., Haripriya, H., Rani, R. A., & Aravind, S. Automated irrigation with advanced seed germination and pest control. In *Proceedings of the 2017 IEEE Technological Innovations in ICT for Agriculture and Rural Development (TIAR)*, pp. 64–67, Chennai, India, April 2017.

Ashoka, P., Devi, B. R., Sharma, N., Behera, M., Gautam, A., Jha, A., & Sinha, G. Artificial intelligence in water management for sustainable farming: a review. *Journal of Scientific Research and Reports*, 2024, 30(6), 511–525.

Baek, S. S., Pyo, J., & Chun, J. A. Prediction of water level and water quality using a CNN-LSTM combined deep learning approach. *Water*, 2020, 12, 3399.

Balasooriya, B. K., Rajapakse, J., & Gallage, C. A review of drinking water quality issues in remote and indigenous communities in rich nations with special emphasis on Australia. *Science of the Total Environment,* 2023, 903, 166559.

Bharambe, K. P., Shimizu, Y., Kantoush, S. A., Sumi, T., & Saber, M. Impacts of climate change on drought and its consequences on the agricultural crop under worst-case scenario over the Godavari River Basin, India. *Climate Services*, 2023, 32, 100415.

Bibri, S. E., Krogstie, J., Kaboli, A., & Alahi, A. Smarter eco-cities and their leading-edge artificial intelligence of things solutions for environmental sustainability: a comprehensive systematic review. *Environmental Science and Ecotechnology,* 2024, 19, 100330.

Bilalova, S., Newig, J., Tremblay-Lévesque, L. C., Roux, J., Herron, C., & Crane, S. Pathways to water sustainability? A global study assessing the benefits of integrated water resources management. *Journal of Environmental Management*, 2023, 343, 118179.

Biswas, S., Sharma, L. K., Ranjan, R., Saha, S., Chakraborty, A., & Banerjee, J. S. Chapter 20- Smart farming and water saving-based intelligent irrigation system implementation using the Internet of Things. In *Recent Trends in Computational Intelligence Enabled Research*, Elsevier Inc., p. 339–354, 2021. https://doi.org/10.1016/B978-0-12-822844-9.00043-8

Bogataj, D., Bogataj, M., & Hudoklin, D. Mitigating risks of perishable products in the cyber-physical systems based on the extended MRP model. *International Journal of Production Economics*, 2017, 193, 51–62.

Bolan, S., Padhye, L. P., Jasemizad, T., Govarthanan, M., Karmegam, N., Wijesekara, H., … & Bolan, N. Impacts of climate change on the fate of contaminants through extreme weather events. *Science of the Total Environment,* 2023, 909, 168388.

Bouteska, A., Sharif, T., Bhuiyan, F., & Abedin, M. Z. Impacts of the changing climate on agricultural productivity and food security: evidence from Ethiopia. *Journal of Cleaner Production*, 2024, 449, 141793.

Bratsberg, B., & Rogeberg, O. Flynn effect and its reversal are both environmentally caused. *Proceedings of the National Academy of Sciences of the United States of America,* 2018, 115(26), 6674–6678. https://doi.org/10.1073/pnas.1718793115

Cosgrove, W. J., & Loucks, D. P. Water management: current and future challenges and research directions. *Water Resources Research*, 2015, 51(6), 4823–4839.

Dao, P. U., Heuzard, A. G., Le, T. X. H., Zhao, J., Yin, R., Shang, C., & Fan, C. The impacts of climate change on groundwater quality: a review. *Science of the Total Environment*, 2024, 912, 169241.

Dela Cruz, J. R., Baldovino, R. G., Bandala, A. A., & Dadios, E. P. Water usage optimization of Smart Farm Automated Irrigation System using artificial neural network. In *2017 5th International Conference on Information and Communication Technology (ICoIC7)*, Malacca, Malaysia, IEEE, on May 17 – 19, 2017, pp. 1–5.

Dimensions. 2024. https://app.dimensions.ai/discover/publication.

Ditria, E. M., Buelow, C. A., Gonzalez-Rivero, M., & Connolly, R. M. Artificial intelligence and automated monitoring for assisting conservation of marine ecosystems: a perspective. *Frontiers in Marine Science*, 2022, 9, 918104.

Drogkoula, M., Kokkinos, K., & Samaras, N. A comprehensive survey of machine learning methodologies with emphasis in water resources management. *Applied Sciences*, 2023, 13, 12147. https://doi.org/10.3390/app132212147

Dursun, M., & Özden, S. An efficient improved photovoltaic irrigation system with artificial neural network based modeling of soil moisture distribution – a case study in Turkey. *Computers and Electronics in Agriculture*, 2014, 102, 120–126. https://doi.org/10.1016/j.compag.2014.01.008

Elkiran, G., Nourani, V., Abba, S., & Abdullahi, J. Artificial intelligence-based approaches for multi-station modelling of dissolve oxygen in river. *Global Journal of Environmental Science and Management*, 2018, 4, 439–450.

Elkiran, G., Nourani, V., & Abba, S. I. Multi-step ahead modelling of river water quality parameters using ensemble artificial intelligence-based approach. *Journal of Hydrology*, 2019, 577, 123962.

FAO. *Water Scarcity, Land and Water.* 2024. https://www.fao.org/land-water/water/water-scarcity/en/

Gao, X., Zeng, W., Shen, Y., Guo, Z., Yang, J., Cheng, X., Hua, Q., & Yu, K. Integrated deep neural networks-based complex system for urban water management. *Complexity*, 2020, 2020, 8848324.

Gaya, M. S., Abba, S. I., Abdu, A. M., Tukur, A. I., Saleh, M. A., Esmaili, P., & Wahab, N. A. Estimation of water quality index using artificial intelligence approaches and multi-linear regression. *IAES International Journal of Artificial Intelligence*, 2020, 9, 126.

Ghandar, A., Ahmed, A., Zulfiqar, S., Hua, Z., Hanai, M., & Theodoropoulos, G. A decision support system for urban agriculture using digital twin: a case study with aquaponics. *IEEE Access*, 2021, 9, 35691–35708.

Giordano, M., Barron, J., & Ünver, O. Water scarcity and challenges for smallholder agriculture. In Campanhola, C., Pandey, S. (eds.), *Sustainable Food and Agriculture*. Elsevier-Academic Press-FAO, 2019, pp. 75–94.

Goap, A., Sharma, D., Shukla, A. K., & Rama Krishna, C. An IoT based smart irrigation management system using Machine learning and open source technologies. *Computers and Electronics in Agriculture*, 2018, 155, 41–49.

Gonçalves, R., Soares, J. J. M., & Lima, R. M. F. An IoT-based framework for smart water supply systems management. *Future Internet*, 2020, 12, 114.

Goralski, M. A., & Tan, T. K. Artificial intelligence and sustainable development. *The International Journal of Management Education*, 2020, 18(1), 100330.

Gu, Z., Zhu, T., Jiao, X., Xu, J., & Qi, Z. Neural network soil moisture model for irrigation scheduling. *Computers and Electronics in Agriculture*, 2021, 180, 105801. https://doi.org/10.1016/J.COMPAG.2020.105801

Gupta, A. D., Pandey, P., Feijóo, A., Yaseen, Z. M., & Bokde, N. D. Smart water technology for efficient water resource management: a review. *Energies*, 2020, 13, 6268. https://doi.org/10.3390/en13236268

Hallaji, S. M., Fang, Y., & Winfrey, B. K. Predictive maintenance of pumps in civil infrastructure: state-of-the-art, challenges and future directions. *Automation in Construction*, 2022, 134, 104049.

Hayder, G., Mustafa, H. M., & Kurniawan, I. Implementation of machine learning methods for monitoring and predicting water quality parameters. *Biointerface Research in Applied Chemistry*, 2020, 11, 9285–9295.

Iglesias, A., & Garrote, L. Adaptation strategies for agricultural water management under climate change in Europe. *Agricultural Water Management*, 2015, 155, 113–124.

Jain, A. K., Mao, J., & Mohiuddin, K. M. Artificial neural networks: a tutorial. *Computer*, 1996, 29 (3), 31–44. https://doi.org/10.1109/2.485891

Javaid, M., Haleem, A., Khan, I. H., & Suman, R. Understanding the potential applications of Artificial Intelligence in Agriculture Sector. *Advanced Agrochem*, 2023, 2(1), 15–30.

Jayaraman, P., Nagarajan, K. K., & Partheeban, P. A Review on artificial intelligence algorithms and machine learning to predict the quality of groundwater for irrigation purposes. In *2022 International Conference on Data Science, Agents & Artificial Intelligence (ICDSAAI)*, vol. 1, Chennai, India, IEEE, 2022, pp. 1–8. https://doi.org/10.1109/ICDSAAI55433.2022.10028857

Jha, K., Doshi, A., Patel, P., & Shah, M. A comprehensive review on automation in agriculture using artificial intelligence. *Artificial Intelligence in Agricultue*, 2019, 2, 1–12. https://doi.org/10.1016/j.aiia.2019.05.004

Karamoutsou, L., & Psilovikos, A. Deep learning in water resources management: the case study of Kastoria Lake in Greece. *Water*, 2021, 13, 3364.

Keswani, B., Mohapatra, A., Mohanty, A., Khanna, A., Rodrigues, J., & Albuquerque, V. H. C. Adapting weather conditions based IoT enabled smart irrigation technique in precision agriculture mechanisms. *Neural Computing and Applications*, 2019, 31(1), 277–292.

Kim, T. H., & AlZubi, A. A. AI-enhanced precision irrigation in legume farming: optimizing water use efficiency. *Legume Research: An International Journal*, 2024, 47(8), 1382–1389.

Kim, Y., Evans, R. G., & Iversen, W. M. Remote sensing and control of an irrigation system using a distributed wireless sensor network. *IEEE Transactions on Instrumentation and measurement,* 2008, 57(7), 1379–1387. https://doi.org/10.1109/TIM.2008.917198

Kisi, O., Alizamir, M., & Docheshmeh Gorgij, A. R. Dissolved oxygen prediction using a new ensemble method. *Environmental Science and Pollution Research*, 2020, 27(9), 9589–9603.

Knoll, L., Breuer, L., & Bach, M. Large scale prediction of groundwater nitrate concentrations from spatial data using machine learning. *Science of the Total Environment*, 2019, 668, 1317–1327.

Koren, O., Bagozzi, B. E., & Benson, T. S. Food and water insecurity as causes of social unrest: evidence from geolocated Twitter data. *Journal of Peace Research*, 2021, 58(1), 67–82.

Krishnan, S. R., Nallakaruppan, M. K., Chengoden, R., Koppu, S., Iyapparaja, M., Sadhasivam, J., & Sethuraman, S. Smart water resource management using artificial intelligence – a review. *Sustainability*, 2022. https://doi.org/10.3390/su142013384

Kumar, P., Udayakumar, A., Anbarasa Kumar, A., Senthamarai Kannan, K., & Krishnan, N. Multiparameter optimization system with DCNN in precision agriculture for advanced irrigation planning and scheduling based on soil moisture estimation. *Environmental Monitoring and Assessment*, 2023. https://doi.org/10.1007/s10661-022-10529-3

Kurtulmus, E., Arslan, B., & Kurtulmus, F. Deep learning for proximal soil sensor development towards smart irrigation. *Expert Systems with Applications*, 2022, 198, 116812.

Lauguico, S. C., Concepcion, R. S., Alejandrino, J. D., Tobias, R. R., Macasaet, D. D., & Dadios, E. P. A comparative analysis of machine learning algorithms modeled from machine vision-based lettuce growth stage classification in smart aquaponics. *International Journal of Environmental Science and Development*, 2020, 11(9), 442–449.

Lebu, S., Lee, A., Salzberg, A., & Bauza, V. Adaptive strategies to enhance water security and resilience in low-and middle-income countries: a critical review. *Science of the Total Environment*, 2024, 925, 171520.

Massarelli, C., Campanale, C., & Uricchio, V. F. Artificial intelligence and water cycle management. In Singh, I., Gao, Z., & Massarelli, C. (eds.), *IoT Applications Computing*, (pp. 1–11), London, UK: IntechOpen, 2021.

Melese, S. M. Effect of climate change on water resources. *Journal of Water Resources and Ocean Science*, 2016, 5(1), 14–21.

Mezni, H., Driss, M., Boulila, W., Atitallah, S. B., Sellami, M., & Alharbi, N. Smartwater: a service-oriented and sensor cloud-based framework for smart monitoring of water environments. *Remote Sensing*, 2022, 14(4), 922.

Mishra, B. K., Kumar, P., Saraswat, C., Chakraborty, S., & Gautam, A. Water security in a changing environment: concept, challenges and solutions. *Water*, 2021. https://doi.org/10.3390/w13040490

Morain, A., Ilangovan, N., Delhom, C., & Anandhi, A. Artificial intelligence for water consumption assessment: state of the art review. *Water Resources Management*, 2024, 38(9), 3113–3134.

Morellos, A., Pantazi, X. E., Moshou, D., Alexandridis, T., Whetton, R., Tziotzios, G., et al. Machine learning-based prediction of soil total nitrogen, organic carbon and moisture content by using VIS-NIR spectroscopy. *Biosystems Engineering*, 2016. https://doi.org/10.1016/j.biosystemseng.2016.04.018

Moreno-Rodenas, A. M., Duinmeijer, A., & Clemens, F. H. Deep-learning based monitoring of FOG layer dynamics in wastewater pumping stations. *Water Resources*, 2021, 202, 117482.

Nasir, N., Kansal, A., Alshaltone, O., Barneih, F., Sameer, M., Shanableh, A., & Al-Shamma'a, A. Water quality classification using machine learning algorithms. *Journal of Water Process Engineering*, 2022, 48, 102920.

Nourani, V., Asghari, P., & Sharghi, E. Artificial intelligence based ensemble modeling of wastewater treatment plant using jittered data. *Journal of Cleaner Production*, 2021, 291, 125772.

Nourani, V., Baghanam, A. H., Adamowski, J., & Kisi, O. Applications of hybrid wavelet-artificial intelligence models in hydrology: a review. *Journal of Hydrology*, 2014. https://doi.org/10.1016/j.jhydrol.2014.03.057

Obaideen, K., Yousef, B. A., AlMallahi, M. N., Tan, Y. C., Mahmoud, M., Jaber, H., & Ramadan, M. An overview of smart irrigation systems using IoT. *Energy Nexus*, 2022, 7, 100124.

Oberascher, M., Kinzel, C., Kastlunger, U., Kleidorfer, M., Zingerle, C., Rauch, W., & Sitzenfrei, R. Integrated urban water management with micro storages developed as an IoT-based solution – the smart rain barrel. *Environmental Modelling & Software*, 2021, 139, 105028.

Olawade, D. B., Wada, O. Z., Ige, A. O., Egbewole, B. I., Olojo, A., & Oladapo, B. I. Artificial intelligence in environmental monitoring: advancements, challenges, and future directions. *Hygiene and Environmental Health Advances*, 2024, 12, 100114.

Oliveira, F., Costa, D. G., Assis, F., & Silva, I. Internet of Intelligent Things: a convergence of embedded systems, edge computing, and machine learning. *Internet of Things*, 2024, 26, 101153.

Pandey, D. K., & Mishra, R. Towards sustainable agriculture: harnessing ai for global food security. *Artificial Intelligence in Agriculture*, 2024, 12, 72–84.

Parra, L., Botella-Campos, M., Puerto, H., Roig-Merino, B., & Lloret, J. Evaluating irrigation efficiency with performance indicators: a case study of citrus in the East of Spain. *Agronomy*, 2020, 10(9), 1359. https://doi.org/10.3390/agronomy10091359

Patel, G. S., Rai, A., Das, N. N., & Singh, R. P., Eds. *Smart Agriculture: Emerging Pedagogies of Deep Learning, Machine Learning and Internet of Things*. CRC Press, Boca Raton, FL, 1st edition, 2021.

Pigola, A., da Costa, P. R., Carvalho, L. C., Silva, L. F. D., Kniess, C. T., & Maccari, E. A. Artificial Intelligence-driven digital technologies for implementing sustainable development goals: a perspective from Brazil and Portugal. *Sustainability*, 2021, 13(24), 13669. https://doi.org/10.3390/su132413669

Raei, E., Asanjan, A. A., Nikoo, M. R., Sadegh, M., Pourshahabi, S., & Adamowski, J. F. A deep learning image segmentation model for agricultural irrigation system classification. *Computers and Electronics in Agriculture*, 2022, 198, 106977.

Ramadan, M. N., Ali, M. A., Khoo, S. Y., Alkhedher, M., & Alherbawi, M. Real-time IoT-powered AI system for monitoring and forecasting air pollution in industrial environments. *Ecotoxicology and Environmental Safety*, 2024, 283, 116856.

Rejeb, A., Abdollahi, A., Rejeb, K., & Treiblmaier, H. Drones in agriculture: a review and bibliometric analysis. *Computers and Electronics in Agriculture*, 2022, 198, 107017.

Rivas-Sánchez, Y. A., Moreno-Pérez, M. F., & Roldán-Cañas, J. Environment control with low-cost microcontrollers and microprocessors: application for green walls. *Sustainability*, 2019, 11(3), 782.

Rojek, I., Jasiulewicz-Kaczmarek, M., Piechowski, M., & Mikołajewski, D. An artificial intelligence approach for improving maintenance to supervise machine failures and support their repair. *Applied Sciences*, 2023, 13, 4971. https://doi.org/10.3390/app13084971.

Saboe, D., Ghasemi, H., Gao, M. M., Samardzic, M., Hristovski, K. D., Boscovic, D., Burge, S. R., Burge, R. G., & Hoffman, D. A. Real-time monitoring and prediction of water quality parameters and algae concentrations using microbial potentiometric sensor signals and machine learning tools. *Science of the Total Environment*, 2021, 764, 142876.

Saggi, M. K., & Jain, S. A survey towards decision support system on smart irrigation scheduling using machine learning approaches. *Archives of Computational Methods in Engineering*, 2022, 29(6), 4455–4478.

Saggi, M. K., & Jain, S. Reference evapotranspiration estimation and modeling of the Punjab Northern India using deep learning. *Computers and Electronics in Agriculture*, 2019, 156, 387–398.

Said, S., Ihab, R., Hesham, O., Abou Tabl, I., Maged, N., Youssef, S., & Elagamy, M. AIOT-Arch: furthering artificial intelligence in big data IoT applications. *IOP Conference Series: Materials Science and Engineering*, 2021, 1051(1), 012008.

Saleh, M. A., Rasel, H. M., & Ray, B. A comprehensive review towards resilient rainfall forecasting models using artificial intelligence techniques. *Green Technologies and Sustainability*, 2024, 2(3), 100104.

Saraiva, M., Protas, É., Salgado, M., & Souza, C. Automatic mapping of center pivot irrigation systems from satellite images using deep learning. *Remote Sensing*, 2020, 12, 558.

Sarkar, S. K., Talukdar, S., Rahman, A., & Roy, S. K. groundwater potentiality mapping using ensemble machine learning algorithms for sustainable groundwater management. *Frontiers in Engineering and Built Environment*, 2022, 2(1), 43–54.

Sasmita, M., & Mohanty, B. P. Global climate change: a cause of concern. *National Academy Science Letters*, 2009, 32(5–6), 149–156.

Saudi University-Computer and Information Sciences. Artificial intelligence in decision-making: a case study. *Journal of King Saud University – Computer and Information Sciences*, 2021, 34(8), 5602–5618. https://doi.org/10.1016/j.jksuci.2021.05.013

Shankar, P., Werner, N., Selinger, S., & Janssen, O. Artificial intelligence-driven crop protection optimization for sustainable agriculture. In *2020 IEEE/ITU International Conference on Artificial Intelligence for Good (AI4G)*, 2020. https://doi.org/10.1109/AI4G50087.2020.9311082

Sillberg, C., Kullavanijaya, P., & Chavalparit, O. Water quality classification by integration of attribute-realization and support vector machine for the Chao Phraya River. *Journal of Ecological Engineering*, 2021, 22, 70–86.

Silva, A. F., Löfkvist, K., Gilbertsson, M., Os, E. V., Franken, G., Balendonck, J., Pinho, T. M., Boaventura-Cunha, J., Coelho, L., Jorge, P., et al. Hydroponics monitoring through UV-VIS spectroscopy and artificial intelligence: quantification of nitrogen, phosphorus, and potassium. *Chemistry Proceedings*, 2021, 5, 88.

Soori, M., Dastres, R., & Arezoo, B. AI-powered blockchain technology in industry 4.0: A review. *Journal of Economy and Technology*, 2024, 1, 222–241. https://doi.org/10.1016/j.ject.2024.01.001

Sulistyaningsih, T., Sunarto, S., Saiman, S., & Oktaviari, V. Artificial intelligence and decision support system to determine policies for controlling river pollution from industrial sectors. *IOP Conference Series: Earth and Environmental Science*, 2021, 717, 012026. https://doi.org/10.1088/1755-1315/717/1/012026

Talaviya, T., Shah, D., Patel, N., Yagnik, H., & Shah, M. Implementation of artificial intelligence in agriculture for optimization of irrigation and application of pesticides and herbicides. *Artificial Intelligence in Agriculture*, 2020, 4, 58–73.

Tchonkouang, R. D., Onyeaka, H., & Nkoutchou, H. Assessing the vulnerability of food supply chains to climate change-induced disruptions. *Science of the Total Environment*, 2024, 920, 171047.

Thai-Nghe, N., Thanh-Hai, N., & Chi Ngon, N. Deep learning approach for forecasting water Quality in IoT systems. *International Journal of Advanced Computer Science and Applications*, 2020, 11, 686–693.

Thakur, A. K., Singh, R., Gehlot, A., Kaviti, A. K., Aseer, R., Suraparaju, S. K., & Sikarwar, V. S. Advancements in solar technologies for sustainable development of the agricultural sector in India: a comprehensive review on challenges and opportunities. *Environmental Science and Pollution Research*, 2022, 29(29), 43607–43634.

Tsang, S. W., & Jim, C. Y. Applying artificial intelligence modeling to optimize green roof irrigation. *Energy and Buildings*, 2016, 127, 360–369.

Upreti, A. Machine learning application in G.I.S. and remote sensing: an overview. *International Journal of Multidisciplinary Research and Growth Evaluation*, 2022. https://doi.org/10.20944/preprints202207.0302.v1

Van Eck, N. J., & Waltman, L. CitNetExplorer: a new software tool for analyzing and visualizing citation networks. *Journal of Informetrics*, 2014, 8(4), 802–823. https://doi.org/10.1016/j.joi.2014.07.006

Vanipriya, C. H., Maruyi, M. S., & Gupta, G. Artificial Intelligence-Enabled Plant Emotion Expresser in the Development of Hydroponics System. *Materials Today Proceedings*, 2021, 45, 5034–5040.

Vij, A., Vijendra, S., Jain, A., Bajaj, S., Bassi, A., & Sharma, A. IoT and machine learning approaches for automation of farm irrigation system. *Procedia Computer Science*, 2020, 167, 1250–1257.

Vyas, S., Shabaz, M., Pandit, P., Parvathy, L. R., & Ofori, I. Integration of artificial intelligence and blockchain technology in healthcare and agriculture. *Journal of Food Quality*, 2022. https://doi.org/10.1155/2022/4228448

Wei, H., Xu, W., Kang, B., Eisner, R., Muleke, A., Rodriguez, D., & Harrison, M. T. Irrigation with artificial intelligence: problems, premises, promises. *Human-Centric Intelligent Systems*, 2024, 4, 1–19.

World Bank Open Data. *Renewable Internal Freshwater Resources per Capita (Cubic Meters)*. 2024. https://data.worldbank.org/indicator/ER.H2O.INTR.PC

Zakaria, M. N. A., Malek, M. A., Zolkepli, M., & Ahmed, A. N. Application of artificial intelligence algorithms for hourly river level forecast: a case study of Muda River, Malaysia. *Alexandria Engineering Journal*, 2021, 60, 4015–4028.

Zhang, J., Yu, Y., Yan, J., & Chen, J. Data-driven parameter prediction of water pumping station. *Water*, 2023, 15(6), 1128.

Zhao, L., Dai, T., Qiao, Z., Sun, P., Hao, J., & Yang, Y. Application of artificial intelligence to wastewater treatment: a bibliometric analysis and systematic review of technology, economy, management, and wastewater reuse. *Process Safety and Environmental Protection*, 2020, 133, 169–182.

Zhou, Z., Majeed, Y., Diverres Naranjo, G., & Gambacorta, E. M. T. Assessment for crop water stress with infrared thermal imagery in precision agriculture: a review and future prospects for deep learning applications. *Computers and Electronics in Agriculture*, 2021, 182, 106019. https://doi.org/10.1016/J.COMPAG.2021.106019

8 Smart Water Infrastructure

Chunchu Suchith Kumar, Neroju Manasa, Guda Bhargavi, and Arunabha Pal

8.1 INTRODUCTION

8.1.1 The Need for Smart Water Infrastructure

8.1.1.1 Global Water Challenges

Water scarcity has now become one of the prime global issues caused by several associated factors like population growth, urbanization, and the growing demand of industries. According to the recent estimate, the world population is expected to be nearly 9.7 billion in the year 2050, further straining water resources (UN DESA, 2022). The growth accelerates the demand for water not only for drinking and sanitation but also in agriculture, industry, and energy production, which are necessary to support growing communities and economies.

The primary cause of water scarcity due to compound growth is urbanization. While in excess of 55% of the global population resides within cities, currently, it is projected that by the year 2050, it will reach 68% (UN, 2019). The shift of population from the countryside to the urban centers increases the demand for water in the cities, thereby straining the existing water supply systems, a great deal of which is old and/or ineffective. The pressure on metropolitan water systems results in recurring water shortages, concerns about the supply quality, and a greater dependence on underground sources of water, exhausting the regional aquifers quickly than they can regenerate.

Apart from the issues related to headcount and urban spread, water requirements for industrial purposes are on the rise as well. Water-intensive industries like agriculture, manufacturing, and energy generation are the leading users of water in the world today, making up about 70% of total water withdrawals (FAO, 2021). Water consumption in industries has a sizable degree of usage as well as pollution to the environment through emissions of wastes to rivers, lakes, and oceans, hence shortening the supply of clean and healthy water to the people and the environment (Jensen, 2023).

The interplay of these elements has elevated the risk of water scarcity to a fundamental danger to the health of mankind and the environment. In its estimation, the World Resources Institute claims that almost a quarter of the people across the globe reside in highly water-stressed, where water consumption overwhelms the available supply demands of the population (WRI, 2020). Global warming has made the

DOI: 10.1201/9781003547099-9

problem much worse because it has increased occurrences of dry seasons, changed patterns of rainfall, and increased the rate of glacier melting, all of which have negatively impacted water availability and worsened the existing water challenge (Gosling & Arnell, 2016).

In addressing these pressing challenges, the design and establishment of smart water infrastructure (SWI) provides a revolutionary solution. With the fond adoption of technologies such as the Internet of Things (IoT) and artificial intelligence (AI), it is conceivable that SWI could transform water management, cut-down wastages, and promote appropriate water distribution, all in the long run contributing toward the mitigation of the water scarcity challenges that the world faces (Sinha et al., 2023).

8.1.1.2 Current Limitations of Traditional Water Systems

The traditional water supply systems put in place several decades ago are inadequate to meet the needs of the present society. Such systems were constructed for less populated areas and uncomplicated urban and industrial setups, thus rendering them wasteful and ineffective as the global water demand increases (Harris Lovett et al., 2019). Dated systems, mostly depending on physical controls and inadequate data gathering, find it hard to cope with the modern demand in operating water infrastructures, causing constant problems in the form of water loss, energy wastage, and deterioration in the quality of service.

An important drawback of conventional water supply systems is the amount of water that gets wasted, mainly because of the old pipes, leaks, and poor maintenance practices. Research suggests that in older cities, the water losses from leakages and inefficient appliances integrated into the networks are referred to as non-revenue water (NRW) and stand in the range of 30%–50%. NRW is of utmost importance, particularly in water-scarce regions where every drop counts. Water utilities experience challenges in the enhancement of leak detection and repair response factors due to the dominant configuration of infrastructure underground, which increases the costly time taken to detect and fix these issues (Nguyen et al., 2022).

Additionally, a further challenge is presented by the energy-demanding aspect of conventional water systems. The steps of pumping, treating, and further delivering water require a lot of energy, which increases operating expenses while adding to the greenhouse gases emitted, hence heightening environmental issues. Most of the existing water facilities do not incorporate the need for optimizing energy consumption in real time since most of their operations depend on old regimens rather than on the changing demand (Kim et al., 2021). Such inefficiency increases energy use, causing increased expenditure as well as decreasing the sustainability of water services.

Water quality control in distribution systems is also restricted by the constraints of age-old methods. Such events can include the aging of the constitutive materials, the buildup of sediments, surface ingress contamination, and liquidity traveling through long pipelines (Thompson, 2021). Such systems do, however, have some drawbacks. The absence of continuous monitoring allows a lapse in control of water quality that may translate to health threats, especially in areas with high levels of pollution. It is also clear that unless there are quick, intelligent reactions, containing the dispersal of toxins in the environment, and degrading quality of water, it is hard for these systems to guarantee the provision of clean and safe water to all populations (Anderson et al., 2023).

The limited scalability of the conventional approaches to water resources management increases the constraint on this system of urbanization expansion. As metropolitan areas grow in size, completing all the required facilities for the treatment and importation of water over larger territories becomes more and more expensive and difficult. Capital-intensive pipeline expansion projects involving replacing older existing pipelines or extending the networks are frequently postponed because of the costs involved, regulations concerning such activities, or the sheer size of the upgrades required. It is therefore not surprising that in many urban centers, where efficient scaling does not take place, there are only intermittent supplies of water, and the quality of the service offered is very poor, adding to the already existing water crisis.

These deficiencies emphasize the critical and immediate need for the advancement of the existing systems in the water sector, Rehabilitation and replacement of existing wastewater systems with SWI utilization implies a way of overcoming these problems through the use of technology that improves the functionality of the system, prevention of wastage and allows the water system to be able to adapt to present-day demands on water.

8.1.1.3 Transition to Smart Infrastructure

Modern water distribution systems can no longer afford to remain stagnant, and thus there is a need for a modification in the structural design and operational paradigms of traditional water systems to SWI (Rodríguez-Alonso et al., 2021). SWI employs state-of-the-art technologies, including the IoT, AI, and incorporates management by big data to facilitate optimized means of provision, management, and use of the water resource. In stark contrast to traditional water systems that are mostly dependent on manual controls and suffer delays in data transmission, SWI presents a solution that entails automation and access to real-time data, hence enabling the water utilities to accurately monitor, control, and manage the water resources.

The IoT is one of the main technologies that support SWI. Operating water systems with SWI, water systems can reach out and touch; they can sense and send the data all the time on the monitored objects, such as flow rates, pressure, and analysis of water quality. Leaks, for example, are equipped with real-time sensors that can inform the team of the start of a leak immediately, minimizing water losses and preventing expensive damage. Such a real-time feedback system not only aids in the conservation of water but also in enhancing the health of the metrics by anticipating and fixing problems even long before they develop into bigger ones.

The enhancement of SWI with the use of AI involves both prediction and decision-making by the users of the system. For instance, AI systems can use historic maintenance records and sensor-related data from equipment in the water system to identify patterns, predict water usage, and even when to service specific equipment components. For instance, fans of entertainment subscription services such as Netflix are able to predict the seasons of the shows that they have previously aired and stop the water pumping in advance to avoid a wastage of resources. Furthermore, these systems can focus on the areas at most risk and help to distribute the resources in a way that is efficient, and most importantly, budget-friendly. This likelihood is also very favorable in a region which is generally known to have insufficient water supply as it allows for the effective use of resources (Álvarez Díez et al., 2024).

Managerial effectiveness based on empirical evidence is also a significant factor in SWI because it helps utilities to take actions that are backed by extensive data. This can be achieved by combining the concepts of big data and cloud computing, whereby SWI systems can integrate numerous data including but not limited to, sensor data, weather forecast data and even historical consumption data for processing and analysis. With cloud technology, all the data is within a single database, enabling the interaction between the global water stakeholders, water utilities, as well as relevant policy makers. This reduces the time needed for decision-making, hence improving the management of the water resources available while at the same time building up a stronger system, as all the decisions made are backed with timely and historical data.

Similarly, the adoption of SWI practices further enhances environmental sustainability. SWI employs real-time monitoring and automation to control the water systems' energy use by operating pumps and valves on demand. Research indicates that costs associated with energy pumps used in SWI technologies are reduced by about 25%, thereby addressing the issue of carbon emissions levels relative to water distribution. In addition, systems equipped with SWI may also include alternative energy options, such as operation on solar-powered sensors, hence adverse effects on the environment through fossil fuels are greatly mitigated (Alkaisi et al., 2017).

The prospect of water infrastructure revolutionization by SWI implementation in terms of efficiency and reliability remains a great addition to the economic, social, and environmental challenges experienced. While the demand for water increases globally, and the effects of climate change become pronounced, such a concept is appropriate, and SWI can be implemented for securing water access for urban and rural populations without fear of changes. SWI, which incorporates IoT, AI, and data management, is a welcome advancement in the attempt to overcome the weaknesses of the current water systems and move us closer to sustainable management of water resources.

8.2 UNDERSTANDING SWI

8.2.1 Smart Water Infrastructure

SWI is a sophisticated system of water management that utilizes advanced technologies including the IoT, computerization, data analysis, and cloud network, for real-time monitoring, controlling, and optimizing the water distribution system and water usage. SWI is seen to move away from the traditional and passive methods of water management systems to capable and intelligent network systems that are able to deal with problems effectively and in an environmentally friendly manner. More importantly, the components that comprise the SWI, which are the IoT sensors, automation and control systems, data analytics, and cloud computing, are integrated in such a way as to enhance the efficiency of water systems, minimize wastage, increase the dependability of services, and help in managing water resources safely.

8.2.2 Core Technologies

Every core technology defined in SWI is vital for making water systems smart, efficient, innovative, and capable of meeting the present and future needs. Following this, the functions as well as the roles of each technology are described elaborately based on SWI.

8.2.2.1 IoT Sensors for Real-Time Monitoring

IoT sensors are one of SWI's key technologies based on collecting information about the flow, pressure, temperature, and water quality in the systems. It is placed in strategic positions within the infrastructure where, consequently, leaks, contamination, and irregularities can be monitored continuously as well as identified promptly. Real-time data from the IoT sensors enables utilities to identify problems early and address them before they become expensive problems that cause extensive water loss and compromise the quality of water delivered to consumers. IoT sensors in urban water systems, for example, can help minimize non-revenue water by way of promptly flagging a leakage or burst pipe, an aspect which older systems might not detect for hours or even days of substantial water and financial loss.

8.2.2.2 Data Analytics for Pattern Identification and Demand Forecasting

The SWI focuses on data analytics at its core, which helps to condense the cloud of data generated from the IoT sensors into useful information. Data analytics applications help in the analysis of the historical and current data on water consumption in order to reveal trends, predict consumption tomorrow, and other similar consumption insights including seasonal variations and consumer trends. More sophisticated mathematical models allow for predictive analysis, which can assess things such as weather, past use, and demographics of a region in order to estimate how much demand there will be. This enables the utilities to plan wastage, peak water supply periods, and of course when, or if, resources will be available or in high demand. The application of such insights also helps to promote the efficiency of the entire system, which ensures that the available water resources are utilized in a proper and purposeful manner.

8.2.2.3 Automation and Control Systems for Optimizing Water Flow

The use of automation and control systems allows SWI to function without supervision, that is, to balance the distribution and the flow of water depending on the prevailing conditions. Such systems allow utility companies to control the flow of water from afar, by turning off the supply or substantially withdrawing it, to control the pressure and consequently save on energy. For example, automation is capable of controlling the opening and the closing of valves, thus reducing the chances of creating high pressure in specific regions, enhancing stability by reducing the chances of burst pipes and loss of water. In regions with a limited supply of water, automation assists in the utilization of the resource by doing better allocation of the available water and at times even favoring water distribution to the areas that are in need the most like hospitals (Olsson, 2021). Automation helps to cut down the need for human involvement in the systems, which in turn leads to improved flexibility and agility of water systems when there are unforeseen changes or disasters (Sweeney & Kabouris 2020).

8.2.2.4 Cloud Computing for Data Storage, Sharing, and Analysis

Cloud computing offers the benefits of easy access and abstraction of IT resources, making it an ideal environment for the management of the large amounts of data generated by SWI. Water companies take advantage of cloud infrastructure by being able to employ the concept of data warehousing where all the data is stored offshore and is available to everyone from local authorities to operational water managers and engineers (Moran et al., 2020). Furthermore, cloud-based systems enable coordination even when the processes are spread out across several players because they allow them to access, evaluate, and act on information simultaneously, which is critical in incidences where resources have to be allocated or emergencies managed. In addition, the cloud also makes it easier to incorporate high-level data analysis, which allows for fast and smooth execution of intricate algorithms and forecasting models. Such a technology goes a long way in providing flexibility in managing resources such as data, which are essential for SWI because of the enormous scale of system operations that would be difficult to achieve through conventional data storage and processing approaches.

Comprehensively, all these primary components make the SWI an innovative water management solution by improving efficiency and conservation of the systems themselves and their flexibility. By intra-connecting IoT, data, automation, and cloud performance, SWI aims to overcome the challenges posed by conventional water infrastructure as well as the current societal requirements of having water resources that are both clean and available at all times (Mahmood et al., 2024)

8.3 SWI'S CONTRIBUTION TO SUSTAINABLE DEVELOPMENT GOALS

SWI provides a solid basis for sustainable development objectives, particularly the sixth Sustainable Development Goal (SDG 6) which relates to the provision of clean water and sanitation for all. It contributes to the improvement of access, quality, efficiency, and integrated management of water resources, all of which help in the realization of the SDG 6 water targets. Outside the scope of SDG 6, SWI also helps in furthering the goals of other SDGs, such as SDG 11 (Sustainable Cities and Communities) and SDG 13 (Climate Action), which seek to promote efficient and resilient urban water systems and climate-resilient infrastructure, respectively.

8.3.1 Alignment with SDG 6 (Clean Water and Sanitation)

8.3.1.1 SDG 6.1: Accessibility and Affordability of Clean Drinking Water

One connecting thread of the SDG 6 targets is the aim of realizing universal and equal access to safe and affordable drinking water. Water Smart Infrastructure (SWI) systems are in support of achieving this aim by ensuring the availability and accessibility of water through proper functioning of the distribution network and monitoring of the supply systems in real time. For example, in SWI, IoT sensors can check water pressure and flow rate, which helps utility companies locate places of low service area or high disruptions. By demand management in the underserved regions, SWI bridges the service gaps and provides a steady, low-cost water supply, whenever needed, to the rural as well as urban populations. Besides, efficient monitoring and forecasting of SWI costs can help the utility companies optimize their operation costs, and they might even be able to lower prices of the services provided to customers.

8.3.1.2 SDG 6.3: Maintaining Water Quality by Detecting Contaminants Early

Sustainable Development Goal 6.3 aims at improving water quality by reducing pollution, eliminating hazardous materials, and controlling the release of chemicals. SWI's IoT sensors are built for the purpose of detection in advance of contamination, processes of which include the continuous observation of water quality parameters such as pH, turbidity, and microbial content. These parameters are reported instantly. Advanced data analytics aids in the quick pinpointing of pollution sources and the occurrences of contamination so that measures can be taken promptly for the sake of public health and the safety of drinking water. Dealing with quality-related problems beforehand, SOI makes sure that standard levels of water quality are met on all occasions, thus enhancing the safety and efficiency of water systems for consumption.

8.3.1.3 SDG 6.4: Improving Water-Use Efficiency across Sectors

SDG 6.4 concerns the enhancement of water-use efficiency in sectors such as agriculture, industry, and urban utilities. SWI promotes better water management through such initiatives as demand forecasting and predictive analytics with automation to cut waste and improve resource allocation. On the other hand, SWI encourages the use of 'smart' irrigation systems for agricultural purposes, which, due to optimization of water usage, are able to reschedule watering of crops depending on real-time soil moisture availability, climatic conditions, and the crops being grown (Bwambale et al., 2022). Urban application of SWI helps to curb non-revenue water losses by facilitating rapid response to leaks and/or illegal connections, which is efficient in many areas where there is a risk of domestic and industrial supply of water being interrupted due to scarcity.

8.3.1.4 SDG 6.5: Supporting Integrated Water Resources Management

As a part of Integrated Water Resources Management (IWRM), Supports Water Information (SWI) systematically manages and controls the water, land, and all related resources. It is implied above in SDG 6.5 on IWRM, necessitated by the multi-stakeholder processes, which implies that information and resources are cross-sectorial for the reason of proper IWRM planning and management practices. SWI improves the participatory nature of decision-making and management by making available databases on water-related materials and by offering opportunities to interested parties and especially the water management authorities, unions, and NGOs to access the databases within the management objectives. In addition, the SWI systems developed include the so-called modeling, or forecasting systems, which allow the stakeholders to consider not only the current situation with water management, but the predictions so that the good management of water resources is seen and practiced in time (Martinez, 2019).

8.3.2 Connection to Other SDGs

8.3.2.1 SDG 11: Sustainable Cities and Communities

SWI advances Sustainable Development Goal 11 by building resiliency focused on the water infrastructure of sustainable cities and communities. Smart infrastructure in urban water systems guarantees reliable water supply and also takes care of storm and flooding management through monitoring and control system using real-time techniques. SWI improves the resilience of urban areas, making it possible to build better cities that are safer and more sustainable by managing water stoppages and minimizing damage to infrastructure. Because of SWI, urban areas face less threat of resource exhaustion as there are fewer water losses and a more effective water supply sickle, thus addressing the problem of overgrowing metropolises.

8.3.2.2 SDG 13: Climate Action

SWI promotes SDG 13 through the provision of instruments aimed at adaptation and mitigation of climate change in the sphere of water management. Water scarcity tends to be worsened by climate change, which also enhances the occurrence of extreme weather events, such as floods and droughts. Thanks to real-time data analysis, SWI has improved service delivery in predicting and mitigating the effects of such occurrences, thus allowing delivery even when the weather is extreme. Built-in control systems reduce energy consumption and hence air emissions of water supply and distribution systems by optimizing the use of water using demand-based water distribution schedules. SWI also facilitates the use of other clean energy like solar sensors, which minimizes the use of non-renewable sources of energy and promotes climate change mitigation. As demonstrated, the approach taken by SWI to adduce and actualize its benefits to the current state of water provision cuts across the entire spectrum of SDG 6 and even goes further to cover SDG 11 and SDG 13, which water and climate are the contemporary challenges of the world.

8.4 SWI AND CLIMATE CHANGE

Scientific Water Section (SWS) has become one of the necessary tools in the fight against global warming because it provides means for both mitigation and adaptation. The water sector is the most sensitive to climate change as changes in temperature, precipitation patterns and increases in the occurrences of extreme weather events can compromise how much water there is and at what quality. Therefore, SWS helps to improve climate resilience and sustainability by reducing greenhouse gases emission, reducing inefficient water distribution and water resource management.

8.4.1 Mitigation Strategies

8.4.1.1 Reducing Greenhouse Gas (GHG) Emissions through Optimized Water Distribution

Evidently, conventional water management techniques have low performance levels as they rely on obsolete approaches and high energy consumption for water sourcing, cleansing, and supplying, which all enhance their contribution toward GHG emissions. Innovative water management systems (SWI) significantly reduce such emissions by dwelling on improved distribution of water using modern technology, robotics, and predictive analytics (Patel & Zhang, 2022). This technology facilitates real-time monitoring and control whereby the water systems can modify the flow depending on the current need, minimizing waste of water, and utilizing the water pumps only when strictly required. In this way, a greater efficiency will help to lower both the water loss and energy use, thereby lowering the carbon footprints for both residential and business water systems.

For instance, automated control systems built into SWI systems can regulate bulk water producer pumps and vary the outlet flow rate in accordance with the incoming demand. Pumps and other equipment can be scheduled for use during certain predictive cycles of demand using created tools of analytic models. This helps avoid peak energy costs and emissions while enhancing the efficiency of operations. Taken together, these functions assist in lessening the effects of global warming with regard to the energy used in transporting and treating water.

8.4.1.2 Incorporating Renewable Energy Sources

The compatibility of SWI with other sources of energy, such as wind energy and solar energy, is another concern within the mitigation strategy of SWI. Most SWI components from IoT sensors to automated control systems and data analytics can be run on renewable energy sources with little or no use of fossil fuels. For example, water quality and flow monitoring solar sensors can be mounted in far-flung areas where haze and moisture levels may affect other power sources. SWI aids in further reductions of GHG emissions by encouraging the use of alternative energy sources in the water sector and hence providing an effective and sustainable infrastructure, which promotes healthy climatic conditions.

In some instances, water system integration has been implemented along with hybrid energy systems, where the water system operators can utilize either renewable energy sources or traditional energy sources, depending upon which one is in higher demand within the system. This hybrid approach promotes access to energy while alleviating the risks of creating emissions, thus facilitating the energy transition for the concerned populations.

8.4.2 Adaptation Strategies

8.4.2.1 Managing the Effects of Droughts through Controlled Water Distribution

Of the many extents to which SWI can shift dynamics within a community, rainfall management remains one of the most important, especially in dry periods. Climate change is a global phenomenon that has resulted in many parts of the world experiencing more and more dry seasons each year, coupled with water regulations. Such management systems are assisted by SWI's approach to database management, which controls the drought threats through levels of water and the use of the water supply by regions. Based on weather predictions, current soil moisture contents, and past consumption trends, SWI assists the utility in rationing water supplies to the customers in more efficient ways, such that a drought would not easily deplete the resources.

In smart climatic environments, for instance, advanced localized agricultural practices with the IoT sensors can assess the soil water storage status and turn on irrigation access only when a certain threshold is reached, thus saving a lot of freshwater use in agricultural practice. Such measures save water yet still assure that crops are sufficiently watered in times of drought. It is also worth noting that the SWI systems of systems have predictive analytics capabilities that anticipate water demands of the population in the future. This gives time for the community's water management system to allocate resources accordingly and avoid wasting water or experiencing drastic water scarcity.

8.4.2.2 Flood Management through Real-Time Monitoring and Resource Conservation

Besides this, SWI also plays an important role in flood management during excessive rain or bad weather by controlling the water flow. Climate change has affected human society as it has increased the occurrence and severity of floods, which threaten both structures and human life. SWI systems allow water utilities to monitor hydrometric data, reservoirs, and radar estimates of precipitation in real time and thus make the decision to operate warning systems and take flood risk alleviation actions.

In the automated control system of SWI, for example, a water release from the storage reservoirs can be done prior to the forecasted heavy rains in order to create extra space and avoid spill over. In addition, the areas where floods are common can apply SWI and generate flooded area models for ensuring preparedness actions. Similar SWI technologies can also be applied in the dry season urban stormwater management systems for regulating water flow to the catchment areas so as to minimize the effects of flash floods and pipe excess water into reservoirs for storage and use later. This strategy helps not only in reducing losses due to floods but also preserves the captured water because the excess water available during floods is stored and treated for use in dry seasons.

8.4.2.3 Increasing Climate Resilience through Integrated Resource Management

In conclusion, the climate resilience building aspects of SWI are enhanced through its resource management ability in promoting water governance within and among different sectors. SWI aims to achieve socio-economic balance by integrating data from agricultural, municipal, and industrial water users. This coordination prevents over-exploitation of water resources by one sector to the detriment of other sectors, which is critical given that climate change has escalated the 'water wars' within nations. Integrating water resources also helps adaptation in the long run through the development of systems that can penetrate the infrastructure and respond to climate changes in the future. For instance, SWI systems can be utilized to create water management systems that adapt to the increasing frequency of droughts or floods along with other changes, such as the environment, population, and economic development, for those communities that foresee this happening. SWI therefore provides an integrated water resource management strategy that promotes conservation of natural resources and adaptation to the effects of climate change.

8.5 TECHNOLOGICAL INNOVATIONS AND PRACTICAL APPLICATIONS

8.5.1 Case Studies

8.5.1.1 Singapore's Water Management System

Singapore is an island city-state where the population is very high as well as the freshwater resources. The government is, however, implementing the use of smart water solutions to assure the availability, if not the sustainability, of the water resource. Given the little precipitation and the nonexistence of large natural water mass, the water management policies for Singapore are largely technological efficiency and self-sustenance. The national water agency of Singapore, the Public Utilities Board (PUB), has also developed and promoted the use of SWIs, which incorporate IoT devices and predictive information in the prevention and management of water quality and water resources distribution.

The main element of Singapore's SWI is its extensive network of smart sensors strategically positioned across the entire water infrastructure. As a result, the system enables the constant monitoring and recording of water quality, water pressure, and water flow rates in order to identify leakages or contamination as soon as possible. Such a system also possesses predictive analytics, where internal regression is run using historical information and future water demand is estimated, therefore allowing PUB to regulate the supply of water in between extremes of oversupply and undersupply. That way, energy costs are also lowered since water pumps are powered when only required, owing to reliable demand forecasts. This network of smart below water quality sensors is parallel to Singapore's "NEWater" program, which manufactures high-purity water from effluent using advanced membrane filtration and UV disinfection technologies, where the treated recycled water begins to re-enter the system. Such practice to incorporate employed packages of treated wastewater is very resourceful in that it enables Singapore to cut down on fresh water sources from outside.

8.5.1.2 India's Hi-Tech Irrigation

Water shortage is one of the major problems faced in India, especially in dry and semi-dry areas where agriculture primarily relies on watering the crops. With the traditional methods of irrigation, excessive amounts of water are consumed resulting in wastage as well, which in turn causes a more severe shortage of water in the area. To this end, smart irrigation technologies have been adopted by the country that employs SWI principles to reduce the use of water as well as promote sustainable agricultural practices.

High-tech irrigation systems in India have integrated soil moisture sensors that are enabled with the IoT and help monitor the hydrating level of the soil instantly. So, the system will only allow for irrigation to be done when the moisture content dips to a certain level in the soil, hence ensuring that crops are provided with adequate water without over-irrigating the crops (Kumar et al., 2022). The information gathered from these sensors is then uploaded to a system where all the irrigated zones are stored, and the farmers are able to rotate irrigation in different areas remotely, hence saving water as well as ensuring optimum use of water supply available.

In addition, predictive and descriptive analytics are employed to help understand the changes in climate and predict rainfall, thereby allowing farmers to adjust their irrigation periods on the basis of expected rainfall and efficiently use water resources. Through its smart irrigation systems, India is able to support enhanced productivity for farmers while curtailing the abuse of water resources, thus promoting both food security and ecological balance.

8.5.2 Barcelona's Smart Water Network

Barcelona has a historical problem with regard to urban water management in conserving water within the limits of the city as well as the population of the city increases day by day. This initiative to use the smart water network (SWN), which is a smart meter network within the city of Barcelona, is indicative of the effort that the city has put toward environmentally friendly activities and optimal utilization of resources. By integrating modern technologies in monitoring and analysis of data, the SWN of Barcelona decreases wastage of water, enhances recovery from damages to the system after a failure and helps to distribute water efficiently to the metropolitan areas.

Moreover, the SWN of Barcelona possesses innovative flow meters and pressure measuring devices strategically located in the water distribution system for effective 'real-time' monitoring. These devices are able to identify various irregularities in the system including leaks and burst pipes which are quickly picked up and dealt with by the repair crews and therefore water loss is very much kept at a minimum (Mendez et al., 2021). Moreover, smart meters are installed at certain places and these smart meters also provide data segregation neighborhood wise facilitating each and every area consumption pattern analysis for proper resource distribution.

On the other hand, the SWN applies predictive maintenance models based on the historical as well as present performance of the system in predicting the likelihood of a failure of the system and the best period to carry out maintenance. With this element in the SWN, wastage of water through interruptions of supply during turns is few and less severe as such needs are cautioned on time before they arise. Such a strategy, along with the culture of sharing information with the public, has established Barcelona as one of the most advanced cities in managing urban water.

8.5.3 Challenges and Opportunities for Scaling SWI

8.5.3.1 Challenges for Scaling SWI

8.5.3.1.1 Capital Costs for Setting Up SWI Systems

It is often the case that the advent of smart water systems entails a considerable capital outlay on technology, infrastructure development, and capacity building. Most of the municipalities and water service providers, especially in the less developed areas, have budget limitations which make it difficult to set aside monetary resources for SWI projects (Gupta et al., 2020).

8.5.3.1.2 Integration Issues with Existing Infrastructures

A great number of water systems in use today depend on obsolete technologies that, on the contrary, may not be suitable for modern SWI solutions. Merging state-of-the-art technologies with a legacy system can be a very complicated affair that necessitates heavyweight enhancements and changes to the current system of operations. Moreover, the presence of varying standards across the different technologies may further hinder integration processes.

8.5.3.1.3 Digital Divide and Limited Access to Technology

The gap in access to technology, especially in rural and underserved places, is a major hindrance to the realization of SWI. Most of the regions do not have good internet access and technical know-how as well as the capital to effectively implement the smart technologies. This divide can worsen the already prevailing disparities in access to and quality of water resources.

8.5.3.1.4 Data Privacy and Security Concerns

Due to the significance of collecting and analyzing data for SWI, the issues of data privacy and security come into play. It is essential to safeguard such information from Internet-based risks, and probably, many of the utilities do not have either the means or the know-how to deal with such threats.

8.5.3.1.5 Regulatory and Governance Challenges

The absence of well-defined regulatory frameworks concerning SWI can affect its operationalization. In the absence of conducive policies and governance structures, the utility of such technologies by the companies may remain a distant dream in the near future. Additionally, varying regulations in different geographical locations may also limit the wide scale application of SWI solutions.

8.5.3.1.6 Public Resistance to Change

The shift toward Water Smart Management practices can be confronted by the opposition from various stakeholders such as utility staff as well as the public. It could be due to apprehension regarding job losses, uncertainties regarding the new technologies to be used, or possibly higher water tariffs, among other reasons that come in opposition to the use of SWI solutions (Farrelly & Brown 2011).

8.5.3.2 Opportunities for Scaling SWI

8.5.3.2.1 Collaborative Partnerships for Resource Mobilization

Working in synergy with one another creates a strong impetus for maximizing available resources for supporting SWI programs. For instance, these synergies help in the sharing of resources and expertise, lower costs, thereby attracting the required investments for smart water Technologies implementation.

8.5.3.2.2 Innovative Financing Solutions

Specific tailored financing mechanisms like green bonds, performing investing, and performance-based contracts can furnish the capital needed for the implementation of SWI. It is possible to attract private investments by explaining the benefits of sustainable water management to the society and the environment.

8.5.3.2.3 *Advancements in Technology*

The ongoing development of technology, especially in relation to IoT, AI, and data analysis, creates a significant possibility for the improvement of SWI. Given the lower costs of cloud computing as well as sensors, it is possible for even the least developed municipalities to smarten up (Kumar & Choudhury 2024). These improvements can enhance operational efficiencies and also improve water management practices.

8.5.3.2.4 *Capacity Building and Knowledge Transfer*

Capacity-building programs enhance the level of technical competency and skills in water resources management programs. Such training and knowledge-sharing activities can help local communities or utility personnel operate SWI technologies more efficiently and enhance their ownership as well as sustainability.

8.5.3.2.5 *Regulatory Support for Innovation*

When augmented with favorable regulatory conditions that stimulate innovative and investment activities in SWI, its broad-based acceptance can be achieved. Governments may provide certain financial gadgets to encourage utilities to embrace smart technologies, for instance, tax holidays or subsidies, thereby encouraging the growth of water management technologies.

8.5.3.2.6 *Enhancing Resilience to Climate Change*

Introducing SWI would help enhance resilience against the impacts of climate change through better water management practices. Communities are also less prone to droughts and floods as they could better predict the patterns of extreme weather and manage water resources properly (Tiwari et al., 2024).

8.6 POLICY AND GOVERNANCE FOR SWI EXPANSION

8.6.1 Need for Regulatory Frameworks

The ability to develop appropriate regulation systems is one of the critical determinants that will affect the growth of SWI. Such systems create a healthy context for the provision of new water management technologies and their acceptance. Regulation will always have many roles: it ensures consistency and predictability; stresses the need to take care of the sensitive parties and encourages environmental care.

To begin with, regulatory frameworks have the potential to ease the burden of investments in SWI infrastructure by adopting the principle whereby a fence is put with clear-cut rules and regulations in which the investors can stay. This clarity lessens the risk perceptions in funding such investments and therefore opens avenues for private sector investment in SWI projects. By way of illustration, broad ranging prohibitions that include tax credits or even subsidies for businesses using advanced technology can motivate city or utility companies to invest in new, modern technology.

Secondly, adequate regulations encourage the sharing of information between different parties such as government, businesses, and civil society, within the context of the water management system. For instance, in addition to water availability and external water piping system, access to information on water quality and usage patterns respectively, can help for better decisions (Trebilcock & Rosenstock 2015). Policymakers can seek appropriate regulatory measures to allow open access to data and information so that these stakeholders are well informed and planning is more effective.

For instance, IWRM, which aims at the proper management of water resources in all aspects, can be implemented thanks to a strong set of regulations. When policies adhere to IWRM principles, they enable the water sector and its related activities to be coordinated together vertically and sustained over time. Such measures are more critical, especially in areas where there is little or no water available as there are demands on the already stretched resources.

8.6.2 International Partnerships and Support

The smooth integration of SWI, especially in regions facing high water scarcity, is significantly helped by the establishment of international partnerships. Coaction among states, international organizations, and private businesses enables the pool of different skills, capabilities, and technologies necessary to tackle potential water obstacles. When it comes to transnational water management initiatives, for instance, the United Nations and any other such institutions have preset modalities. This facilitates the exchange of information and experiences pertaining to best practices among countries with regard to heating up the implementation of SWI strategies. For example, collaboration with foreign aid organizations assists in getting the much-needed funds and expertise to support developing countries in 'leapfrogging' to advanced technologies.

The private sector has its share of contributions in this process, too. Thanks to their know-how and financial might, private business practitioners can work together with the public sector and develop SWI solutions that fit certain regional problems. This is the case as technology companies can join efforts with a municipal authority to design a fitting intake system that helps in such a community's monitoring of water quality and distribution.

Moreover, such partnerships can support local actors by equipping them with the education and the tools that would allow them to run smart water systems efficiently. Such capacity-building approaches give a sense of belonging to the local communities, making them an integral part of the rationalization of the water sources they have. It encourages not only the improvement of water management in terms of its technical and operational features but even the incorporation of climate change adaptation strategies in the water systems.

Finally, in regard to the former, where one of them can be represented by the tendencies in the regulatory environment and the international partnerships, both of them need to complement each other if scaling SWI is to be achieved. Countries can devise water management strategies that enable them to increase their investment and focus on achieving the Sustainable Development Goals (SDGs) by developing regulatory instruments to streamline investment and data sharing among stakeholders.

8.7 CONCLUSION

Due to the present-day circumstances of persistent water availability challenges aggravated by climate change, SWI has become a necessity. SWI employs smart technologies such as the IoT, AI, and data analytics to improve water management systems by supporting real-time management and improved distribution of the resources. SWI, therefore, does not only respond to short-term water scarcity issues, but also plays an important role in the realization of SDG 6 on water and sanitation: ensuring availability and sustainable management of water and sanitation for all.

Specifically, it encourages access and affordability of water (SDG 6.1), promotes quality of water (SDG 6.3), enhances efficiencies in the use of water (SDG 6.4), and encourages IWRM (SDG 6.5). These benefits are important because they enhance the efforts to ensure clean water — a basic necessity for the health and well-being of the underprivileged population. Additionally, such technology would allow for cutbacks in wasting water, while working toward the needs of the several conflicting sectors: urban, agriculture, and industry demand.

SDG 6 interlinkage with other global goals focuses on the broader significance of implementing SWI. Here, SWI has been shown to support sustainable cities (SDG 11) and climate action (SDG 13), further extending the respect that water resources management, for example has to other dimensions of development.

8.8 WAY FORWARD

Going forward, it is crucial to employ a multilayered strategy that encompasses technology, policy, and people to optimize the sustainable water management benefits of SWI. The establishment of viable regulatory practices to stimulate the adoption of SWI investment strategies together with support for nested public-private community partnerships are necessity for governments and policy makers.

Public sector agencies, private sector companies, and non-profit organizations can engage in active partnerships for the purpose of knowledge transfer and rolling out improved practices of use in water management. Such studies, for example, have been shown to be very important in adverse geographical locations, when, for instance, climates are dry and there is a scarcity of water. There are International help and support, hence the countries can implement SWI in a way that its positive spread and effect is simultaneous among the country residents.

In addition to that, there is a need for constant training and up-skilling of the local community and other stakeholders for the proper use of SWI technologies. It is this ownership that supports and motivates the local people to embrace and engage with environmental sustainability practices.

To wind up, the power of SWI in transforming the water management status quo is very promising. A strategy that steers all of these factors simultaneously is viable; technology, policy and willing partners' efforts will go a great way in attaining water sustainability for all. There is no disputing that this multifaceted approach is necessary in addressing the challenging water issues of the present day and ensuring that sustainable access to clean and safe water is achieved for ages to come.

REFERENCES

Alkaisi, A., Mossad, R., & Sharifian-Barforoush, A. (2017). A review of the water desalination systems integrated with renewable energy. *Energy Procedia*, 110, 268–274.

Álvarez Díez, A., Pena Rois, R., Mocanu, I., Orzan, C., Brebenel, C., Stere, J., … & Fernández Montenegro, J. M. (2024). Reinforcement learning-based DSS for coagulant and disinfectant dosage selection on drinking water treatment plants. *Water Supply*, 24(1), 86–102.

Anderson, L. E., DeMont, I., Dunnington, D. D., Bjorndahl, P., Redden, D. J., Brophy, M. J., & Gagnon, G. A. (2023). A review of long-term change in surface water natural organic matter concentration in the northern hemisphere and the implications for drinking water treatment. *Science of the Total Environment*, 858, 159699.

Bwambale, E., Abagale, F. K., &Anornu, G. K. (2022). Smart irrigation monitoring and control strategies for improving water use efficiency in precision agriculture: a review. *Agricultural Water Management*, 260, 107324.

FAO. (2021). *Water Use in Industry and Agriculture*. Food and Agriculture Organization of the United Nations, Rome, 2021.

Farrelly, M., & Brown, R. (2011). Rethinking urban water management: experimentation as a way forward?. *Global Environmental Change*, 21(2), 721–732.

Gosling, S. N., & Arnell, N. W. (2016). A global assessment of the impact of climate change on water scarcity. *Climatic Change*, 134(3), 371–385.

Gupta, A. D., Pandey, P., Feijóo, A., Yaseen, Z. M., &Bokde, N. D. (2020). Smart water technology for efficient water resource management: a review. *Energies*, 13(23), 6268.

Harris-Lovett, S., Lienert, J., & Sedlak, D. (2019). A mixed-methods approach to strategic planning for multi-benefit regional water infrastructure. *Journal of Environmental Management*, 233, 218–237.

Jensen, A. S., Coffman, V. R., Schullehner, J., Trabjerg, B. B., Pedersen, C. B., Hansen, B., … & Sigsgaard, T. (2023). Prenatal exposure to tap water containing nitrate and the risk of small-for-gestational-age: a nationwide register-based study of Danish births, 1991–2015. *Environment International*, 174, 107883.

Kim, J., Kweon, S. J., & Hwang, S. W. (2021). Spatial equality and equity for an effective emergency water distribution system: points of distribution. *Journal of Water Resources Planning and Management*, 147(3), 04020111.

Kumar, A., Dubey, S. K., Sendhil, R., Mishra, A. K., Sah, U., Suna, T., & Chand, R. (2022). Smart and sustainable food production technologies. In *Smart and Sustainable Food Technologies* (pp. 3–24). Singapore: Springer Nature.

Kumar, P., & Choudhury, D. (2024). Innovative technologies for effective water resources management. In *Water Crises and Sustainable Management in the Global South* (pp. 555–594). Springer, DOI:10.1007/978-981-97-4966-9_18

Mahmood, H. S., Abdulqader, D. M., Abdullah, R. M., Rasheed, H., Ismael, Z. N. R., & Sami, T. M. G. (2024). Conducting in-depth analysis of AI, IoT, Web Technology, Cloud Computing, and Enterprise Systems integration for enhancing data security and governance to promote sustainable business practices. *Journal of Information Technology and Informatics*, 3(2), 297–332.

Martinez, A. A. (2019). Protecting sources of urban water supply in Haiti: an institutional analysis. *Journal of Water, Sanitation and Hygiene for Development*, 9(2), 237–246.

Moran, T., Saracino, A., Sugg, Z., & Thompson, B. (2020). *Evaluating the Use of Data Platforms for Water Management Decisions. Water in the West*. California: Stanford Digital Repository.

Nguyen, M. D., Thomas, M., Surapaneni, A., Moon, E. M., & Milne, N. A. (2022). Beneficial reuse of water treatment sludge in the context of circular economy. *Environmental Technology & Innovation*, 28, 102651.

Olsson, G. (2021). Urban water supply automation–today and tomorrow. *AQUA – Water Infrastructure, Ecosystems and Society*, 70(4), 420–437.

Rodríguez-Alonso, C., Pena-Regueiro, I., & García, Ó. (2024). Digital twin platform for water treatment plants using microservices architecture. *Sensors*, 24(5), 1568.

Sinha, S. K., Babbar-Sebens, M., Dzombak, D., Gardoni, P., Watford, B., Scales, G., … & Meeker, M. (2023). Smart one water: an integrated approach for the next generation of sustainable and resilient water systems. In *Oxford Research Encyclopedia of Environmental Science*. Oxford University Press.

Sweeney, M., & Kabouris, J. (2020). Modeling, instrumentation, automation, and optimization of water resource recovery facilities (Review of 2018 Literature) DIRECT. *Water Environment Research*, 92(10), 1618–1624.

Thompson, D. D. (2021). Hazardous environmental effects. In *Encyclopedia of Security and Emergency Management* (pp. 403–408). Cham: Springer International Publishing. DOI: 10.1007/978-3-319-70488-3_146

Tiwari, S., Chidambaram, N., & Tyagi, A. K. (2024). Need of 21st century: a sustainable environment together with a smart environment. In *Operational Research for Renewable Energy and Sustainable Environments* (pp. 199–224). United States, Hershey, Pennsylvania: IGI Global.

Trebilcock, M., & Rosenstock, M. (2015). Infrastructure public–private partnerships in the developing world: lessons from recent experience. *The Journal of Development Studies*, 51(4), 335–354.

UN DESA. (2022). *World Population Prospects 2022*. New York, USA: United Nations Department of Economic and Social Affairs.

UN. (2019). *Urbanization Trends and Implications*. New York, USA: United Nations Department of Economic and Social Affairs, Population Division.

WRI. (2020). *Global Water Stress and Demand Projections*. Washington, D.C.: World Resources Institute.

Part II

Applications in Agriculture

9 Smart Technologies to Achieve Net Zero Emissions in Food Production Systems

Joan Cecilia C. Casila, Kanta Kobayashi, Joseph Conrad M. Li, and Eric Jay F. Bulalacao

9.1 OVERVIEW

The net-zero emissions target in Sustainable Development Goal 13 and the Paris Agreement long-term goals requires commitment from various industry sectors to reduce greenhouse gases (GHG) emissions, which are notable contributors to climate change. Economic growth is paramount for a country's development, but it typically entails environmental tradeoffs in the form of GHG emissions. The water sector plays a significant role in direct and indirect GHG production. In particular, the agriculture and aquaculture food production sectors rely heavily on water (Aryal et al., 2024), leading to the production of harmful gases (Casila et al., 2022) such as methane (CH_4), nitrous oxide (N_2O), and carbon dioxide (CO_2) (Lynch et al. 2021; Myhre et al., 2013). In this chapter, food production systems are referred to as agriculture, aquaculture, and livestock.

Emissions from these production systems are difficult to quantify due to the distinct processes through which food is generated, along with challenges to uniformly account for methods and sectoral boundaries. In response, several studies of AI-based carbon capturing (Priya et al., 2023), carbon sequestration (Wang et al., 2024), and carbon footprint (Miao et al., 2023; Ozlu et al., 2022) have become widespread initiatives to evaluate the environmental impact of key agriculture and aquaculture products.

Rice production is an example of a water-intensive agricultural industry that releases GHG emissions. For instance, the application of urea and lime, the use of tractors, and the transportation of produce lead to CO_2 emissions. Methane is produced due to the anaerobic conditions of the flooded field. Additionally, the use of synthetic fertilizers and animal manure emits N_2O (Shakoor et al., 2021). Several studies have been conducted by the International Rice Research Institute about the effect of reduced CO_2 in the soil and automated methane emission monitoring (Wassman et al., 2000). While two cropping seasons in a year reduce CO_2 in the soil due to a rest period, three croppings in a year lead to higher yield but higher CO_2 is generated. Methane emissions increase with the effect of the continuous

DOI: 10.1201/9781003547099-11

submergence of lowland rice in water. CO_2 is further generated, especially in set-ups with higher fertilizer input. This highlights the need for research on the potential of leveraging smart technologies, particularly artificial intelligence (AI).

For irrigation and hydropower resources, interest is currently placed in organic matter that settles in deeper water impoundments or reservoir areas that transform under low-oxygen conditions, leading to methane production (Lessman et al., 2023). The methane escapes into the atmosphere either through diffusion in the reservoir water or through water movements downstream (e.g., degassing when water is released through turbines).

The aquaculture sector also contributes to global warming. Aquaculture practices could be significantly improved through advancements in AI. By employing sensors to monitor real-time water quality parameters (e.g., pH, turbidity, electrical conductivity, salinity, ammonium ion, and dissolved oxygen) and fish migration, valuable insights could be gained into fish behavior and migration patterns. Data analytics combined with artificial neural network (ANN), the Internet of Things (IoT), and machine learning (ML) allows us to optimize feed management, reduce feed wastage, monitor fish populations, adjust fishing quota, and determine the efficient feed formulation and feeding schedule to improve growth rates. Further innovations could be explored, such as using smarter sensors and IoT to monitor water resources, which can help preserve and protect marine ecosystems, for cultural preservation and sustainability, and to effectively identify and remove specific contaminants in wastewater treatment systems.

The continued innovation of AI in handling large amounts of data using algorithms provides fast and efficient solutions for sustainable food production. ML could predict and analyze future trends and events of weather to optimize the use of water for systems. Automation and robotics powered by AI use sensors to improve farming practices in monitoring crop, seafood, and livestock health, maximize the use of water, and minimize environmental impacts.

This book chapter covers the role of water use and withdrawal in relation to climate change. It gathers existing AI-based research, analyzes the effectiveness and sustainability of various solutions, and identifies gaps in net-zero emissions through water conservation in agriculture, aquaculture, and livestock. With the availability of AI sensors that transmit real-time data on soil moisture levels, methane emissions in rice fields, available soil nutrients such as nitrogen, phosphorus, and potassium, as well as physicochemical water quality parameters in impoundments and aquaculture systems, to name some, it is possible to significantly monitor and reduce the amounts of water applied, water replacement needs, GHG emissions, and nutrient inputs.

9.2 WATER AS A VITAL SOURCE

Navigating the pressing challenges of water use, water withdrawals must be approached with care and foresight. Ritchie et al. (2022) and Ritchie and Roser (2015) mentioned that water withdrawals—often referred to as "water abstractions"—involve the extraction of freshwater from surface and groundwater sources, such as lakes and rivers. Quantifying water usage across various industries is essential to pinpoint sectors that can enhance their water management practices, leading

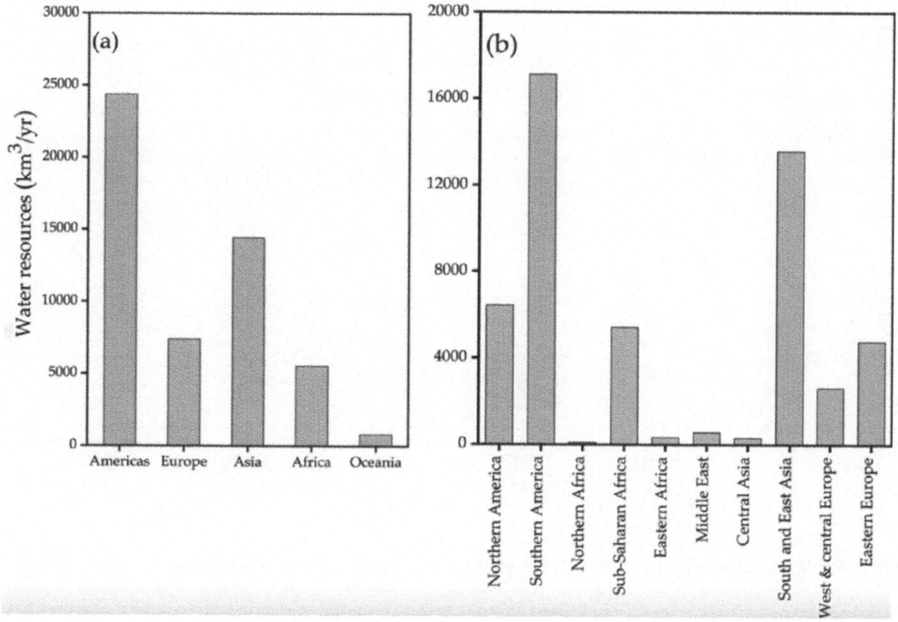

FIGURE 9.1 Water resources distribution (a) continental and (b) regional. (Musie & Gonfa, 2023.)

to sustainable resource use that benefits both the environment and society. Positive impact could be made by improving efficiency and conservation methods to ensure that this vital resource remains available for future generations (FAO, 2022).

The U.S. Geological Survey (2024, as cited in National Ocean Service) estimates that if all of Earth's water were collected into a sphere, it would be about 1,385 km (860 m) in diameter, containing approximately 332.5 million cubic miles (1,386 million cubic kilometers) of water—equivalent to over a trillion liters per cubic kilometer (or 264 billion gallons). However, despite this abundance, less than 1% of Earth's water is readily accessible for human use, because 99% is found in groundwater and only 1% is on surface water (Musie & Gonfa, 2023).

Research shows that freshwater resources are unevenly distributed around the globe. Some regions receive abundant water supplies, while others grapple with significant shortages (Musie & Gonfa, 2023). This imbalance presents considerable challenges, particularly for areas that heavily rely on freshwater for agriculture and daily life. This situation is illustrated in Figure 9.1, which highlights both the continental and regional distributions of water resources.

9.2.1 Rising Demand for Freshwater

The increasing demand for freshwater faces challenges and opportunities (Hubbart & Ross, 2024). According to the International Water Management Institute in their Comprehensive Assessment of Water Management in Agriculture, the increase in water withdrawal can be traced to two key factors: supply-push and demand-pull.

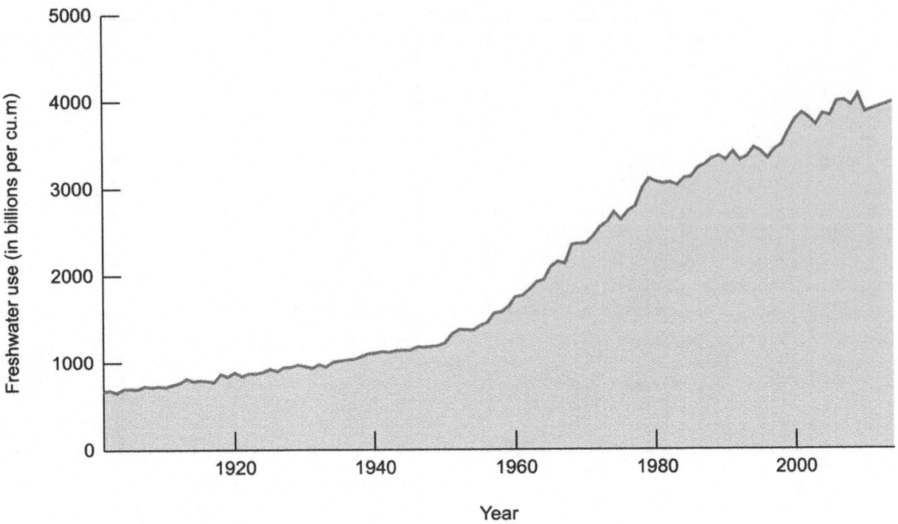

FIGURE 9.2 Historical trend of freshwater use from 1901–2014. (Ritchie & Roser, 2024.)

Supply-push factors, such as generous government subsidies and the widespread availability of affordable pumps and drilling technologies, have made accessing water resources more attractive than ever. Meanwhile, demand-pull factors underscore the rising need for food, driving an expansion in food production systems to meet the needs of growing populations.

The trend in freshwater use, as shown in Figure 9.2, from 1901 to 2014, shows significant changes with a general increase over time. A major factor was the development of new groundwater technologies, like drilling rigs, which allowed more water extraction (Gun, 2019). The data shows a sharp rise in water use during the 1950s and 1960s when freshwater use increased from about 520 million cubic meters in 1950 to over 1.4 billion cubic meters by 1968. This increase highlights how technology helped meet the growing need for water due to factors like agricultural growth, industrial development, and population growth (SMS Foundation, 2023).

9.2.2 Role of Water in Agriculture and Food Production

Understanding the critical issue of water withdrawals across various sectors is vital. Particularly as global demand continues to escalate. This includes how agriculture, industry, and domestic use influence water withdrawal and consumption. The agricultural sector is responsible for approximately 70% of global water withdrawals. Figure 9.3 illustrates the central role agriculture plays in the discourse on water management (FAO, 2022). This data reflects the intricate processes within the agricultural sector, including pre-planting activities, planting, post-harvest handling, waste management, production, and delivery to consumers. Therefore, this section focuses on the food production system within the agricultural sector, highlighting its essential contributions and challenges.

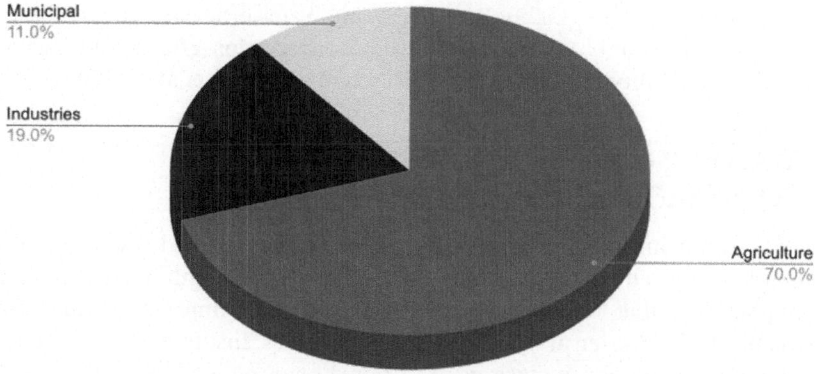

FIGURE 9.3 Comparison of water withdrawals among different sectors. (FAO, 2024.)

Water is crucial in various farming practices like irrigating crops, providing drinking water for livestock, or maintaining water quality in aquaculture (Apure, 2023). These elements are vital for food security and sustainability. For instance, efficient irrigation helps maximize crop yields, while clean water is essential for keeping fish and livestock healthy.

Fish is a vital source of animal protein in the human diet. With population increase and economic advancements, an increase in fish consumption and global demand for food fish is projected (IAHR, 2020). Asia accounts for roughly 90% of global aquaculture output and has emerged as the leading provider in the international seafood market (FAO, 2016), with 60% of the produce coming from aquaculture. In aquaculture, maintaining the right water quality is key to ensuring healthy and safe fish and seafood (Fujs & Kashiwase, 2023).

Meat production, particularly beef farming, tops the ranking in water use. For every pound of beef, around 6,800 L (1,800 gallons) of water is required (The Water Footprint, 2018). This includes the substantial amounts of water needed to grow feed crops like corn and soy—often irrigated intensively—and the daily water intake of each cow (Wagner & Engle, 2021). Beyond feeding and drinking water, beef processing itself demands significant water for hygiene and maintenance, which increases pressure on local water resources and highlights critical environmental concerns (Ritchie et al., 2022).

Most cereals consume around 1,644 liters of water per kilogram. PhilRice explains that rice requires approximately 2,500 L of water per kilogram, and this can escalate to 4,000 L if water management practices are not optimized (Pascual, 2021). Given its heavy water usage, the rice industry must embrace efficient water practices to meet growing demand while conserving resources.

Rice and beef represent only a fraction of the agricultural products that demand substantial water resources. Other crops such as nuts, avocados, and sugarcane also rank high in water consumption, along with various types of livestock (Hoekstra, 2012). Each of these products possesses a distinct water footprint influenced by factors such as cultivation practices, regional climate conditions, and specific processing requirements (Wang et al., 2023). By prioritizing water efficiency across these sectors, we can significantly contribute to water conservation efforts and ensure the long-term sustainability of food production systems (Ritchie et al., 2022).

9.3 WATER USE AND GHG EMISSIONS IN FOOD PRODUCTION SYSTEMS

There is a significant relationship between GHG emissions and water withdrawal, particularly in agriculture. Numerous researchers have highlighted this connection, emphasizing that water usage in farming directly impacts global warming (Silverman, 2022; Huo et al., 2022; World Resources Institute, 2014). Figure 9.4 shows the GHG emissions in gigatons of carbon dioxide equivalents ($GtCO_2eq$) from different sectors (Filonchyk et al., 2024). Interestingly, while agriculture accounts for a large share of freshwater use, it emits relatively low levels of GHGs compared to other sectors. By conserving water, communities can play a part in reducing the agricultural sector's carbon footprint by 0.4% which is a significant move toward achieving net-zero emissions goals.

9.3.1 GHG Emissions across Agricultural Farming Practices

Agriculture is essential for feeding our expanding global population, yet it contributes to climate change through the "Heat-Trapping Trio" of GHGs: methane, nitrous oxide, and carbon dioxide (Dinoy, 2024). These gases are released through various farming practices, from livestock production to fertilizer application, and their

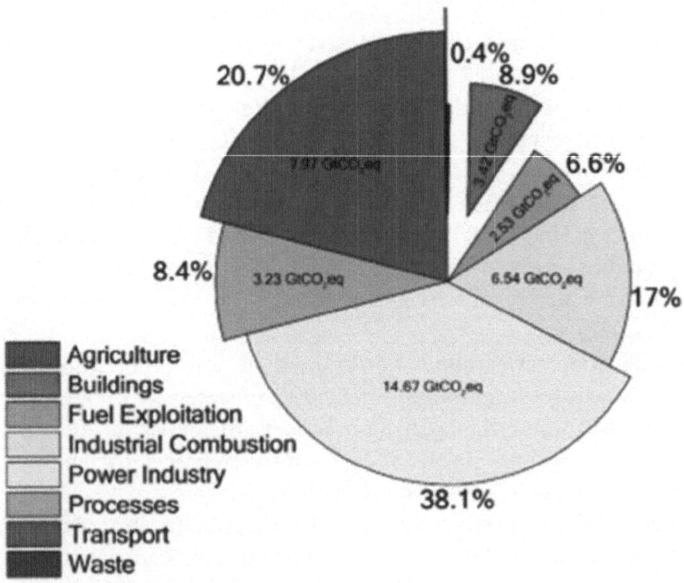

FIGURE 9.4 GHG emissions in gigatons of carbon dioxide equivalents ($GtCO_2eq$) in various sectors. (Filonchyk et al., 2024.)

effects on the climate are profound. Recognizing the origins of these emissions in our farming practices is essential for crafting sustainable solutions that can lessen their effects and promote a healthier environment for generations to come.

Methane, a greenhouse gas with significant warming potential, contributes about 16% of global emissions with a heat-trapping ability of over 25 times greater than carbon dioxide over a century (Garthwaite, 2021). One of the main agricultural sources of methane is livestock farming, particularly cattle, due to enteric fermentation, which releases methane during digestion (Zhang et al., 2023a). Manure management also adds to these emissions; anaerobic conditions in manure lagoons lead to methane production as organic matter decomposes without oxygen (Agga et al., 2022). Beyond livestock, rice paddies are substantial methane contributors, as flooded, low-oxygen fields create ideal conditions for methanogens— methane-producing microbes— that thrive with the added nitrogen from synthetic fertilizers (Li et al., 2023). In major rice-producing countries, like the Southeast and South Asia regions, rice production produces methane accounting for up to 20% of the national GHG inventories (IRRI, 2024). Furthermore, aquaculture operations, particularly shrimp farming, generate methane as organic waste decomposes in anaerobic pond sediments, driven by acetoclastic methanogenesis (Tan et al., 2023). This indicates that methane is primarily driven by aerobic and anaerobic digestion processes.

Nitrous oxide contributes roughly 6% of global GHG emissions and has a heat-trapping ability nearly 300 times that of carbon dioxide (Chrobak, 2021). Livestock farming contributes to these emissions through manure management when manure is stored or treated anaerobically, releasing N_2O into the atmosphere from the breakdown of nitrogen compounds (Zhu et al., 2021). A significant portion of N_2O emissions in agriculture also originates from nitrogen-based fertilizer application when soil microbes convert some of the nitrogen into N_2O (Lu et al., 2022), particularly in warm and moist conditions that enhance microbial activity (Cruz-Paredes et al., 2021). This accounts for approximately 60% of the total N_2O emissions from agricultural lands (Li et al., 2020). Aquaculture further contributes to N_2O emissions through the accumulation of organic waste in anaerobic pond sediments, where uneaten feed and waste decompose, converting nitrogen compounds into N_2O (Yang et al., 2023).

Carbon dioxide emissions are responsible for approximately 76% of total GHG emissions, primarily resulting from the burning of fossil fuels, deforestation, and industrial processes (Nunes, 2023). In agriculture, fossil fuel combustion for machinery, transportation, and processing significantly contributes to CO_2 levels (Bahman et al., 2023). Moreover, land clearing for crops and livestock not only releases stored carbon in soil and vegetation but also contributes to soil disturbances during tillage, exposing carbon to the atmosphere (Derpsch et al., 2024). The energy-intensive processes involved in producing synthetic fertilizers, heavily reliant on fossil fuels, add another layer of carbon emissions. Additionally, aquaculture contributes to CO_2 emissions indirectly through the energy used for aeration, pumping, and maintaining water quality in aquaculture systems that rely on fossil fuels (Zhang et al., 2024). Furthermore, the production and transportation of feed for aquaculture can exacerbate CO_2 emissions, particularly if sourced from land-clearing activities or involving significant processing. Collectively, these practices highlight the urgent need for sustainable approaches in both agriculture and aquaculture to mitigate carbon emissions and combat climate change.

9.3.2 Aquaculture's Carbon Footprint

There has been growing concern regarding GHG emissions from aquaculture. Small-scale aquaculture ponds, which are particularly widespread in developing countries, are suggested to emit the largest amount of GHGs per unit area (Holgerson, 2015). GHG generation, migration, transformation, and release in aquaculture are intricate and linked to biogeochemical processes that involve carbon and nitrogen (Chen et al., 2023). These processes are influenced by the physicochemical water quality properties, agricultural management practices, substrate quality, water depth, and presence of aquatic plants. A study on dissolved concentrations and fluxes of CO_2 and CH_4 was conducted in small coastal earthen shrimp ponds (Tong et al., 2021). Results showed that subtropical earthen aquaculture ponds have more CH_4 emission per square meter than most inland water bodies.

Excessive feeding to boost production could also lead to eutrophication and water quality deterioration, as uneaten feed contributes to the accumulation of dead algae, which, when decomposed, increases GHG emissions (Pu et al., 2022). To quantify these emissions and come up with GHG inventories, smart technologies such as water sensors and the use of AI must be leveraged.

9.3.3 Livestock Water Use and GHG Emissions

Livestock products play a crucial role in global food security, contributing 17% of the world's caloric intake and 33% of protein intake (Rojas-Downing et al, 2016; Rosegrant et al., 2009). The global demand for these products is anticipated to double by 2050, largely driven by rising living standards around the world (Rojas-Downing et al., 2016). Additionally, livestock is one of the typical water-intensive products. The production of animal products consumes more than 30% of global agricultural water (Mekonnen & Hoekstra, 2012). The water usage required for animal products is higher compared to crop products with similar nutritional value (Lin et al., 2021).

Traditional livestock management practices face several obstacles. Resource inefficiency is an example, especially in terms of feed and water management. Conventional farming methods often depend on extensive grazing, which can lead to overgrazing and soil degradation. Moreover, inefficient feeding practices result in waste and may provide inadequate nutrition for cattle, negatively affecting their health and productivity. The environmental impact of traditional cattle farming is also a major concern, as unsustainable practices like deforestation for pasture expansion and poor waste management contribute to habitat loss, soil erosion, and increased greenhouse gas emissions (Pawar et al., 2024). Livestock farming emits a lot of GHG and contributes to global warming. The most important GHGs emitted by the livestock industry are methane and nitrous oxide, which contribute more to global warming than carbon dioxide. Methane is a gas mainly produced by enteric fermentation and manure storage. The process of enteric fermentation in ruminants such as cattle represents over a quarter of the GHG emissions from the agriculture sector (US EPA, n.d.). Nitrous oxide is a molecule that is produced by manure storage and the use of organic and inorganic fertilizers. Additionally, the production of feed, along with the associated emissions of CO_2 and N_2O from soil, represents another significant source of environmental impact in the livestock industry (Grossi et al., 2019).

9.3.4 Socioeconomic Ripple Effects

The socioeconomic ripple effects of GHGs, especially the heat-trapping trio CH_4, N_2O, and CO_2 are deep and widespread, affecting communities and economies around the world. CH_4 presents significant threats to agricultural stability, contributing to extreme weather events that disrupt food production and inflate prices, ultimately leading to food insecurity (Anderson, 2022). The economic burden mostly affects vulnerable groups, particularly in developing countries, where many people depend on farming and have limited adaptation resources. Another gas emission primarily emitted from different agricultural practices is nitrous oxide, as it not only affects air quality and human health but also reduces the productivity in agriculture, straining farmer incomes and threatening livelihoods (Wozniacka, 2019). Carbon dioxide, the most important GHG, fuels climate change, leading to more frequent and severe natural disasters that may harm local economies and force communities to relocate (Lindsey, 2024). These gases contribute to ongoing social and economic challenges, worsening inequalities, and slowing down sustainable development. It is important to address these emissions to create strong communities and promote fair economic growth and a sustainable future for all.

9.4 INNOVATIONS IN AGRICULTURE FOR WATER MANAGEMENT

Farmers traditionally relied on simple techniques and manual labor, digging irrigation channels by hand and monitoring rainfall based on personal observations. Technological advancements allowed the use of pumps, sensors, and data collection for more efficient water management. Today, AI, poised to transform water management, making it smarter, more efficient, and sustainable (Bibri et al., 2024). This discussion will explore historical progression, the benefits of AI, and its contributions to achieving net-zero emissions.

In the 21st century, smart irrigation systems emerged, utilizing real-time weather data and soil moisture sensors to optimize irrigation practices. These systems conserve water and improve crop health, reducing the need for excessive fertilizers that contribute to GHG emissions (Lakhiar et al., 2024). Advancements in sensor technology also enhanced water quality monitoring, highlighting the importance of responsible water management for achieving net-zero emissions (Essamlali et al., 2024).

Techniques like Alternate Wetting and Drying (AWD) in rice cultivation have proven effective in conserving water and reducing methane emissions by allowing fields to dry intermittently, thereby minimizing anaerobic conditions (Lampayan et al., 2014). However, other methods, such as mulching and zero-tillage practices, have seen limited adoption despite their potential benefits for soil health and moisture retention (Tao et al., 2006).

Several systems for water conservation are continuously being studied. The aerobic rice system, which involves growing rice in well-drained, non-flooded, and unsaturated soils without ponded water, was found responsive to inputs and optimized water management techniques (Garg et al., 2022; Mallareddy et al., 2023). Similarly, raised-bed systems (Choudhury et al. 2007) and saturated soil culture (Kima et al., 2014) are continuously studied because they offer promising water-saving techniques to optimize water use and enhance crop yields.

9.5 AI APPLICATIONS IN FOOD PRODUCTION AND WATER CONSERVATION

Global population growth and climate change have led to water scarcity. Innovative solutions using AI can be used to manage and conserve water effectively. AI is an innovative tool for water management, especially within food production systems that are both resource-intensive and environmentally impactful. The application of AI in water conservation spreads widely, from agriculture, smart irrigation, aquaculture, livestock, and water supply management. This section covers the role of AI in water conservation and GHG emissions reduction through advanced monitoring techniques, predicting, and optimizing water use, highlighting successful implementations and their tangible impacts.

9.5.1 AGRICULTURE

The integration of AI technology in agriculture has great potential to optimize water use and improve efficiency across the sector as a whole. For instance, AI can help optimize irrigation and water management by analyzing factors such as crop type, weather, health of the crops, and soil characteristics. ML uses historical data to create customized irrigation schedules that improve water efficiency and crop productivity while reducing energy use and promoting agricultural sustainability (Nova. 2023).

AI could play an important role in reducing GHG emissions from agriculture in various ways. The appropriate use and monitoring of water resources using AI indirectly leads to a reduction in GHG emissions during withdrawal from reservoirs and optimizes fertilizer use, leading to lower emissions from fertilizers (Xu et al., 2024). AI can contribute to a reduction of CH_4 produced due to the anaerobic condition of the paddy field (Abbasi et al., 2019). Currently, CH_4 emissions from rice paddy fields in the Sanjiang Plain, China, were estimated based on remote sensing data, and the results provided reliable estimates of the spatial distribution of CH_4 emissions at the regional scale when compared with the Intergovernmental Panel on Climate Change (IPCC) approach and previous studies (Sun et al., 2017). Low-cost sensor-based automated in-situ GHG measurement systems were also developed and implemented. The results obtained from this work demonstrated its reliability and showed that controlled intermittent flooding (CIF) treatment reduced CH_4 emissions and water usage compared to intermittent flooding (IF) and continuous flooding (CF) treatments (Rajasekar and Selvi, 2022). It suggests that AI-driven systems can provide precise control over irrigation, using real-time data from sensors to monitor parameters. These systems enable farmers to optimize water use, improving yield while minimizing waste.

Several studies utilized remote sensing and ML algorithms. AI allows real-time monitoring of soil moisture, crop health, and water requirements, supporting informed irrigation decisions (Efremova et al., 2019; Inoue and Yokoyama, 2017; Yokoyama, 2017). It prevents over- and under-irrigation through optimal soil moisture management and facilitates early detection of crop stress and disease, allowing farmers to supply water precisely when needed (Wei et al., 2024). This approach reduces water waste and curtails energy consumption associated with irrigation systems. Additionally, AI analyzes various data sources, such as weather and crop growth information, to optimize irrigation schedules and water management strategies (Saikai et al., 2023). This not only conserves water but also improves agricultural productivity and sustainable water use (Nova, 2023). By integrating AI with remote sensing data and sensor-based GHG measurement systems, effective irrigation control can be achieved, further minimizing water usage and CH_4 emissions. Figure 9.5 shows the components and their interdependencies in an AI-enabled precision agriculture system.

Another AI application in agriculture was conducted by Montazeri et al. (2023), which maximized the use of AI to model movement and chromium (Cr) from the discharge in the river up to absorption of tomatoes in Mashhad, Iran. Here, chromium was measured at three different river depths and accumulation effects. Hoseinzadeh and Astiaso Garcia (2024) also evaluated the integration of AI in greenhouse agriculture for its potential to enhance sustainability and crop efficiency. A detailed approach between traditional farming and AI-based approaches was developed after re-evaluating publicly through noise reduction methods and advanced time series analysis. Akkaş et al. (2024) addressed the issue of extensively using pesticides for pests that affect the crops. AI was used to control, detect, and prevent pests

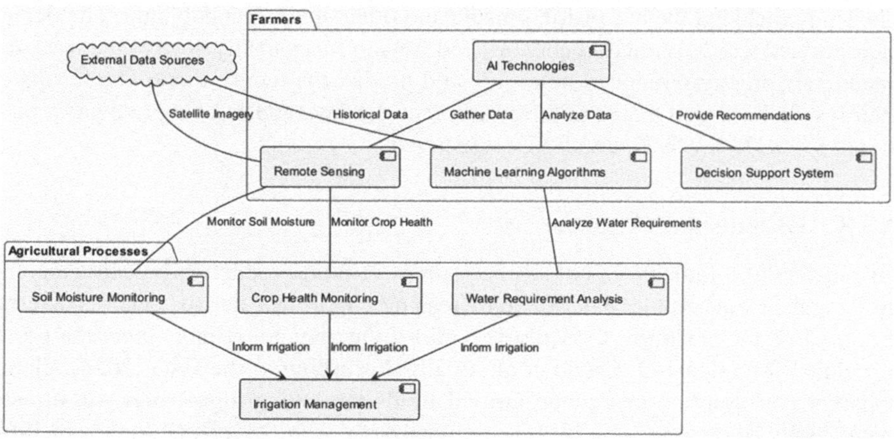

FIGURE 9.5 Components and their interdependencies in an AI-enabled precision agriculture system. (Nova, 2023.)

through advanced analytics technologies and real-time data from IoT with an average efficiency of 98.87% for controlling and preventing the pests at a value.

9.5.2 Smart Irrigation Systems

Kasera and Acharjee (2024) designed a long-range (LoRa) edge IoT for a sustainable smart irrigation system from real-time data of tomato plants. Water usage, demand, and status were monitored through smart sensors. A hybrid algorithm for machine-to-machine communication (M2M) named linked least traversal (LLT) was created, while the proposed reinforcement learning (RL) was based on optimal soil wetness closeness policy (OSWCP). The systems automatically schedule a drip irrigation and predict the water needs of the plants for maximum growth. The results obtained the highest accuracy of 97.88% using real-time data from OSWCP. Manocha et al. (2024) developed a platform using IoT to collect soil and meteorological data for field test. The study provided important information to reduce water consumption and improve farm operations. In Morocco, Bouali et al. (2024) aimed to connect small-scale farmers by developing a cloud-based irrigation system. The system optimized a large amount of data, with storage and analysis to enhance decision-making for water conservation in arid areas. The study also included weather forecasting for irrigation control during inconsistent rainfall using solar power. Similarly, the study of Jain (2023) using IoT through web and Android applications provided solutions for continuous control and monitoring of a smart irrigation system to avoid water waste.

Another agriculture-related AI application is using a 6 G-enabled Internet of Things (6G-IoT) by Sitharthan et al. (2023). It utilized a prediction algorithm for rainfall by analyzing historical data of weather. This supplied the necessary water for crops by measuring the soil moisture. Under a controlled environment, the study demonstrated accuracy, precision, and sensitivity values of 86.34%, 91%, and 89.28%, respectively (Sitharthan et al. 2023). Sangeetha et al. (2022) also studied an intelligent agriculture that enhanced crop production and agricultural benefits in India. The study included the use of IoT sensors and other environmental data. The device incorporated a radial function network and GPS to successfully control the irrigated system, air pressure, reduced humidity, and predict the temperature. The results of their research showed that sensitivity, accuracy, hit rate, and catching rate had values of 0.9834, 0.7734, 0.8955, and 0.77, respectively.

9.5.3 Fisheries and Aquaculture

The aquaculture industry is one of the sources of food production by culturing and growing fish. Aquaculture has gained increasing importance due to concerns that traditional livestock farming, agriculture, and fishing may not reliably meet the growing global food demand (David et al., 2023). According to the FAO (2024), global apparent consumption of aquatic animal foods (excluding algae) reached an estimated 162 million tonnes in 2021, a significant rise from 28 million tonnes in 1961. However, traditional aquaculture presents environmental challenges, limited production capacity, and high labor demands (Vo et al., 2020). These conventional methods also lack real-time data collection, hindering rapid responses to health issues and environmental changes (Wu et al., 2022). High-density aquaculture, implemented to satisfy increasing marine product demand, often deteriorates water quality and facilitates disease spread, thereby reducing production efficiency. Sustainable aquaculture requires the optimization of water, energy, and resource use (Mandal & Ghosh, 2023).

Maintaining optimal water quality, appropriate stocking densities, and effective disease prevention is crucial. AI technology can enhance monitoring and management in aquaculture by providing tools for real-time monitoring, data analysis, predictive modeling, and decision support (Mandal & Ghosh, 2023; Mustapha et al., 2021). AI analyzes historical growth data to identify trends and optimize operations, leading to improved growth performance and resource efficiency (Yue & Shen, 2022). By continuously monitoring parameters like temperature, pH, dissolved oxygen, and nutrient levels, AI systems can detect deviations from ideal conditions and implement corrective actions in real time, reducing fish mortality and enhancing productivity (Kaur et al., 2023; Gladju et al., 2022). In Hong Kong, 12 key fish culture zones have been set up with real-time telemetry data monitoring stations having spatial distances ranging from 2.5 to 20 km (IAHR, 2020). The availability of high-frequency real-time temperature, salinity, dissolved oxygen (DO), and chlorophyll fluorescence data (at 10-minute intervals) allows the possibility of forecasting algal bloom risks on a daily basis.

Effective water quality management through AI not only conserves water but also minimizes costs by customizing feeding based on fish growth stages and nutritional requirements (Glencross et al., 2023; Siad & Bouzid, 2023). Furthermore, AI can help reduce GHG emissions by monitoring and managing organic matter accumulation at the bottom of aquaculture systems. Advanced technologies like image recognition and deep learning models, such as convolutional neural networks (CNN), can automatically detect disease signs in fish by analyzing images and videos (Zhang et al., 2020). These models classify fish images as healthy or diseased, aiding in early disease detection (Mandal & Ghosh, 2023). AI can also accurately estimate biomass, allowing for optimal stocking densities, harvest times, and production plans based on growth rates, feeding patterns, and market demand (Gladju et al., 2022; Mandal & Ghosh, 2023). Overall, AI integration not only boosts aquaculture productivity but also improves energy and water consumption.

Mohd Jais et al. (2024) designed and developed an accurate water quality monitoring system for an aquaculture tank culturing Asian seabass in Malaysia. The device was used to improve the yields and sustainability of fish farming. This is done with the help of IoT and sensors for accurate water quality management. Additionally, a printed circuit board (PCB) that is compact and efficient was developed to accurately read the data using sensors and communicate in an aqua environment in Pune, Maharashtra. It allowed the collection of critical data that reduces the percentage of diseases and mortality with minimal human interaction and maintenance cost (Shete et al. 2024)

The integration of IoT devices with Firebase's cloud data system for reliable and accurate data in culturing a catfish *(Siluriformes)* was studied in India to develop successful monitoring of fishpond conditions in real time, using sensors such as pH, temperature, and turbidity (Sung et al. 2023). Meanwhile, Zheng et al. (2023b) used a spatiotemporal attention network (STAN) to analyze the feeding behavior of golden pompano. Algorithms and techniques for spatial image and optical image were used to evaluate the performance of their methods. The STAN achieved an accuracy level of 97.97%. Lastly, Chen et al. (2022) conducted a case study aquaculture fish farm in Taiwan, introducing a deep learning model with a Convolutional Neural Network (CNN) that classifies three types of appearance of grouper. The experimental result of their research reveals that the model for classifying different types of abnormal appearance of grouper can successfully reach an average value of 98.94% accuracy during the phase II task.

9.5.4 Livestock

The integration of AI into livestock has the potential to transform the industry, fostering innovation, improving operational efficiency, and promoting sustainability (Pawar et al., 2024). AI can contribute to farmers' decision-making in various ways using data obtained through remote sensing technology, such as sensors. The sensors can provide information on water consumption, water quality data, and livestock behavior (Tang et al., 2020). AI can integrate this information and optimize water usage. In addition, AI can be a powerful tool for enhancing health monitoring and disease detection in cattle farming, enabling farmers to identify potential health issues early and implement timely interventions (Pawar et al., 2024). In terms of animal welfare, AI can process the vast amounts of data obtained from wearable devices and environmental sensors, identify patterns in behavior and physiological changes, and detect signs of stress and illness at an early stage (Neethirajan, 2023a). AI can also integrate multiple data sets to assess the overall welfare of animals, as well as detect problems in real time and support proactive management. AI-driven automated feeding systems in livestock enhance efficiency, precision, and resource management by reducing labor and optimizing feed usage. These systems can customize feed rations for each animal based on factors such as age, weight, health, and nutritional needs, ensuring more precise and individualized care (Pawar et al., 2024). Additionally, AI can play a key role in reducing GHG emissions in livestock farming by supporting precision feeding systems that minimize feed waste and methane emissions, while promoting the use of renewable energy to power digital tools and farm processes (Neethirajan, 2023b).

9.6 POLICIES ON AI REGULATION AND DATA SECURITY

While AI significantly impacts food production, ethical issues such as data privacy and security must be addressed. The vast amounts of data generated by AI can be beneficial but also pose risks, necessitating strong regulations to ensure ethical use. The rapid development of AI often outpaces the establishment of ethical frameworks, highlighting the need for careful regulation and understanding of its long-term effects on sustainable food production (Neethirajan, 2023a). AI regulations in various countries are presented in the succeeding paragraphs.

The EU AI Act, effective in 2024, is the first comprehensive legal framework for AI, employing a risk-based classification system. It bans systems with "unacceptable risks" (e.g., social scoring) and heavily regulates "high-risk" systems in critical sectors like healthcare, while imposing lighter requirements on "limited-risk" systems like chatbots. The Act emphasizes transparency, fairness, and accountability, particularly for high-risk AI, requiring ongoing risk assessments and compliance with EU standards. Its extraterritorial scope mandates compliance from non-EU companies targeting the EU market, setting a global benchmark for AI regulation (Upmann, 2024; Global Legal Insights, 2024).

In 2023, the U.S. government began implementing new regulations to ensure responsible AI use. Key areas of concern include intellectual property, competition law, and governance, with new guidance from the U.S. Patent & Trademark Office and increased scrutiny from antitrust regulators. While no comprehensive federal AI regulation exists, an executive order promoting safe AI development was issued in 2023, and state-level regulations are evolving, particularly regarding privacy and fairness in sectors like insurance and employment (Global Legal Insights, 2024).

In 2023, China advanced its AI regulation with laws like the Interim Measures for the Management of Generative AI Services, focusing on data security and ethical oversight. In 2024, it established a robust legal framework mandating state oversight of AI systems, especially those posing security risks, and requiring mandatory reporting of AI data usage. China's regulations balance rapid innovation with strict state control, ensuring oversight of sensitive technologies such as biometric data and surveillance tools (Upmann, 2024; Global Legal Insights, 2024).

India is advancing in AI adoption across sectors like healthcare, agriculture, and finance. While specific AI legislation is currently lacking, the forthcoming Digital India Act is expected to promote ethical and responsible AI use, particularly for high-risk applications like deepfakes. The government has introduced advisories emphasizing transparency, accountability, and fairness (Global Legal Insights, 2024).

AI regulation varies by region, balancing innovation, human rights, and security. Although global principles like transparency and accountability may converge, enforcement and scope will differ, underscoring the need for international cooperation (Upmann, 2024).

9.7 WASTEWATER MANAGEMENT POLICIES AND ITS IMPLICATIONS TO AGRICULTURE

Effective wastewater management is crucial for reducing water and GHG footprints, significantly impacting sustainable agriculture. Untreated runoff from farms can harm ecosystems and exacerbate climate change (Hughes et al., 2020). Wastewater treatment facilities play a vital role in cleaning water waste, removing pollutants, and lowering GHG emissions (Silva, 2022). Examining global wastewater management practices reveals their importance in environmental protection and resource conservation, especially as we pursue net-zero emissions.

Germany has implemented the "Wastewater Ordinance," mandating municipalities to invest in advanced treatment technologies that recycle nutrients like phosphorus for fertilizers (Federal Ministry for the Environment, Nature Conservation, Nuclear Safety, and Consumer Protection, n.d.). This circular economy approach enhances water quality and supports sustainable agriculture.

Singapore's "Four National Taps" framework integrates wastewater treatment into its water supply system, with NEWater facilities ensuring that treated wastewater meets drinking water standards (Susantono & Li, 2023). This innovative strategy exemplifies Singapore's commitment to water sustainability amid resource limitations.

China's "Water Pollution Prevention and Control Action Plan" aims to treat at least 95% of urban sewage by 2020, addressing water quality issues through advanced technologies (Schmid & Xiong, 2023). Policies promoting treated wastewater reuse in agriculture are crucial for sustainable farming, particularly in water-scarce regions (Al Hamedi et al., 2022).

In the United States, the Clean Water Act governs wastewater management, with the EPA establishing national pollutant discharge standards (US EPA, 2024). A recent shift toward sustainable practices includes green infrastructure and innovative treatment technologies like membrane bioreactors (National Association of Clean Water Agencies, n.d.).

Previous studies have used the application of AI and ML in sewer management. ANN and decision trees have been utilized to forecast rates of sewer collapses and blockages, which are essential for proactive management of sewer infrastructure. The research involved data from sensors measuring levels, flow, and water quality, in addition to information from closed-circuit television (CCTV) inspection footage, along with details about sewer characteristics, environmental factors, and maintenance history. ML techniques have also been employed to identify and anticipate blockages as well as to create models that can forecast threshold flow conditions necessary for achieving self-cleansing in sewer systems (IAHR, 2020).

Incorporating AI into wastewater management could enhance efficiency by optimizing treatment processes in real time, identifying waste patterns, and rapidly detecting pollution sources. This advancement would not only protect the environment but also lead us toward a cleaner, more sustainable future.

9.8 AI-DRIVEN PATHWAYS TO WATER CONSERVATION AND SUSTAINABLE FOOD PRODUCTION

The global food system is facing a great challenge to feed the growing population while mitigating the environmental impacts and carbon footprints. Excessive water withdrawal, which is linked to GHG emissions and water scarcity in some regions, further complicates this challenge (Tian et al. 2017). However, the new wave of innovations powered by AI holds the potential to revolutionize food production systems while achieving net-zero emissions.

9.8.1 AI as a Catalyst for Sustainable Water Management

AI is transforming water management through efficient resource allocation, waste reduction, and optimized irrigation practices (Mana et al., 2024). For instance, precision irrigation utilizes AI-powered sensors and drones to gather real-time data on soil moisture, crop health, and weather patterns, creating precise irrigation schedules that minimize water waste and maximize efficiency (Lakhiar et al., 2023). Additionally,

AI algorithms in smart water meters help detect leaks and anomalies in water distribution, enabling timely repairs and reducing water and energy loss (Pochepskiy, 2024). In aquaculture, AI-powered sensors monitor water quality parameters like temperature, pH, dissolved oxygen, salinity, and ammonia, allowing for automated responses to maintain optimal conditions and prevent potential issues (Nabik, 2023). Data-driven methods, such as ANN and early warning systems to predict harmful algal blooms (IAHR, 2020; Dabrowski et al., 2016; Coad et al., 2014), have also been studied to protect the fish environment. ANN models are further assimilated to sea surface temperature and vertical density stratification, validated against extensive field data.

9.8.2 Beyond Water Conservation: AI for Net-Zero Emissions

Beyond water conservation, AI plays a crucial role in achieving net-zero emissions in food production systems. AI algorithms analyze data to optimize crop yields, predicting ideal planting times, fertilization strategies, and harvest schedules, thereby maximizing output and minimizing the need for land expansion, a major driver of deforestation (Shang, 2023). AI-powered systems also track food from farm to plate, predicting demand, optimizing logistics, and monitoring shelf life, thus minimizing waste at all stages of the supply chain and reducing the environmental footprint of food production (Buss, 2023). Furthermore, AI can help farmers transition to sustainable practices like organic farming by analyzing data on soil health, pest management, and crop rotation, reducing reliance on synthetic fertilizers and pesticides with their significant environmental impacts (Sen, 2023).

9.8.3 The Future of AI in Food Production

The future of AI in food production looks promising, driven by ongoing research and development in several key areas. Advanced predictive analytics are enhancing the accuracy of AI models, allowing for better forecasts of weather patterns, crop yields, and market demand, which leads to more proactive and efficient resource management (Pandey & Mishra, 2024). Additionally, AI-powered robots are being designed to automate tasks such as planting, harvesting, and packaging, further streamlining agricultural processes while reducing labor costs (Lika, 2024). Moreover, AI is pivotal in the advancement of vertical farming systems that grow crops in stacked layers indoors, offering significant benefits in water conservation, energy efficiency, and a decreased reliance on pesticides (Shang, 2023).

9.9 CHALLENGES AND LIMITATIONS

A study by Sparrow et al. (2022) shows that the integration of AI in agriculture presents both benefits and significant risks that must be carefully considered. While AI has the potential to enhance productivity, it may also lead to unintended consequences such as (1) job displacement across various roles in agriculture, as AI systems increasingly perform tasks traditionally done by humans. (2) The transformation of agricultural work is expected, with fewer workers needed to oversee machines, shifting the nature of jobs toward more managerial roles, although many remaining jobs may require less skill. Additionally, there is a risk of (3) increased concentration of ownership in the agricultural sector, as larger entities can leverage AI and big data more effectively than smaller farms, exacerbating inequalities and potentially threatening biodiversity and food security.

Another important thing that needs to be addressed is data availability and access, as the wide-scale adoption of AI relies on the collection of vast amounts of accurate and reliable data, which must be accessible to all stakeholders (Kelly et al., 2023). Additionally, ethical considerations pose a significant concern, as biased AI algorithms can lead to unintended consequences; thus, it is vital to establish ethical guidelines and promote transparency in AI development and deployment (Osasona et al., 2024). Thus, while the benefits of AI are compelling, the associated risks warrant serious attention from governments and stakeholders to ensure the public good is prioritized.

9.10 CONCLUSION

The integration of AI presents a significant opportunity to address the challenges of achieving net-zero emissions in food production systems in the context of agriculture, aquaculture, and livestock sectors. The utilization of AI technologies, such as smart IoT sensors, data analytics, ANN, and ML algorithms, allows precise monitoring of water quality, crop growth, seafood, fish, and livestock health, and environmental impacts, leading to optimized resource management and reduced GHG emissions.

Leveraging smart technologies powered by AI, the data will be automatically sent to the cloud. Data can be shared among farm managers allowing adjustments to irrigation schedules and immediate attention to crop rot or produce mortality, solving the sources of possible yield reduction. AI-based smart and sustainable food production systems could provide solutions to problems of water shortage and water quality and increase productivity while conserving water and achieving net-zero emissions. To promote and achieve climate change adaptation, mitigation, and resilience while pursuing renewed paths of socioeconomic activities, AI must be incorporated to monitor and conserve water, particularly in the agriculture, aquaculture, and livestock sectors.

REFERENCES

Abbasi, T., Luithui, C., & Abbasi, S. A. (2019). Modelling methane and nitrous oxide emissions from rice paddy wetlands in India using Artificial Neural Networks (ANNs). *Water*, 11(10), 2169. https://doi.org/10.3390/w11102169

Agga, G. E., Couch, M., Parekh, R. R., Mahmoudi, F., Appala, K., Kasumba, J., Loughrin, J. H., & Conte, E. D. (2022). Lagoon, anaerobic digestion, and composting of animal manure treatments impact on tetracycline resistance genes. *Antibiotics*, 11(3), 391. https://doi.org/10.3390/antibiotics11030391

Akkaş, Ö. M., Tosun, C., & Gökçearslan, Ş. (2024). Artificial Intelligence (AI) and cheating: The concept of generative artificial intelligence (GenAI). In *Transforming Education with Generative AI: Prompt Engineering and Synthetic Content Creation* (pp. 18). IGI Global. https://doi.org/10.4018/979-8-3693-1351-0.ch009

Al Hamedi, F. H., Kandhan, K., Liu, Y., Ren, M., Jaleel, A., & Alyafei, M. A. (2022). Wastewater irrigation: A promising way for future sustainable agriculture and food security in the United Arab Emirates. *Water*, 15(12), 2284. https://doi.org/10.3390/w15122284

Anderson, K. (2022). *Methane is the Climate Opportunity Food Companies Can't Afford to Miss*. Environmental Defense Fund. https://business.edf.org/insights/methane-is-the-climate-opportunity-food-companies-cant-afford-to-miss/

Apure. (2023). *Agricultural Uses of Water*. https://apureinstrument.com/blogs/agricultural-uses-of-water/#what-is-agricultural-uses-of-water

Aryal, J. P., Rahut, D. B., Sonobe, T., Tortajada, C., & Chhay, P. (2024). Water resource management in agriculture for achieving food and water security in Asia. *International Journal of Water Resources Development*, 40(3), 319–322. https://doi.org/10.1080/07900627.2024.2332222

Bahman, N., Alalaiwat, D., Abdulmohsen, Z., Al Khalifa, M., Al Baharna, S., Al-Mannai, M., & Younis, A. (2023). A critical review on global CO_2 emission: where do industries stand?. *Reviews on Environmental Health*, 38(4), 681–696.

Bibri, S. E., Krogstie, J., Kaboli, A., & Alahi, A. (2024). Smarter eco-cities and their leading-edge artificial intelligence of things solutions for environmental sustainability: a comprehensive systematic review. *Environmental Science and Ecotechnology*, 19, 100330. https://doi.org/10.1016/j.ese.2023.100330

Bouali, E.-T., Abid, M. R., Boufounas, E.-M., & Morchid, A. (2024). Enhancing water management in smart agriculture: A cloud and IoT-based smart irrigation system. *Results in Engineering*, 10, 102283. https://doi.org/10.1016/j.rineng.2024.102283

Buss, D. (2023). *Leveraging AI on the Road to Net Zero: Can Artificial Intelligence Move Us toward a More Sustainable Food Industry?* Food Technology Magazine. https://www.ift.org/news-and-publications/food-technology-magazine/issues/2023/december/features/leveraging-ai-on-the-road-to-net-zero

Casila, J.C.C., Ureta, J.C., Haddout, S., Priya, K.L. (2022). GHG emissions, environmental technologies, and renewable energy status in the Philippines. In: *Towards Net Zero Emissions: Policy and Practice*, Hanoi, Vietnam, 05 October 2022.

Chen, G., Bai, J., Bi, C., Wang, Y., & Cui, B. (2023). Global greenhouse gas emissions from aquaculture: A bibliometric analysis. *Agriculture, Ecosystems & Environment*, 348, 108405. https://doi.org/10.1016/j.agee.2023.108405

Chen, J. C., Chen, T.-L., Wang, H.-L., & Chang, P.-C. (2022). Underwater abnormal classification system based on deep learning: A case study on aquaculture fish farm in Taiwan. *Aquacultural Engineering*. https://doi.org/10.1016/j.aquaeng.2022.102290

Choudhury, B.U., Bouman, B.A.M. and Singh, A.K. (2007) Yield and water productivity of rice—Wheat on raised beds at New Delhi, India. *Field Crops Research*, 100, 229–239. https://doi.org/10.1016/j.fcr.2006.07.009

Chrobak, U. (2021). *The World's Forgotten Greenhouse Gas*. BBC Future. https://www.bbc.com/future/article/20210603-nitrous-oxide-the-worlds-forgotten-greenhouse-gas

Coad, P., Cathers, B., Ball, J. E., & Kadluczka, R. (2014). Proactive management of estuarine algal blooms using an automated monitoring buoy coupled with an artificial neural network. *Environmental Modelling and Software*, 61, 393–409

Cruz-Paredes, C., Tájmel, D., & Rousk, J. (2021). Can moisture affect temperature dependences of microbial growth and respiration? *Soil Biology and Biochemistry*, 156, 108223. https://doi.org/10.1016/j.soilbio.2021.108223

Dabrowski, T., Lyons, K., Nolan, G., Berry, A., Cusack, C., & Silke, J. (2016). Harmful algal bloom forecast system for SW Ireland. Part I: Description and validation of an operational forecasting model. *Harmful Algae*, 53, 64–76

David, F., Antonio, M., & Manuel, J. (2023). Artificial intelligence-based aquaculture system for optimizing the quality of water: a systematic analysis. *Journal of Marine Science and Engineering*, 12(1), 161. https://doi.org/10.3390/jmse12010161

Derpsch, R., Kassam, A., Reicosky, D., Friedrich, T., Calegari, A., Basch, G., Gonzalez-Sanchez, E., & Dos Santos, D. R. (2024). Nature's laws of declining soil productivity and conservation agriculture. *Soil Security*, 14, 100127. https://doi.org/10.1016/j.soisec.2024.100127

Dinoy, W. (2024). *3 Heat-Trapping Gases Soar to Unprecedented Levels, Scientists Confirm*. Clean Energy Revolution. https://cleanenergyrevolution.co/2024/04/07/3-heat-trapping-gases-soar-to-unprecedented-levels-scientists-confirm/

Efremova, N., Zausaev, D., & Antipov, G. (2019). Prediction of soil moisture content based on satellite data and sequence-to-sequence networks. ArXiv. https://arxiv.org/abs/1907.03697

Essamlali, I., Nhaila, H., & El Khaili, M. (2024). Advances in machine learning and IoT for water quality monitoring: A comprehensive review. *Heliyon*, 10(6), e27920. https://doi.org/10.1016/j.heliyon.2024.e27920

FAO. (2016) Sustainable intensification of aquaculture in the Asia-Pacific region. Documentation of successful practices. Miao, W. and Lal, K. K. (Ed) Bangkok Thailand. Food and Agriculture Organization, United Nations

FAO. (2022). *FAO Strategy on Climate Change 2022–2031*. Rome. Food and Agriculture Organization, United Nations

FAO. (2024). *The State of World Fisheries and Aquaculture 2024 – Blue Transformation in Action*. Rome. Food and Agriculture Organization, United Nations. https://doi.org/10.4060/cd0683en

Federal Ministry for the Environment, Nature Conservation, Nuclear Safety and Consumer Protection (BMUV). (n.d.). *National Water Strategy: Challenges and goals for water management by 2050*. https://www.bmuv.de/DL2717-1

Filonchyk, M., Peterson, M. P., Zhang, L., Hurynovich, V., & He, Y. (2024). Greenhouse gases emissions and global climate change: examining the influence of CO_2, CH_4, and N_2O. *Science of the Total Environment*, 935, 173359. https://doi.org/10.1016/j.scitotenv.2024.173359

Fujs, T., & Kashiwase, H. (2023). *Strains on Freshwater Resources: The Impact of Food Production on Water Consumption*. World Economic Forum. https://www.weforum.org/agenda/2021/06/water-footprint-food-sustainability/

Garg, K. K., Anantha, K., Dixit, S., Nune, R., Venkataradha, A., Wable, P., Budama, N., & Singh, R. (2022). Impact of raised beds on surface runoff and soil loss in Alfisols and Vertisols. *CATENA*, 211, 105972. https://doi.org/10.1016/j.catena.2021.105972

Garthwaite, J. (2021). *Methane and Climate Change*. Stanford University. https://sustainability.stanford.edu/news/methane-and-climate-change-0

Gladju, J., Kamalam, B. S., & Kanagaraj, A. (2022). Applications of data mining and machine learning framework in aquaculture and fisheries: A review. *Smart Agricultural Technology*, 2, 100061. https://doi.org/10.1016/j.atech.2022.100061

Glencross, B., Fracalossi, D. M., Hua, K., Izquierdo, M., Mai, K., Øverland, M., Robb, D., Roubach, R., Schrama, J., Small, B., Tacon, A., P. Valente, L. M., Viana, T., Xie, S., & Yakupityage, A. (2023). Harvesting the benefits of nutritional research to address global challenges in the 21st century. *Journal of the World Aquaculture Society*, 54(2), 343–363. https://doi.org/10.1111/jwas.12948

Global Legal Insights. (2024). *AI, Machine Learning & Big Data Laws and Regulations 2024*. GLI. https://www.globallegalinsights.com/practice-areas/ai-machine-learning-and-big-data-laws-and-regulations/

Grossi, G., Goglio, P., Vitali, A., & Williams, A. G. (2019). Livestock and climate change: Impact of livestock on climate and mitigation strategies. *Animal Frontiers*, 9(1), 69–76. https://doi.org/10.1093/af/vfy034

Gun, J. van der. (2019). *The Global Groundwater Revolution*. https://doi.org/10.1093/acrefore/9780199389414.013.632

Hoekstra, A. Y. (2012). *The Hidden Water Resource Use Behind Meat and Dairy*. Twente Water Centre, University of Twente, Enschede, the Netherlands.

Holgerson, M.A. Drivers of carbon dioxide and methane supersaturation in small, temporary ponds. *Biogeochemistry* 124, 305–318 (2015). https://doi.org/10.1007/s10533-015-0099-y

Hoseinzadeh, S., & Astiaso Garcia, D. (2024). AI-driven innovations in greenhouse agriculture: Reanalysis of sustainability and energy efficiency impacts. *Ecological Modelling and eXperiments*, 26, 100701. https://doi.org/10.1016/j.ecmx.2024.100701

Hubbart, S., & Ross, L. (2024). *The Increasing Demand and Decreasing Supply of Water*. The National Environmental Education Foundation (NEEF). https://www.neefusa.org/story/water/increasing-demand-and-decreasing-supply-water

Hughes, J., Cowper-Heays, K., Olesson, E., Bell, R., & Stroombergen, A. (2020). Impacts and implications of climate change on wastewater systems: A New Zealand perspective. *Climate Risk Management*, 31, 100262. https://doi.org/10.1016/j.crm.2020.100262

Huo, P., Li, H., Huang, X., Ma, X., Liu, L., Ji, W., Liu, Y., & Gao, P. (2022). Dissolved greenhouse gas emissions from agricultural groundwater irrigation in the Guanzhong Basin of China. *Environmental Pollution*, 309, 119714. https://doi.org/10.1016/j.envpol.2022.119714

Inoue, Y., & Yokoyama M. (2017). Drone-based remote sensing of crops and soils and its application to smart agriculture. *Journal of the Remote Sensing Society of Japan*, 37(3), 224–235. https://ci.nii.ac.jp/naid/40021281502

International Association for Hydro-Environment Engineering and Research (IAHR) (Hg.) (2020): Hydrolink 2020/4. Artificial intelligent. Madrid: IAHR (Hydrolink, 2020/4). https://www.iahr.org/library/infor?pid=20138

International Rice Research Institute (IRRI). (2024). *GHG Emissions in Rice Systems: Basics of Mechanisms and Standards for Measurements*. https://education.irri.org/technology-transfer/ghg-emission-in-rice-system-basic-of-mechanisms-and-standard-for-measurements/

Jain, R. K. (2023). Experimental performance of smart IoT-enabled drip irrigation system using and controlled through web-based applications. *Applied Technologies*, 11, 100215. https://doi.org/10.1016/j.atech.2023.100215

Kasera, R. K., & Acharjee, T. (2024). A comprehensive IoT edge-based smart irrigation system for tomato cultivation. *Internet of Things*, 19, 101356. https://doi.org/10.1016/j.iot.2024.101356

Kaur, G., Adhikari, N., Krishnapriya, S., Wawale, S. G., Malik, R. Q., Zamani, A. S., ... & Osei-Owusu, J. (2023). Research article recent advancements in deep learning frameworks for precision fish farming opportunities, challenges, and applications. *Journals of Food Quality*, 1, 4399512.

Kelly, S., Kaye, S., & Oviedo-Trespalacios, O. (2023). What factors contribute to the acceptance of artificial intelligence? A systematic review. *Telematics and Informatics*, 77, 101925. https://doi.org/10.1016/j.tele.2022.101925

Kima, A. S., Chung, W. G., & Wang, Y. -M. (2014). Improving irrigated lowland rice water use efficiency under saturated soil culture for adoption in tropical climate conditions. *Water*, 6(9), 2830–2846. https://doi.org/10.3390/w6092830

Lakhiar, I. A., Yan, H., Zhang, C., Wang, G., He, B., Hao, B., Han, Y., Wang, B., Bao, R., Syed, T. N., Chauhdary, J. N., & Rakibuzzaman, M. (2024). A review of precision irrigation water-saving technology under changing climate for enhancing water use efficiency, crop yield, and environmental footprints. *Agriculture*, 14(7), 1141. https://doi.org/10.3390/agriculture14071141

Lampayan, R. M., Rejesus, R. M., Singleton, G. R., & Bouman, B. A. (2014). Adoption and economics of alternate wetting and drying water management for irrigated lowland rice. *Field Crops Research*, 170, 95–108.

Lessman, O., Fernandez, J.E., Cruz, K., Peeters, F. (2023). Methane emissions due to reservoir flushing: A significant emission pathway?. *Biogeosciences*, 20(11), 4057–4076. https://doi.org/10.5194/bg-20-4057-2023

Li, C., Fan, M., Wang, X., Li, X., Zhao, G., Liu, G., & Zhao, J. (2023). Effects of nitrogen fertilizers on the bacterial community diversity and the weathering of purple mudstone in Southwest China. *Frontiers in Microbiology*, 14, 1164826. https://doi.org/10.3389/fmicb.2023.1164826

Li, Z., Xia, S., Zhang, R., Zhang, R., Chen, F., & Liu, Y. (2020). N_2O emissions and product ratios of nitrification and denitrification are altered by K fertilizer in acidic agricultural soils. *Environmental Pollution*, 265, 115065. https://doi.org/10.1016/j.envpol.2020.115065

Lika, K. (2024). *Field to Farm: How Robots in Agriculture Revolutionize Farming*. Bota Systems. https://www.agritechtomorrow.com/article/2024/06/field-to-farm-how-robots-in-agriculture-revolutionize-farming/15585

Lin, L., Gao, X., Zhao, Y., Wang, L., An, T., Liu, C., Qiao, Y., & Wu, P. (2021). Evaluation of the water consumption of animal products and the virtual water flow pattern associated with interprovincial trade in China. *Journal of Cleaner Production*, 328, 129599. https://doi.org/10.1016/j.jclepro.2021.129599

Lindsey, R. (2024). *Climate Change: Atmospheric Carbon Dioxide*. Climate.gov. Reviewed by Ed Dlugokencky. https://www.climate.gov/news-features/understanding-climate/climate-change-atmospheric-carbon-dioxide

Lu, L., Chen, C., Ke, T., Wang, M., Sima, M., & Huang, S. (2022). Long-term metal pollution shifts microbial functional profiles of nitrification and denitrification in agricultural soils. *Science of the Total Environment*, 830, 154732. https://doi.org/10.1016/j.scitotenv.2022.154732

Lynch, J., Cain, M., Frame, D., & Pierrehumbert, R. (2021). Agriculture's contribution to climate change and role in mitigation is distinct from predominantly fossil CO_2-emitting sectors. *Frontiers in Sustainable Food Systems*, 4, 518039. https://doi.org/10.3389/fsufs.2020.518039

Mallareddy, M., Thirumalaikumar, R., Balasubramanian, P., Naseeruddin, R., Nithya, N., Mariadoss, A., Eazhilkrishna, N., Choudhary, A. K., Deiveegan, M., Subramanian, E., Padmaja, B., & Vijayakumar, S. (2023). Maximizing water use efficiency in rice farming: A comprehensive review of innovative irrigation management technologies. *Water*, 15(10), 1802. https://doi.org/10.3390/w15101802

Mana, A., Allouhi, A., Hamrani, A., Rehman, S., El Jamaoui, I., & Jayachandran, K. (2024). Sustainable AI-based production agriculture: Exploring AI applications and implications in agricultural practices. *Smart Agricultural Technology*, 7, 100416. https://doi.org/10.1016/j.atech.2024.100416

Mandal, A., & Ghosh, A. R. (2023). Role of artificial intelligence (AI) in fish growth and health status monitoring: A review on sustainable aquaculture. *Aquaculture International*, 32(3), 2791–2820. https://doi.org/10.1007/s10499-023-01297-z

Manocha, A., Sood, S. K., & Bhatia, M. (2024). IoT-digital twin-inspired smart irrigation approach for optimal water utilization. *Sustainable Computing: Informatics and Systems*, 38, 100947. https://doi.org/10.1016/j.suscom.2023.100947

Mekonnen, M., & Hoekstra, A. (2012). A global assessment of the water footprint of farm animal products. *Ecosystems*. 15. https://doi.org/10.1007/s10021-011-9517-8.

Myhre, G., Shindell, D., Bréon, F.-M., Collins, W., Fuglestvedt, J., Huang, J., Koch, D., Lamarque, J.-F., Lee, D., Mendoza, B., Nakajima, T., Robock, A., Stephens, G., Takemura, T., & Zhang, H. (2013). Anthropogenic and natural radiative forcing. In T. F. Stocker, D. Qin, G.-K. Plattner, M. Tignor, S. K. Allen, J. Doschung, A. Nauels, Y. Xia, V. Bex, & P. M. Midgley (Eds.), *Climate Change 2013: The Physical Science Basis. Contribution of Working Group I to the Fifth Assessment Report of the Intergovernmental Panel on Climate Change* (pp. 659–740). Cambridge University Press. https://doi.org/10.1017/CBO9781107415324.018

Miao, Z., Zhao Z., Long T., Chen, X. (2023). Carbon footprint in agriculture sector: A literature review. *Carbon Footprints*, 2(13). https://dx. doi.org/10.20517/cf.2023.29

Mohd Jais, N. A., Abdullah, A. F., Mohd Kassim, M. S., Abd Karim, M. M., & Muhadi, N. ' A. (2024). Improved accuracy in IoT-based water quality monitoring for aquaculture tanks using low-cost sensors: Asian seabass fish farming. *Heliyon*, 10, e29022. https://doi.org/10.1016/j.heliyon.2024.e29022

Montazeri, A., Chahkandi, B., Gheibi, M., Eftekhari, M., Wacławek, S., Behzadian, K., & Campos, L. C. (2023). A novel AI-based approach for modelling the fate, transportation, and prediction of chromium in rivers and agricultural crops: A case study in Iran. *Ecotoxicology and Environmental Safety*, 263, 115269. https://doi.org/10.1016/j.ecoenv.2023.115269

Musie, W., & Gonfa, G. (2023). Fresh water resource, scarcity, water salinity challenges and possible remedies: A review. *Heliyon*, 9(8), e18685. https://doi.org/10.1016/j.heliyon.2023.e18685

Mustapha, U. F., Alhassan, W., Jiang, N., & Li, L. (2021). Sustainable aquaculture development: A review on the roles of cloud computing, internet of things and artificial intelligence (CIA). *Reviews in Aquaculture*, 13(4), 2076–2091. https://doi.org/10.1111/raq.12559

Nabik, N. (2023). *Sustainable Aquaculture with AI: A Leap Forward in Pond Farming*. Medium. https://medium.com/@nabikautomations/sustainable-aquaculture-with-ai-a-leap-forward-in-pond-farming-14e3fad03981

National Association of Clean Water Agencies. (n.d.). *Homepage*. https://www.nacwa.org/

National Ocean Service. (n.d.). *Why is the Ocean Salty?* NOAA. https://oceanservice.noaa.gov/facts/oceanwater.html

Neethirajan, S. (2023a). Artificial intelligence and sensor innovations: Enhancing livestock welfare with a human-centric approach. *Human-Centric Intelligent Systems*, 4(1), 77–92. https://doi.org/10.1007/s44230-023-00050-2

Neethirajan, S. (2023b). The significance and ethics of digital livestock farming. *AgriEngineering*, 5(1), 488–505. https://doi.org/10.3390/agriengineering5010032

Nova, K. (2023). AI-enabled water management systems: An analysis of system components and interdependencies for water conservation. *Eigenpub Review of Science and Technology*, 7(1), 105–124.

Nunes, L. J. (2023). The rising threat of atmospheric CO_2: A review on the causes, impacts, and mitigation strategies. *Environments*, 10(4), 66. https://doi.org/10.3390/environments10040066

Osasona, F., Amoo, O. O., Atadoga, A., Abrahams, T. O., Farayola, O. A., & Ayinla, B. S. (2024). Reviewing the ethical implications of AI in decision-making processes. *International Journal of Management & Entrepreneurship Research*, 6(2), 322–335. https://doi.org/10.51594/ijmer.v6i2.773

Ozlu, E., Arriaga, F. J., Bilen, S., Gozukara, G., & Babur, E. (2022). Carbon footprint management by agricultural practices. *Biology*, 11(10). https://doi.org/10.3390/biology11101453

Pandey, D. K., & Mishra, R. (2024). Towards sustainable agriculture: Harnessing AI for global food security. *Artificial Intelligence in Agriculture*, 12, 72–84. https://doi.org/10.1016/j.aiia.2024.04.003

Pascual, K. S. (2021). *Conserve Water to Manage Crisis, Farmers Told*. Philippine Rice Research Institute. https://www.philrice.gov.ph/conserve-water-to-manage-crisis-farmers-told/

Pawar, D.J., Sonavale, D.R., & Sarkale, M.P. (2024). transforming cattle farming with artificial intelligence: Innovations, applications, and implications for precision livestock management and sustainable agriculture practices. *Revista Electronica De Veterinaria,* 25(1), 525–537. https://veterinaria.org/index.php/REDVET/article/view/541

Pochepskiy, O. (2024). *Sustainable Practices in Food and Water Management: Leveraging AI for Success*. Business Analyst. https://www.cleverence.com/articles/business-blogs/sustainable-practices-in-food-and-water-management-leveraging-ai-for-success/

Priya, A., Devarajan, B., Alagumalai, A., & Song, H. (2023). Artificial intelligence enabled carbon capture: A review. *Science of the Total Environment*, 886, 163913. https://doi.org/10.1016/j.scitotenv.2023.163913

Pu, Y., Zhang, M., Jia, L., Zhang, Z., Xiao, W., Liu, S., Zhao, J., Xie, Y., & Lee, X. (2022). Methane emission of a lake aquaculture farm and its response to ecological restoration. *Agriculture, Ecosystems & Environment*, 330, 107883. https://doi.org/10.1016/j.agee.2022.107883

Rajasekar, P., & Selvi, J. A. V. (2022). Sensing and analysis of greenhouse gas emissions from rice fields to the near field atmosphere. *Sensors*, 22, 4141. https://doi.org/10.3390/ s22114141

Ritchie, H., & Roser, M. (2015). *Water Use and Stress: How Much Water Do We Use? How Did It Change Over Time?* Our World in Data. https://ourworldindata.org/water-use-stress#water-withdrawals

Ritchie, H., Rosado, P., & Roser, M. (2022). *Environmental Impacts of Food Production*. Our World in Data. https://ourworldindata.org/environmental-impacts-of-food

Rojas-Downing, M. M., Nejadhashemi, A. P., Harrigan, T., & Woznicki, S. A. (2016). Climate change and livestock: Impacts, adaptation, and mitigation. *Climate Risk Management*, 16, 145–163. https://doi.org/10.1016/j.crm.2017.02.001

Rosegrant, M. W., Fernandez, M., Sinha, A., Alder, J., Ahammad, H., de Fraiture, Charlotte, Eickhour, B., Fonseca, J., Huang, J., Koyama, O., Omezzine, A. M., Pingali, P., Ramirez, R., Ringler, C., Robinson, S., Thornton, P., van Vuuren, D., Yana-Shapiro, H. (2009). Looking into the future for agriculture and AKST. In: McIntyre, B. D., Herren, H. R., Wakhungu, J., Watson, R. T. (Eds.). *International Assessment of Agricultural Knowledge, Science and Technology for Development (IAASTD): Agriculture at a Crossroads, global report*. Washington, DC, USA: Island Press. pp. 307–376. https://hdl.handle.net/10568/37336

Saikai, Y., Peake, A., & Chenu, K. (2023). Deep reinforcement learning for irrigation scheduling using high-dimensional sensor feedback. *PLOS Water*, 1(9), e0000169. https://doi.org/10.1371/journal.pwat.0000169

Sangeetha, P., Kumar, N., Ambalgi, A. P., Haleem, S. L. A., Thilagam, K., & Vijayakumar, P. (2022). IoT-based smart irrigation management system for environmental sustainability in India. *Sustainable Energy Technologies and Assessments*, 55, 101973. https://doi.org/10.1016/j.seta.2022.101973

Schmid, R., & Xiong, X. (2023). China's environmental solutions. *Applied Microbiology and Biotechnology*, 107(3), 987–1002. https://doi.org/10.1007/s00253-022-12340-z

Sen, S. (2023). *Smart Farming and Water Conservation: The Role of AI*. LinkedIn. https://www.linkedin.com/pulse/smart-farming-water-conservation-role-ai-shilpi-sen

Shakoor, A., Shahzad, S. M., Chatterjee, N., Arif, M. S., Farooq, T. H., Altaf, M. M., Tufail, M. A., Dar, A. A., & Mehmood, T. (2021). Nitrous oxide emission from agricultural soils: Application of animal manure or biochar? A global meta-analysis. *Journal of Environmental Management*, 285, 112170. https://doi.org/10.1016/j.jenvman.2021.112170

Shang, B. (2023). *The Role of AI in Creating a More Sustainable Food System*. Forbes Technology Council. https://www.forbes.com/councils/forbestechcouncil/2023/07/20/the-role-of-ai-in-creating-a-more-sustainable-food-system/

Shete, R. P., Bongale, A. M., & Dharrao, D. (2024). IoT-enabled effective real-time water quality monitoring method for aquaculture. *MethodsX*, 11, 102906. https://doi.org/10.1016/j.mex.2024.102906

Siad, O., & Bouzid, C. (2023). Biotech meets artificial intelligence to enhance the value of by-products in animal nutrition. *Biological Sciences*, 3(1), 353–365.

Silva, J. A. (2022). Wastewater treatment and reuse for sustainable water resources management: A systematic literature review. *Sustainability*, 15(14), 10940. https://doi.org/10.3390/su151410940

Silverman, F. (2022). *Exploring the Depths of Water's Role in Climate Change*. Yale School of the Environment. https://environment.yale.edu/canopy/2022/feature/exploring-depths-waters-role-climate-change

Sitharthan, R., Rajesh, M., Vimal, S., Saravana Kumar, E., Yuvaraj, S., Kumar, A., Raglend, J. I., & Vengatesan, K. (2023). A novel autonomous irrigation system for smart agriculture using AI and 6G-enabled IoT network. *Microprocessors and Microsystems*, 104, 104905. https://doi.org/10.1016/j.micpro.2023.104905

SMS Foundation. (2023). *The Role of Technology in Water Management*. https://www.smsfoundation.org/the-role-of-technology-in-water-management/

Sparrow, R., Howard, M., & Degeling, C. (2021). Managing the risks of artificial intelligence in agriculture. *NJAS: Impact in Agricultural and Life Sciences*, 93(1), 172–196. https://doi.org/10.1080/27685241.2021.2008777

Sun M, Zhang Y, Ma J, Yuan W, Li X, Cheng X (2017) Satellite data based estimation of methane emissions from rice paddies in the Sanjiang Plain in northeast China. *PLoS ONE* 12(6), e0176765. https://doi.org/10.1371/journal.pone.0176765

Sung, W.-T., Isa, I. G. T., & Hsiao, S.-J. (2023). An IoT-based aquaculture monitoring system using Firebase. *Computers, Materials & Continua*, 74(2), 41022. https://doi.org/10.32604/cmc.2023.041022

Susantono, B., & Li, S. H. (2023). Urban water future: What can we learn from the Singapore experience?. *CSID Journal of Infrastructure Development*, 4(1), 4–20. https://doi.org/10.32783/csid-jid.v4i1.220

Tan, J., Lichtfouse, E., Luo, M., Liu, Y., Tan, F., Zhang, C., Chen, X., Huang, J., & Xiao, L. (2023). Aquaculture drastically increases methane production by favoring acetoclastic rather than hydrogenotrophic methanogenesis in shrimp pond sediments. *Aquaculture*, 563, 738999. https://doi.org/10.1016/j.aquaculture.2022.738999

Tang, W., Biglari, A., Ebarb, R., Pickett, T., Smallidge, S., & Ward, M. (2020). A smart sensing system of water quality and intake monitoring for livestock and wild animals. *Sensors*, 21(8), 2885. https://doi.org/10.3390/s21082885

Tao, H., Brueck, H., Dittert, K., Kreye, C., Lin, S., & Sattelmacher, B. (2006). Growth and yield formation of rice (Oryza sativa L.) in the water-saving ground cover rice production system (GCRPS). *Field Crops Research*, 95(1), 1–12.

The Water Footprint (2018). *In Footprints, Water in Your Food*. Water Calculator. https://watercalculator.org/footprint/water-footprint-beef-industrial-pasture/

Tian, Y., Hu, H., & Zhang, J. (2017). Solution to water resource scarcity: Water reclamation and reuse. *Environmental Science and Pollution Research*. https://doi.org/10.1007/s11356-016-8331-4

Tong, C., Bastviken, D., Tang, K. W., Yang, P., Yang, H., Zhang, Y., Guo, Q., & Lai, D. Y. (2021). Annual CO_2 and CH_4 fluxes in coastal earthen ponds with Litopenaeus vannamei in southeastern China. *Aquaculture*, 545, 737229. https://doi.org/10.1016/j.aquaculture.2021.737229

U.S. Environmental Protection Agency (EPA). (2024). *Summary of the Clean Water Act*. https://www.epa.gov/laws-regulations/summary-clean-water-act

U.S. Environmental Protection Agency (EPA). (n.d.) *Sources of Greenhouse Gas Emissions*. https://www.epa.gov/ghgemissions/sources-greenhouse-gas-emissions#:~:text=Livestock%2C%20especially%20ruminants%20such%20as,emissions%20from%20the%20Agriculture%20sector

Upmann, P. (2024). *Global AI Regulation in 2024: Navigating the Complex Landscape of AI Laws*. Now.Digital | Mind The Ai Data. https://now.digital/world-of-data/global-ai-regulation-in-2024-navigating-the-complex-landscape-of-ai-laws/

Vo, T. T., Ko, H., Huh, J., & Kim, Y. (2020). Overview of smart aquaculture system: Focusing on applications of machine learning and computer vision. *Electronics*, 10(22), 2882. https://doi.org/10.3390/electronics10222882

Wagner, J., & Engle, T. (2021). Invited review: Water consumption and drinking behavior of beef cattle, and effects of water quality. *Applied Animal Science*, 37(4), 418–435. https://doi.org/10.15232/aas.2021-02136

Wang, J., Wu, Q., He, Y., Li, Y., Xu, J., & Jiang, Q. (2024). Maximizing the carbon sink function of paddy systems in China with machine learning. *Science of the Total Environment*, 909, 168542. https://doi.org/10.1016/j.scitotenv.2023.168542

Wang, Q., Huang, K., Liu, H., & Yu, Y. (2023). Factors affecting crop production water footprint: A review and meta-analysis. *Sustainable Production and Consumption*, 36, 207–216. https://doi.org/10.1016/j.spc.2023.01.008

Wassmann, R., Lantin, R. S., Neue, H. U., Buendia, L. V., Corton, T. M., & Lu, Y. (2000). Characterization of methane emissions from rice fields in Asia. III. Mitigation options and future research needs. *Nutrient Cycling in Agroecosystems*, 58(1), 23–36. https://doi.org/10.1023/A:1009874014903

Wei, H., Xu, W., Kang, B., Eisner, R., Muleke, A., Rodriguez, D., deVoil, P., Sadras, V., Monjardino, M., & Harrison, M.T. (2024). Irrigation with artificial intelligence: Problems, premises, promises. *Human-Centric Intelligent* Systems, 4, 187–205. https://doi.org/10.1007/s44230-024-00072-4

World Resources Institute. (2014). *Wetting and Drying: Reducing Greenhouse Gas Emissions and Saving Water in Rice Production*. https://www.wri.org/research/wetting-and-drying-reducing-greenhouse-gas-emissions-and-saving-water-rice-production

Wozniacka, G. (2019). *The Greenhouse Gas No One's Talking About: Nitrous Oxide on Farms, Explained*. Civil Eats. https://civileats.com/2019/09/19/the-greenhouse-gas-no-ones-talking-about-nitrous-oxide-on-farms-explained/

Wu, Y., Duan, Y., Wei, Y., An, D., & Liu, J. (2022). Application of intelligent and unmanned equipment in aquaculture: A review. *Computers and Electronics in Agriculture*, 199, 107201. https://doi.org/10.1016/j.compag.2022.107201

Xu, P., Li, G., Zheng, Y., Fung, J. C., Chen, A., Zeng, Z., Shen, H., Hu, M., Mao, J., Zheng, Y., Cui, X., Guo, Z., Chen, Y., Feng, L., He, S., Zhang, X., Lau, A. K., Tao, S., & Houlton, B. Z. (2024). Fertilizer management for global ammonia emission reduction. *Nature*, 626(8000), 792–798. https://doi.org/10.1038/s41586-024-07020-z

Yang, F., Sun, C., Wang, H., Hu, X., Wang, S., Zhang, M., Zhang, L., & Zhong, J. (2023). Significant spatiotemporal variability of nitrous oxide emissions from a temperate reservoir experiencing intensive aquaculture disturbance. *Agriculture, Ecosystems & Environment*, 348, 108427. https://doi.org/10.1016/j.agee.2023.108427

Yokoyama, T. (2017). A review on the numerical simulation of equatorial plasma bubbles toward scintillation evaluation and forecasting. *Progress in Earth and Planetary Science*, 4(1), 1–13. https://doi.org/10.1186/s40645-017-0153-6

Yue, K., & Shen, Y. (2022). An overview of disruptive technologies for aquaculture. *Aquaculture and Fisheries*, 7(2), 111–120. https://doi.org/10.1016/j.aaf.2021.04.009

Zhang, S., Yang, X., Wang, Y., Zhao, Z., Liu, J., Liu, Y., Sun, C., & Zhou, C. (2020). Automatic fish population counting by machine vision and a hybrid deep neural network model. *Animals*, 10(2), 364. https://doi.org/10.3390/ani10020364

Zhang, J., Chen, L., Chen, Y., & Groisman, P. (2023a). Comparing process-based models with the inventory approach to predict CH_4 emission of livestock enteric fermentation. *Environmental Research Letters*, 18(3), 035002. https://doi.org/10.1088/1748-9326/acb6a8

Zheng, K., Yang, R., Li, R., Guo, P., Yang, L., & Qin, H. (2023b). A spatiotemporal attention network-based analysis of golden pompano school feeding behavior in an aquaculture vessel. *Computers and Electronics in Agriculture*, 203, 107610. https://doi.org/10.1016/j.compag.2022.107610

Zhang, Z., Liu, H., Jin, J., Zhu, X., Han, D., & Xie, S. (2024). Towards a low-carbon footprint: Current status and prospects for aquaculture. *Water Biology and Security*, 3(4), 100290. https://doi.org/10.1016/j.watbs.2024.100290

Zhu, Y., Butterbach-Bahl, K., Merbold, L., Leitner, S., & Pelster, D. E. (2021). Nitrous oxide emission factors for cattle dung and urine deposited onto tropical pastures: A review of field-based studies. *Agriculture, Ecosystems & Environment*, 322, 107637. https://doi.org/10.1016/j.agee.2021.107637

10 AI Integration for Irrigation Optimization

Tri Retnaningsih Soeprobowati and Ovide Decroly Wisnu Ardhi

10.1 EXPLORING HOW AI IS ENHANCING WATER EFFICIENCY IN AGRICULTURE

As global population growth and climate change increase the demand for food production, efficient water management has become essential to sustainable farming practices. The growing global population, projected to reach 9.8 billion by 2050, alongside climate-induced challenges, intensifies the need for sustainable agricultural practices that prioritize practical water usage (United Nations, Department of Economic and Social Affairs, Population Division, 2017). Research indicates that the agricultural sector faces challenges in meeting food demand due to these pressures, which emphasize the critical role of effective water management to support sustainability in agriculture (Kannan & Anandhi, 2020). Agriculture consumes a significant portion of global water resources, yet traditional irrigation techniques often lead to inefficiencies and substantial water wastage, exacerbating the strain on already limited water resources (Pandya, 2019). In response to these challenges, precision agriculture—a technology-driven approach that optimizes agricultural inputs—transforms how water is utilized in farming. This approach integrates advanced technologies to enhance water management, reducing waste and enabling higher crop yields (Karthikeyan et al., 2020).

Key technologies within precision agriculture, such as artificial intelligence (AI), Internet of Things (IoT) sensors, and data-driven farming practices, contribute significantly to improving water efficiency (Pandya, 2019). By employing IoT and AI, farmers can make informed decisions based on real-time data, which ensures that water is applied precisely when and where it is needed, minimizing waste and supporting healthier crops (Lal, 2013). Integrating AI, weather forecasts, and soil analysis further enables precision in water application, allowing farmers to adapt irrigation to real-time environmental needs, thus promoting sustainable water use in agriculture (Karthikeyan et al., 2020). By leveraging these technological advancements, agriculture can better address security while conserving essential resources (Benke & Tomkins, 2017). Figure 10.1 shows the framework for precision agriculture and AI-driven water efficiency, illustrating how water efficiency strategies, precision irrigation techniques, and data collection methods like IoT sensors and weather data are interconnected. The figure also highlights challenges such as technology costs, training needs, and development potential, all of which are crucial in shaping future directions for sustainable water management in agriculture.

DOI: 10.1201/9781003547099-12

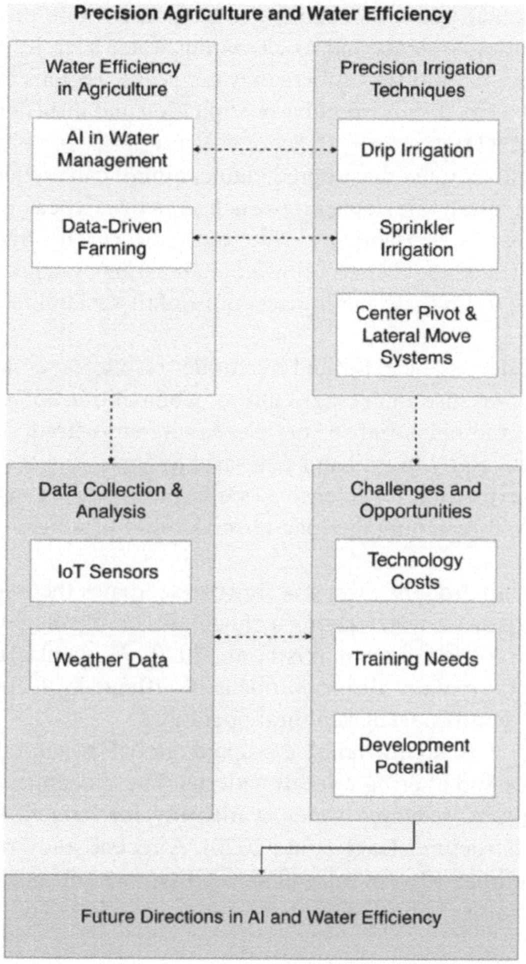

FIGURE 10.1 Precision agriculture and AI-driven water efficiency framework.

10.1.1 Overview of Water Challenges in Agriculture

Agriculture is one of the most water-dependent sectors worldwide. Water requirements for large-scale irrigation are a critical factor in determining the success of food production. According to the FAO, agriculture accounts for up to 70% of the world's total freshwater use, underscoring its reliance on this resource (Food and Agriculture Organization of the United Nations, 2020; Pereira, 2017). With a rapidly growing global population and an increasing demand for food, the strain on freshwater resources is only expected to intensify, exacerbating challenges related to water management in agriculture (Mekonnen & Hoekstra, 2011; Aivazidou, 2021).

Many regions, mainly tropical and subtropical areas, depend heavily on seasonal rainfall for irrigation systems. However, this reliance introduces significant uncertainty due to increasingly unpredictable rainfall patterns caused by climate change.

Research indicates that rainfall variability is becoming more pronounced, disrupting traditional farming systems and exacerbating water scarcity (Feloni & Nastos, 2024). For example, smallholder farmers in water-scarce regions experience reduced agricultural productivity due to irregular precipitation and insufficient adaptive strategies (He & Rosa, 2023).

Studies have demonstrated that unpredictable rainfall can reduce crop yields by as much as 30%, particularly for staple crops such as maize, wheat, and rice, which are highly sensitive to water availability (Achite et al., 2021). This highlights the urgent need for adaptive strategies, such as rainwater harvesting systems and improved soil water management, to mitigate the impacts of rainfall variability (Feloni & Nastos, 2024).

Beyond rainfall dependence, limited freshwater resources remain one of the most critical challenges for sustainable agriculture. Competition for water among agricultural, domestic, and industrial sectors places immense strain on already stressed water systems, particularly in arid and semi-arid regions (Yao et al., 2019). In India, groundwater over-extraction for intensive rice farming has led to severe declines in groundwater levels, threatening the long-term viability of irrigation systems (Feloni & Nastos, 2024).

In regions prone to droughts, water scarcity exacerbates the issue further, forcing farmers to adopt expensive water-saving technologies or rely on government-imposed water allocation strategies (Tzanakakis et al., 2020). Without affordable solutions, small-scale farmers remain disproportionately affected, deepening the divide between large and small-scale agricultural operations.

Climate change has significantly disrupted global water cycles, intensifying droughts and floods and altering rainfall patterns. These changes have direct implications for agriculture, reducing water availability for irrigation while damaging critical water infrastructure (Garg et al., 2020). A recent study revealed that rising temperatures and climate-driven rainfall variability pose substantial risks to regions heavily reliant on rainfed agriculture, particularly in Africa and Asia (He & Rosa, 2023).

Additionally, increasing rainfall variability has led to periodic water stress, causing significant reductions in crop productivity and increasing the frequency of extreme weather events such as floods (Achite et al., 2021). As agricultural expansion continues, additional pressure is placed on already scarce water resources, necessitating integrated water management approaches.

To combat the challenges of water scarcity and climate change, innovative technologies such as AI have emerged as critical tools for improving water management. AI-based systems enable real-time monitoring of water availability, predict irrigation needs based on weather patterns, and optimize water use through advanced analytics (Abuhammad, 2024).

Precision irrigation technologies powered by AI have been shown to reduce water consumption by up to 25% while maintaining or even enhancing crop yields (Tzanakakis et al., 2020). For example, AI-driven systems analyze field data such as soil moisture, temperature, and crop health to dynamically adjust irrigation schedules, ensuring efficient water use and reducing wastage. These advancements are crucial for building resilient agricultural systems that can adapt to climate variability and ensure food security (Feloni & Nastos, 2024).

10.1.2 Machine Learning Models for Irrigation Optimization

Machine learning (ML) models are revolutionizing modern irrigation management by offering data-driven approaches to optimize water use based on historical data, weather patterns, real-time field conditions, and crop-specific requirements (Sayari et al., 2021). Traditional irrigation systems often rely on fixed schedules or manual estimates, which can lead to overwatering or underwatering, resulting in wasted resources and potential crop stress (Eltarabily et al., 2023). ML models, on the other hand, enable precise irrigation tailored to the needs of each crop and environment, improving water efficiency, reducing costs, and promoting sustainable farming practices (El Bilali & Taleb, 2020).

A key way ML enhances irrigation is through the analysis of historical data. By examining records of weather conditions, crop yields, soil moisture levels, and irrigation schedules, ML algorithms identify patterns to inform future water needs (Del-Coco et al., 2024). For example, Ambildhuke and Banik (2022) demonstrated how Long Short-Term Memory (LSTM) models could analyze seasonal trends to predict soil water content and evapotranspiration, achieving a mean squared error (MSE) as low as 0.014. This ensures accuracy and adaptability to seasonal variability, reducing the need for constant manual adjustments.

The integration of IoT sensors and machine learning (ML) technologies has revolutionized irrigation systems by enabling precise prediction of crop water requirements and soil moisture levels, allowing for data-driven irrigation scheduling (Bhuvana, 2024; Kadiyala et al., 2024). Advanced feature selection methods identify critical variables such as evapotranspiration, depletion, and soil water deficit, significantly improving prediction accuracy and system efficiency (Goap et al., 2018). Logistic regression models trained on historical data and expert knowledge enhance the ability to deliver real-time predictions of crop water needs, enabling automated irrigation systems that dynamically adapt to environmental changes (Shahriar et al., 2021). For instance, Jenitha and Rajesh (2024) utilized regression models to analyze soil moisture dynamics and optimize irrigation schedules, reducing water consumption by up to 63.05%. This supports the efficient use of limited water resources, particularly in regions facing water scarcity. These systems leverage IoT sensors to monitor real-time field conditions such as soil moisture, weather, and temperature, transmitting live data to ML models for immediate analysis and decision-making (Kumar & Manjula, 2024).

In addition to regression, decision tree algorithms provide logical, step-by-step irrigation recommendations based on factors such as weather forecasts, current soil moisture, and water availability (Arulraj & Karthikeyan, 2024). These models help determine whether, when, and how much to irrigate, making them especially useful in farms with diverse crops requiring varied watering schedules (Mabasha, 2024). By tailoring irrigation practices to the precise needs of each crop type, decision trees prevent over-irrigation and promote efficient water use (Sayari et al., 2021).

Advanced deep learning (DL) techniques, such as neural networks, further enhance irrigation precision. Neural networks recognize complex patterns in large datasets and adapt dynamically to new information (Patil et al., 2024). For example, Traore et al. (2021) employed Deep Neural Networks (DNNs) to predict equivalent water thickness (EWT) in crops using UAV data, achieving high accuracy and enabling precise irrigation schedules. These models improve over time as they learn from new data, making them increasingly effective in addressing complex irrigation challenges (Trabelsi & Bel Hadj Ali, 2022).

Wireless Sensor Networks (WSNs) integrated with ML models provide real-time soil moisture insights, facilitating immediate irrigation adjustments (Khriji et al., 2021). For example, Singh et al. (2024) combined WSNs with LSTM models to improve soil monitoring accuracy and enable data-driven irrigation planning. Dolaptsis et al. (2024) applied a hybrid LSTM approach for irrigation scheduling in maize fields, achieving high accuracy in predicting soil moisture content reduction. Patrizi et al. (2022) developed a virtual soil moisture sensor using LSTM networks and low-cost WSN nodes, addressing challenges in traditional sensor deployment. Koné et al. (2023) compared various deep learning models for soil moisture prediction, with a hybrid CNN-LSTM model outperforming standalone LSTM and Bi-LSTM models. These studies highlight the potential of combining WSNs with advanced machine learning techniques to enhance precision agriculture practices and optimize water resource management.

The growing integration of ML and DL into irrigation management marks a transformative shift in agricultural practices. By leveraging these technologies, farmers gain access to precise, data-informed strategies for water use tailored to individual crops and environments (Trabelsi & Bel Hadj Ali, 2022). This evolution not only enhances agricultural sustainability but also strengthens food security in the face of increasing environmental challenges (Sayari et al., 2021). As ML and DL models become more accessible, they are poised to redefine the future of farming, enabling efficient, resilient, and sustainable agricultural systems (Patil et al., 2024).

10.2 PRECISION IRRIGATION TECHNIQUES AND DATA-DRIVEN FARMING PRACTICES

As global agriculture evolves to meet the demands of a growing population and changing climate, the need for efficient water use has never been greater (Ghilan et al., 2024). Precision irrigation techniques and data-driven farming practices are key strategies that enable farmers to optimize water use based on the specific needs of crops, soil conditions, and environmental factors (Chauhdary et al., 2023). By tailoring irrigation practices to field conditions, precision irrigation reduces waste, improves crop health, and enhances sustainability in agriculture (Adeyemi et al., 2017; Sharma et al., 2021).

Precision irrigation moves beyond traditional methods, using advanced technologies to apply water more accurately and efficiently (Keswani et al., 2020; González Perea et al., 2018). This section will cover a variety of precision irrigation methods, including drip irrigation, sprinkler systems, and center pivot and lateral move systems (Liang et al., 2020). Each technique is uniquely suited to different crop types, soil compositions, and farm sizes, enabling farmers to choose the best approach for their specific needs (Capraro et al., 2018; Keswani et al., 2019). Additionally, these methods are often enhanced by IoT sensors and AI-driven decision-making, which allow for real-time adjustments based on data collected from the field (Kashyap et al., 2021). IoT facilitates continuous data collection from sensors deployed in the field, while AI algorithms analyze data such as soil moisture, temperature, and rainfall to optimize irrigation schedules (Hussain et al., 2024). This technology has been proven to reduce water consumption by up to 50% while significantly enhancing overall agricultural efficiency (Pongiannan et al., 2023).

Data-driven farming practices further elevate the effectiveness of precision irrigation by integrating information on soil moisture, weather patterns, and crop health (Akhter & Sofi, 2022). With the help of IoT devices, farmers can monitor field conditions continuously, adjusting water distribution and timing based on the latest data (Patil et al., 2024). Predictive analytics can also forecast irrigation needs based on anticipated weather conditions and crop growth stages, allowing for proactive water management (Abdelmoneim et al., 2023).

In this section, we will explore each of these precision irrigation techniques in detail, discuss real-world case studies demonstrating their effectiveness, and examine the challenges and opportunities associated with implementing data-driven and technology-enhanced irrigation practices (Garcia et al., 2023). As we look toward the future, innovations in AI and automation promise even more significant advancements in water efficiency, supporting the development of resilient agricultural systems that can adapt to the complexities of modern farming (Kashyap et al., 2021).

10.2.1 Introduction to Precision Irrigation

Precision irrigation is a highly targeted approach to water management in agriculture that tailors water application according to the specific needs of crops and soil conditions (Adeyemi et al., 2017). This strategy enables farmers to optimize water use, ensuring that each plant receives precisely the right amount of water at the right time, while minimizing waste (Sharma et al., 2021). Unlike traditional irrigation, which often applies water uniformly across a field, precision irrigation employs advanced technologies to monitor and control water distribution based on real-time data (Kamienski et al., 2019). This method has become essential for achieving both productivity and sustainability in modern agriculture (Kashyap et al., 2021).

At its core, precision irrigation involves analyzing soil and crop requirements. Different crops have varying water needs that change with growth stages, climate, and soil type (Patil et al., 2024). For instance, crops like tomatoes and cotton require more water during their flowering and fruiting stages, while leafy greens may need consistent, shallow watering throughout their lifecycle (Abdelmoneim et al., 2023). By aligning irrigation with these unique requirements, precision irrigation ensures that plants receive the water necessary for optimal growth, without overuse. Recent advancements in precision irrigation technologies have demonstrated the ability to reduce water waste while maintaining optimal soil moisture levels and improving crop productivity (Abioye et al., 2021).

One of the key concepts behind precision irrigation is site-specific water management. Soil conditions can vary significantly within a single field, with differences in texture, organic matter, and water-holding capacity (Bodkhe et al., 2020). For example, sandy soils drain faster and require more frequent watering, while clay soils retain moisture for longer periods (Peter et al., 2024). Precision irrigation systems use these soil variations to adjust water delivery, applying more water to areas with high drainage and less to areas that retain water (Goap et al., 2018). This results in a more balanced distribution that conserves water while promoting uniform crop health (Akhter & Sofi, 2022).

Technology plays a critical role in implementing precision irrigation. Systems often rely on a network of IoT sensors, drones, and even satellite imagery to gather data on soil moisture, temperature, and crop conditions (Choudhari et al., 2024). These data points are processed by AI or machine learning algorithms, which analyze the information to generate specific irrigation recommendations (Hussein et al., 2024). For instance, if soil moisture sensors detect dry conditions specific a particular zone of the field, the system can increase water delivery to that area, ensuring that plants receive adequate hydration without over-irrigating other parts of the field (Kamienski et al., 2019).

Types of Precision Irrigation Techniques are tailored to meet the varying needs of crops and fields. Drip irrigation systems, including surface and subsurface methods, have proven to significantly enhance water efficiency by delivering water directly to the root zone while minimizing evaporation losses, making them particularly effective in water-scarce agricultural regions (Kisekka et al., 2017). Sprinkler systems, on the other hand, distribute water in a controlled manner over a larger area and are particularly useful for crops with high foliage (Baiamonte & Baiamonte, 2019). In some cases, more advanced systems like center pivot and lateral move irrigation are used, which can adjust water flow across different parts of a field based on soil moisture readings, improving irrigation uniformity and reducing water losses (Lamm et al., 2019; Kashyap et al., 2021).

The benefits of precision irrigation extend beyond just water savings quality. Precision irrigation enhances crop yields and quality by aligning water use with actual crop needs. Plants that receive the right amount of water at the right time tend to grow more vigorously and are less susceptible to stress-related diseases (Abdelmoneim et al., 2023). Additionally, optimized water use helps maintain soil health by preventing issues such as waterlogging, erosion, and nutrient leaching, which can occur when excessive water is applied (El Chami et al., 2019). Healthier soils support long-term agricultural productivity, making precision irrigation an essential component of sustainable farming (Hussein et al., 2024).

Another advantage of precision irrigation is its positive impact on energy savings. By reducing unnecessary water application, precision systems lower the energy required to pump and distribute water across fields, leading to decreased fuel and electricity costs (Choudhari et al., 2024). This is particularly valuable in large-scale operations where irrigation constitutes a significant portion of operational expenses. Reducing energy consumption also minimizes the environmental footprint of farming, supporting broader sustainability goals in agriculture (Kamienski et al., 2019).

10.2.2 Types of Precision Irrigation Techniques

Precision irrigation encompasses various methods designed to meet the specific needs of crops while maximizing water efficiency. Each technique is suited to different types of crops, soil conditions, and environmental factors, enabling farmers to choose the most effective solution for their particular context (Zhang et al., 2021).

10.2.2.1 Drip Irrigation

Drip irrigation is one of the most water-efficient irrigation techniques available. This system delivers water directly to the root zone of each plant through a network of tubes and emitters, providing a steady, slow release of water close to where it is most needed (Gan et al., 2018). By targeting water application, drip irrigation minimizes evaporation and runoff, which are common issues in traditional irrigation methods. Studies have shown that drip irrigation significantly reduces soil evaporation and crop evapotranspiration compared to border irrigation, particularly in arid regions, where water efficiency improvements can reach up to 7% annually (Wang et al., 2020).

The benefits of drip irrigation extend beyond just water conservation. Limiting water exposure to the soil surface also reduces weed growth, as the areas between plants remain relatively dry. Additionally, drip irrigation can be adapted to apply fertilizers and nutrients directly to the root zone, enhancing nutrient efficiency and plant uptake (Zhang et al., 2021). This is particularly advantageous for high-value crops such as fruits, vegetables, and vineyards, where uniform water and nutrient distribution are essential for quality and yield (Korzekwa, 2024).

10.2.2.2 Sprinkler Irrigation

Sprinkler irrigation systems distribute water over a larger area, simulating rainfall. Water is sprayed into the air and falls onto the plants and soil, covering a wide area efficiently (Gan et al., 2018). Sprinkler systems offer significant water-saving advantages over traditional flood irrigation, as they allow for controlled water application and reduce the risk of overwatering. These systems can be easily adjusted to deliver specific amounts of water based on crop requirements, soil conditions, and climatic factors (Wang et al., 2020).

10.2.2.3 Center Pivot and Lateral Move Systems

Center pivot and lateral move systems are mechanized irrigation techniques that use large, automated structures to irrigate crops in a controlled manner. Center pivot systems move in a circular pattern around a central pivot point, covering large, circular areas of land, while lateral move systems travel in straight lines across rectangular or square plots (Lamm et al., 2019). These systems incorporate sensors to monitor real-time soil moisture and adjust water application as needed, optimizing irrigation efficiency and improving crop yields (Zhang et al., 2021).

10.2.3 Role of Data-Driven Farming in Water Management

Data-driven farming transforms traditional agriculture by leveraging real-time data and advanced analytics to make informed water management decisions. Farmers can continuously monitor crucial indicators such as crop health, soil conditions, and rainfall patterns with sensors and IoT devices now widely used on farms. This wealth of data provides a clearer understanding of each crop's water needs, allowing farmers to manage water more precisely and efficiently (Sivakumar et al., 2023).

One of the most valuable aspects of data-driven farming is crop health monitoring. Through data collected from sensors and drones, farmers can assess the health of their crops on a granular level, identifying areas of the field that may be under stress due to lack of water. By analyzing these insights, they can adjust irrigation to meet specific crop requirements, reducing the risk of under- or over-watering (Peddi et al., 2022). Additionally, by continuously monitoring crop health, data-driven systems can detect early signs of disease or nutrient deficiency, allowing farmers to address issues before they impact productivity (Suhag et al., 2021).

Data-driven farming also enhances soil condition monitoring. Soil sensors collect real-time information on moisture levels, pH, and nutrient composition, enabling farmers to understand the soil's water-holding capacity and adjust irrigation accordingly. For instance, if some regions of a field retain more moisture than others, irrigation can be reduced in those zones, conserving water. Soil data also helps farmers apply water only when it's necessary, as sensors can signal when moisture falls below optimal levels, thus preventing unnecessary watering (Hajdu et al., 2019).

Rainfall pattern analysis is another critical component of data-driven farming events. For instance, machine learning models and high-resolution rainfall predictions have enabled precise irrigation planning, reducing water use while maintaining crop yields (Zhao et al., 2023). If rain is expected, irrigation can be postponed, minimizing unnecessary water application and avoiding potential runoff (OswaltManoj & Ananth, 2020).

Data-driven farming practices provide essential insights that enhance water management, helping farmers make more precise decisions that save water and improve crop health. As technologies for data collection and analytics continue to advance, these practices are expected to play a crucial role in promoting sustainable agriculture, enabling farmers to optimize resource use, adapt to changing environmental conditions, and ensure long-term productivity (Priya et al., 2023).

10.2.4 Case Studies in AI-Driven Precision Irrigation and Water Management

This section presents a series of case studies showcasing the application of AI technologies in precision irrigation and water management. These examples highlight how AI has been successfully integrated into agricultural practices to optimize water use, improve crop productivity, and promote sustainable water management. By focusing on various crops and irrigation systems, the case studies demonstrate the challenges faced, methods applied, results achieved, and the contributions of AI to agricultural water efficiency.

The first case study investigates the application of Convolutional Neural Networks (CNNs) in pattern classification for onion crops (López-Martínez et al., 2024). The primary aim of the research was to improve water use efficiency by monitoring various environmental factors, such as soil moisture, temperature, and humidity. By utilizing CNNs, the study was able to accurately predict water requirements based on environmental data, leading to optimized irrigation schedules. The model achieved an impressive accuracy rate of 99%, indicating the high potential of deep learning technologies in precision irrigation. The key contribution of this study is its innovative use of CNN models for pattern recognition in agriculture, which can be extended beyond onion crops to other types of vegetation that require precise water management. However, challenges were faced during the implementation of the CNN model, primarily due to the need for large, high-quality datasets for training the model. Despite this, the study highlighted significant opportunities to expand CNN applications for other crops, improving water management in various agricultural settings. Future directions for research could involve integrating other AI models, such as deep reinforcement learning, to improve the model's performance further.

The second case study examines using a hybrid Long Short-Term Memory (LSTM) model to optimize irrigation scheduling for maize farming (Dolaptsis et al., 2024). LSTM is a type of recurrent neural network (RNN) that is well-suited for time-series data, such as soil moisture levels, rainfall forecasts, and crop growth stages. In this study, the hybrid LSTM model integrated multiple data sources, including weather data, sensor readings, and historical soil moisture information, to predict the optimal times for irrigation. The model predicts soil moisture with high accuracy, helping farmers schedule irrigation more effectively and reduce water waste. The contributions of this research lie in its ability to improve water efficiency by integrating predictive analytics into irrigation decision-making processes. The hybrid LSTM model proved particularly effective for large-scale maize farms, where timely and accurate irrigation is essential for maximizing yields. One of the main challenges the study faced was data scarcity, particularly in rural areas where sensor networks are not always available. However, the study provides opportunities to enhance the model by incorporating additional sensor types and testing it across different crops. Looking ahead, developing autonomous irrigation systems using LSTM models could lead to further improvements in water efficiency, reducing the need for manual intervention in irrigation practices.

The third case study explores the integration of AI-driven precision irrigation in legume farming (Kim & AlZubi, 2024). This study highlights the use of remote sensing technologies, combined with AI, to monitor soil moisture, weather conditions, and crop growth in real time. By applying AI algorithms to this data, the irrigation system could automatically adjust water application based on real-time environmental conditions, ensuring that water was used efficiently and only when necessary. The results were promising: AI-driven irrigation reduced water consumption by 15% while maintaining or even improving crop yields. This study contributed to the growing body of research on AI's role in improving water efficiency in agriculture also conserves water and helps maintain healthy crop growth, making it particularly relevant for legume farming, which is highly sensitive to water stress. However, one of the challenges faced in the study was the initial high cost of setting up such an advanced irrigation system and the need for continuous data monitoring to ensure optimal performance. Despite these challenges, the study presented numerous opportunities to refine the system and expand it to other crops. Future research should focus on developing fully autonomous irrigation systems that adjust water delivery in real-time without human intervention, thereby enhancing scalability and ease of use.

The fourth case study investigates the use of Ant Colony Optimization (ACO) to estimate the irrigation water requirements of crops based on environmental and soil data (Li et al., 2024). ACO is a nature-inspired optimization algorithm that mimics the foraging behavior of ants to find optimal solutions to complex problems. This study applied ACO to model irrigation scheduling by analyzing factors such as soil moisture content, temperature, and evapotranspiration rates. The results showed that the ACO-based model significantly improved water use efficiency, reducing water wastage by 25% compared to traditional irrigation methods. The research introduced ACO as a novel optimization technique for irrigation management, which could be adapted to a variety of agricultural settings and crops. One of the main challenges of this study was the complexity of implementing ACO algorithms and the need for accurate and reliable data to ensure the optimization process was effective. However, the study highlighted opportunities to scale the use of ACO in large-scale agricultural systems, particularly in regions where water resources are scarce. Future research could focus on integrating ACO with other AI techniques scheduling further to refine the accuracy and efficiency of irrigation scheduling.

This research combined unmanned aerial systems (UAS) with multi-sensor deep learning frameworks to predict crop yields and optimize water usage (Lee & Park, 2024). The framework was designed to capture high-resolution imagery and sensor data, processed through deep-learning models to predict crop yields. Results indicated that the system achieved a high correlation ($R^2=0.82$) and low RMSE, making it a valuable tool for improving irrigation practices based on accurate yield predictions. The key contribution of this study is its integration of multiple sensors with deep learning, providing a holistic approach to water-efficient farming. Challenges included data fusion complexities, but the research opened opportunities for improving crop-specific irrigation strategies. Future research could focus on refining this framework to make it more scalable and adaptable to different crop types and climates (Table 10.1).

The case studies presented in this chapter provide compelling evidence that AI-driven technologies, such as convolutional neural networks (CNNs), long short-term memory (LSTM) networks, and ACO, can significantly enhance water efficiency in agriculture. These technologies offer innovative solutions for optimizing irrigation scheduling, reducing water consumption, and improving crop yields. However, despite the promising results, challenges remain, including high implementation costs, limited access to data, and the need for continuous monitoring to ensure the systems function optimally. Nonetheless, there are significant opportunities for advancing these technologies, mainly through improvements in sensor technology, the development of autonomous irrigation systems, and the integration of multiple AI models. Future research should focus on improving the affordability and scalability of AI-based irrigation systems and exploring AI's use for long-term climate forecasting to better anticipate irrigation needs. By addressing these challenges and expanding the applications of AI in water management, agricultural systems can become more resilient, sustainable, and water-efficient in the face of growing global demands and climate change.

Figure 10.2 shows the IoT and AI-based water management system, which outlines the sequential steps involved in the process, divided into distinct clusters to

TABLE 10.1
Applications of Ai-Driven Approaches in Precision Irrigation and Water Management, Highlighting Key Principles, Components, and Sources

Application in AI-Driven Precision Irrigation and Water Management	Key Principles, Source, and Publisher of the Document
Pattern classification of an onion crop using Convolutional Neural Networks (CNNs) (López-Martínez et al., 2024) • Key components: Convolutional Neural Networks (CNN), soil moisture sensors, humidity sensors	• Key principles: This study utilizes Convolutional Neural Networks (CNNs) for pattern classification in onion crops, focusing on vegetation, soil, and humidity monitoring.
A Hybrid LSTM approach for irrigation scheduling in maize crop (Dolaptsis et al., 2024) • Key component: Long Short-Term Memory (LSTM), weather forecasting, soil moisture sensors	• Key principles: Uses hybrid LSTM models to predict optimal irrigation schedules for maize farms based on soil and weather data.
AI-driven precision irrigation for legume farms (Kim & AlZubi, 2024) • Key components: remote sensing, AI algorithms, soil moisture sensors, crop growth models	• AI-based precision irrigation in legume farming to monitor soil moisture and adjust irrigation based on real-time data.
Estimating irrigation water requirements using Ant Colony Optimization (ACO) (Li et al., 2024) • Key component: Ant Colony Optimization (ACO), environmental data (soil moisture, temperature)	• Uses Ant Colony Optimization (ACO) to estimate irrigation water requirements, optimizing water scheduling.
Development of a UAS-based multi-sensor deep learning framework for crop yield prediction (Lee & Park, 2024) • Unmanned aerial systems (UAS), multi-sensor data, deep learning models	• Combines UAS-based multi-sensor data with deep learning models to predict crop yield and optimize irrigation.

highlight each phase of the system. This flowchart provides a comprehensive overview of how IoT and AI technologies can optimize irrigation and ensure efficient water usage in agriculture.

1. The flowchart starts with Cluster 1: Data collection and acquisition, where IoT sensors are deployed to measure key parameters like soil moisture, temperature, humidity, rainfall, and water flow. The data collected by these sensors is then transmitted in real time to cloud or edge systems, ensuring that it is ready for processing and analysis.
2. Next, Cluster 2: Data processing and storage stores and pre-processes the data, ensuring quality and consistency. AI-driven AI & Data Analytics Engine models analyze this data to provide insights on water needs, crop types, weather forecasts, evapotranspiration, and soil moisture.

3. In Cluster 3: Decision-making and irrigation control, the AI system generates actionable irrigation schedules and recommendations for water management. Based on these insights, automated irrigation systems adjust in real time, controlling sprinklers, valves, and other components for optimal water distribution.
4. The Cluster 4: Monitoring and feedback step continuously tracks the irrigation system's performance, providing feedback, alerts, and detailed reports on water usage, crop health, and system efficiency. The dashboard enables farmers to make informed decisions about irrigation and resource management.
5. Finally, Cluster 5: Continuous improvement highlights the feedback loop where the system learns from past data and refines future irrigation schedules. This ongoing learning process allows the system to adapt to changing environmental conditions, ensuring long-term sustainability and water efficiency.

This integrated system, shown in Figure 10.2, optimizes water management in agriculture. It leverages IoT for real-time data collection, AI for predictive insights, and automated systems for efficient irrigation control.

10.2.5 Challenges and Opportunities

Implementing precision irrigation and data-driven farming technologies offers significant benefits but presents notable challenges. A primary barrier is the substantial upfront investment required for installing IoT sensors, automated systems, and data management platforms. This high cost can be prohibitive for small-scale farmers, particularly in developing regions, potentially hindering the adoption of these technologies. Additionally, the ongoing expenses of maintaining and upgrading these systems can further strain financial resources, making it difficult for some farmers to embrace precision techniques fully (U.S. Government Accountability Office, 2024).

Another significant challenge is the need for technical training. Precision irrigation and data-driven farming rely on complex systems that require a certain level of expertise to operate effectively. Farmers must understand how to interpret sensor data, adjust settings based on real-time conditions, and troubleshoot equipment. Without adequate training, they may not fully benefit from the technology or may even misuse it, leading to inefficiencies or crop damage. Providing accessible training programs and technical support is essential to ensure that farmers can use these tools successfully (Mizik, 2023).

Despite these challenges, there are considerable opportunities for expanding precision irrigation. Advances in sensor technology and IoT connectivity are driving down costs, making these tools more accessible to a broader range of farmers. Additionally, as awareness of climate change and water scarcity grows, demand for sustainable farming practices increases, which could attract public and private funding for precision irrigation projects. Government subsidies, grants, and partnerships

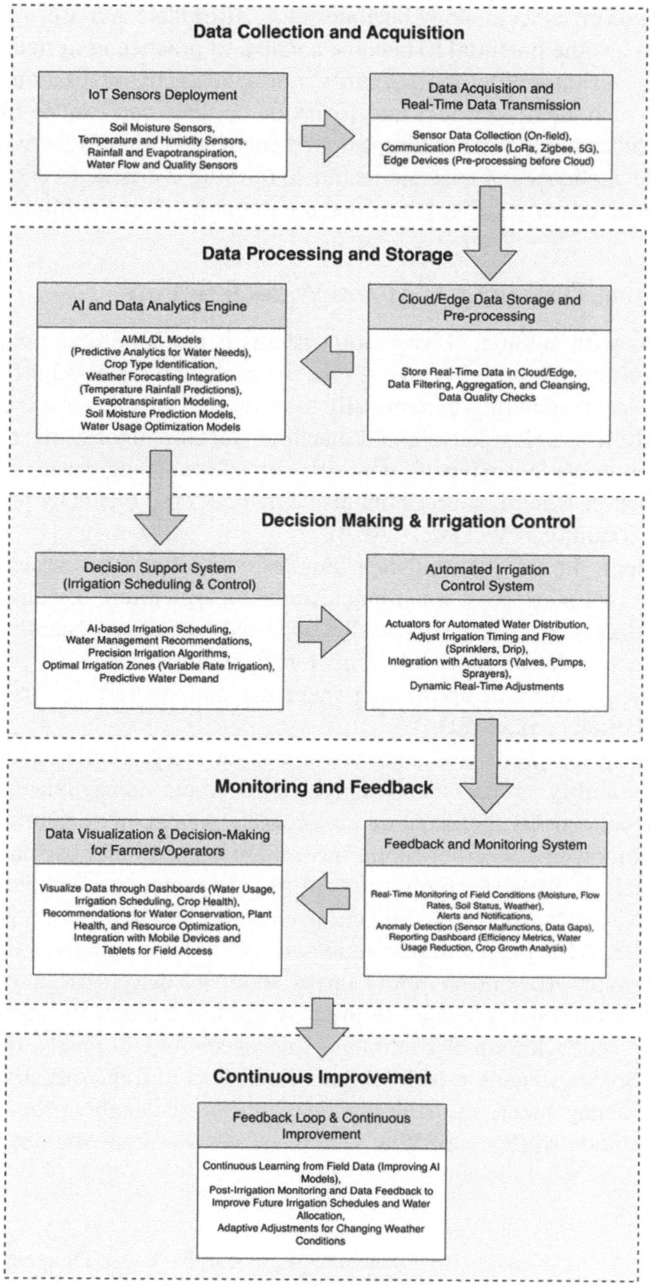

FIGURE 10.2 IoT and AI-based water management systems.

with agricultural technology companies can further support farmers in adopting these methods (Mgendi, 2024).

In the long term, as technology becomes more affordable and user-friendly, precision irrigation has the potential to become a standard practice in agriculture, driving water savings and improving food security worldwide. The integration of advanced technologies, such as AI and IoT, into irrigation systems can lead to more efficient water use, reduced environmental impact, and enhanced crop productivity. However, addressing the challenges of cost and technical training is crucial to ensure that these benefits are realized across diverse farming communities (Karunathilake et al., 2023).

10.2.6 Future Directions in AI and Water Efficiency

Integrating AI with automated irrigation systems is revolutionizing water management in agriculture. These AI-powered systems enable real-time adjustments to irrigation schedules, responding dynamically to changing weather, soil conditions, and crop needs. By leveraging sensors and machine learning algorithms, these systems optimize water usage, ensuring efficient delivery and reducing waste. This approach not only conserves water resources but also enhances crop yields by providing optimal growing conditions (Wei et al., 2024).

Advancements in AI also enhance long-term weather forecasting capabilities, which are crucial for effective water management in agriculture. By analyzing extensive climate data, AI models can predict seasonal variations, enabling farmers to plan irrigation schedules proactively. This foresight allows for better preparation for droughts or heavy rainfall, optimizing water use and supporting resilient farming practices (Khalifeh et al., 2021).

AI's capacity for predictive analytics and data integration is also expanding, enhancing its ability to provide comprehensive water management insights. AI systems can collectively generate more accurate irrigation recommendations by analyzing soil moisture, weather conditions, and historical water use data. This comprehensive analysis allows for highly customized water management plans tailored to specific crop varieties, soil types, and climate conditions (Nsoh et al., 2024).

In the future, AI is poised to play a pivotal role in making agriculture more sustainable and water-efficient. As more farms adopt AI-driven irrigation, data from various regions can be aggregated to improve models and refine algorithms, creating a positive feedback loop of continuous improvement. Ultimately, the integration of AI and precision irrigation technologies could lead to transformative changes in global water management, ensuring that farmers can sustainably produce food for a growing population while conserving vital resources (Assimakopoulos et al., 2024).

REFERENCES

Abdelmoneim, A. A., Khadra, R., Elkamouh, A., Derardja, B., & Dragonetti, G. (2023). Towards affordable precision irrigation: An experimental comparison of weather-based and soil water potential-based irrigation using low-cost IoT-tensiometers on drip irrigated lettuce. *Sustainability*, 16(1), 306. https://doi.org/10.3390/su16010306

Abioye, E. A., Abidin, M. S. Z., Aman, M. N., Mahmud, M. S. A., & Buyamin, S. (2021). A model predictive controller for precision irrigation using discrete lagurre networks. *Computers and Electronics in Agriculture*, 181, 105953. https://doi.org/10.1016/j.compag.2020.105953

Abuhammad, H. (2024). Rainfall prediction using deep learning algorithms. *Journal of Electrical Systems*, 20(7s), 1696–1704. https://doi.org/10.52783/jes.3765

Achite, M., Ceribasi, G., Ceyhunlu, A. I., Wałęga, A., & Caloiero, T. (2021). The Innovative Polygon Trend Analysis (IPTA) as a simple qualitative method to detect changes in environment—Example detecting trends of the total monthly precipitation in semiarid area. *Sustainability*, 13(22), 12674. https://doi.org/10.3390/su132212674

Adeyemi, O., Grove, I., Peets, S., & Norton, T. (2017). Advanced monitoring and management systems for improving sustainability in precision irrigation. *Sustainability*, 9(3), 353. https://doi.org/10.3390/su9030353

Aivazidou, E. (2021). Water management in agriculture and industry: Challenges, trends, and opportunities. *Sustainability*, 14(1), 66. https://doi.org/10.3390/su14010066

Akhter, R., & Sofi, S. A. (2022). Precision agriculture using IoT data analytics and machine learning. *Journal of King Saud University – Computer and Information Sciences*, 34(8), 5602–5618. https://doi.org/10.1016/j.jksuci.2021.05.013

Ambildhuke, G., & Banik, B. G. (2022). IoT based portable weather station for irrigation management using real-time parameters. *International Journal of Advanced Computer Science and Applications*, 13(5). https://doi.org/10.14569/IJACSA.2022.0130533

Arulraj, S., & Karthikeyan, B. (2024). Prediction of irrigation water for plants using machine learning algorithms. In: *2024 International Conference on Computing and Data Science (ICCDS)* (pp. 1–6). https://doi.org/10.1109/ICCDS60734.2024.10560418

Assimakopoulos, F., Vassilakis, C., Margaris, D., Kotis, K., & Spiliotopoulos, D. (2024). Artificial intelligence tools for the agriculture value chain: Status and prospects. *Electronics*, 13(22), 4362. https://doi.org/10.3390/electronics13224362

Baiamonte, G., & Baiamonte, G. (2019). Using rotating sprinkler guns in centre-pivot irrigation systems. *Irrigation and Drainage*, 68. https://doi.org/10.1002/ird.2391

Benke, K., & Tomkins, B. (2017). Future food-production systems: Vertical farming and controlled-environment agriculture. *Sustainability: Science, Practice and Policy*, 13(1), 13–26. https://doi.org/10.1080/15487733.2017.1394054

Bhuvana, T. V. (2024). AI enabled water conservation for irrigation using IoT. *Interantional Journal of Scientific Research in Engineering and Management*, 8(5), 1–5. https://doi.org/10.55041/IJSREM33920

Bodkhe, U., Tanwar, S., Bhattacharya, P., & Kumar, N. (2020). Blockchain for precision irrigation: Opportunities and challenges. *Transactions on Emerging Telecommunications Technologies*, 33. https://doi.org/10.1002/ett.4059

Capraro, F., Tosetti, S., Rossomando, F., Mut, V., & Vita Serman, F. (2018). Web-based system for the remote monitoring and management of precision irrigation: A case study in an arid region of Argentina. *Sensors*, 18(11), 3847. https://doi.org/10.3390/s18113847

Chauhdary, J. N., Li, H., Jiang, Y., Pan, X., Hussain, Z., Javaid, M., & Rizwan, M. (2023). Advances in sprinkler irrigation: A review in the context of precision irrigation for crop production. *Agronomy*, 14(1), 47. https://doi.org/10.3390/agronomy14010047

Choudhari, A., Bhoyar, D. & Badole W. P. (2024). Design of an IoT based monitoring system utilizing deep reinforcement learning and ensemble learning for precision agriculture. In: *2024 2nd International Conference on Sustainable Computing and Smart Systems (ICSCSS)* (pp. 845–852). https://doi.org/10.1109/ICSCSS60660.2024.10625266

Del-Coco, M., Leo, M., & Carcagnì, P. (2024). Machine learning for smart irrigation in agriculture: How far along are we? *Information*, 15(6), 306. https://doi.org/10.3390/info15060306

Dolaptsis, K., Pantazi, X. E., Paraskevas, C., Arslan, S., Tekin, Y., Bantchina, B. B., Ulusoy, Y., Gündoğdu, K. S., Qaswar, M., Bustan, D., &Mouazen, A. M. (2024). A hybrid LSTM approach for irrigation scheduling in maize crop. *Agriculture*, 14(2), 210. https://doi.org/10.3390/agriculture14020210

El Bilali, A., & Taleb, A. (2020). Prediction of irrigation water quality parameters using machine learning models in a semi-arid environment. *Journal of the Saudi Society of Agricultural Sciences*, 19(7), 439–451. https://doi.org/10.1016/j.jssas.2020.08.001

El Chami, D., Knox, J. W., Daccache, A., &Weatherhead, E. K. (2019). Assessing the financial and environmental impacts of precision irrigation in a humid climate. *Horticultural Science*, 46(1), 43–52. https://doi.org/10.17221/116/2017-HORTSCI

Eltarabily, M. G., Abd-Elhamid, H. F., Zeleňáková, M., Elshaarawy, M. K., Elkiki, M., & Selim, T. (2023). Predicting seepage losses from lined irrigation canals using machine learning models. *Frontiers in Water*, 5, 1287357. https://doi.org/10.3389/frwa.2023.1287357

Feloni, E., &Nastos, P. T. (2024). Evaluating rainwater harvesting systems for water scarcity mitigation in small Greek islands under climate change. *Sustainability*, 16(6), 2592. https://doi.org/10.3390/su16062592

Food and Agriculture Organization of the United Nations. (2020). *The State of Food and Agriculture: Climate Change, Agriculture, and Food Security*. Rome, Italy: Food and Agriculture Organization of the United Nations.

Gan, L., Rad, S., Chen, X., Fang, R., Yan, L., & Su, S. (2018). Clock hand lateral, a new layout for semi-permanent sprinkler irrigation system. *Water*, 10(6), 767. https://doi.org/10.3390/w10060767

Garcia, L. D., Lozoya, C., Favela-Contreras, A., & Giorgi, E. (2023). A comparative analysis between heuristic and data-driven water management control for precision agriculture irrigation. *Sustainability*, 15(14), 11337. https://doi.org/10.3390/su151411337

Garg, K. K., Singh, R., Anantha, K. H., Singh, A. K., Akuraju, V. R., Barron, J., Dev, I., Tewari, R. K., Wani, S. P., Dhyani, S. K., & Dixit, S. (2020). Building climate resilience in degraded agricultural landscapes through water management: A case study of Bundelkhand region, Central India. *Journal of Hydrology*, 591, 125592. https://doi.org/10.1016/j.jhydrol.2020.125592

Ghilan, A., Afou, Y. E., Merras, M., & Akkad, N. E. (2024). Data-driven precision agriculture advanced irrigation system for sustainable smart farming. In: *2024 4th International Conference on Innovative Research in Applied Science, Engineering and Technology (IRASET)* (pp. 1–6). https://doi.org/10.1109/IRASET60544.2024.10548267

Goap, A., Sharma, D., Shukla, A. K., & Rama Krishna, C. (2018). An IoT based smart irrigation management system using machine learning and open source technologies. *Computers and Electronics in Agriculture*, 155, 41–49. https://doi.org/10.1016/j.compag.2018.09.040

González Perea, R., Daccache, A., Rodríguez Díaz, J. A., Camacho Poyato, E., & Knox, J. W. (2018). Modelling impacts of precision irrigation on crop yield and in-field water management. *Precision Agriculture*, 19(3), 497–512. https://doi.org/10.1007/s11119-017-9535-4

Hajdu, I., Yule, I., Bretherton, M., Singh, R., & Hedley, C. (2019). Field performance assessment and calibration of multi-depth AquaCheck capacitance-based soil moisture probes under permanent pasture for hill country soils. *Agricultural Water Management*, 217, 332–345. https://doi.org/10.1016/j.agwat.2019.03.002

He, L., & Rosa, L. (2023). Solutions to agricultural green water scarcity under climate change. *PNAS Nexus*, 2(4), pgad117. https://doi.org/10.1093/pnasnexus/pgad117

Hussain, M., N, K., Maurya, I., & Sinha, S. (2024). AI-optimized irrigation for sustainable agriculture. In: *2024 International Conference on Advances in Data Engineering and Intelligent Computing Systems (ADICS)* (pp. 1–9). https://doi.org/10.1109/ADICS58448.2024.10533562

Hussein, A. H. A., Jabbar, K. A., Mohammed, A., & Jasim, L. (2024). Harvesting the future: AI and IoT in agriculture. In: *E3S Web of Conferences* (vol. 477, p. 00090). https://doi.org/10.1051/e3sconf/202447700090

Jenitha, R., & Rajesh, K. (2024). Intelligent irrigation scheduling scheme based on deep bi-directional LSTM technique. *International Journal of Environmental Science and Technology*, 21(2), 1905–1922. https://doi.org/10.1007/s13762-023-04955-0

Kadiyala, S., Potluri, C. S., Birke, A. H., Mehra, A., Rajput, S., & Padmavathi, K. (2024). An intelligent and cost-effective IoT-based irrigation system using machine learning. In: *2024 2nd International Conference on Disruptive Technologies (ICDT)* (pp. 380–385). https://doi.org/10.1109/ICDT61202.2024.10489218

Kamienski, C., Soininen, J.-P., Taumberger, M., Dantas, R., Toscano, A., Salmon Cinotti, T., Filev Maia, R., & Torre Neto, A. (2019). Smart water management platform: IoT-based precision irrigation for agriculture. *Sensors*, 19(2), 276. https://doi.org/10.3390/s19020276

Kannan, N., &Anandhi, A. (2020). Water management for sustainable food production. *Water*, 12(3), 778. https://doi.org/10.3390/w12030778

Karthikeyan, L., Chawla, I., & Mishra, A. K. (2020). A review of remote sensing applications in agriculture for food security: Crop growth and yield, irrigation, and crop losses. *Journal of Hydrology*, 586, 124905. https://doi.org/10.1016/j.jhydrol.2020.124905

Karunathilake, E. M. B. M., Le, A. T., Heo, S., Chung, Y. S., & Mansoor, S. (2023). The path to smart farming: Innovations and opportunities in precision agriculture. *Agriculture*, 13(8), 1593. https://doi.org/10.3390/agriculture13081593

Kashyap, P. K., Kumar, S., Jaiswal, A., Prasad, M., &Gandomi, A. H. (2021). Towards precision agriculture: IoT-enabled intelligent irrigation systems using deep learning neural network. *IEEE Sensors Journal*, 21(16), 17479–17491. https://doi.org/10.1109/JSEN.2021.3069266

Keswani, B., Mohapatra, A. G., Keswani, P., Khanna, A., Gupta, D., & Rodrigues, J. (2020). Improving weather dependent zone specific irrigation control scheme in IoT and big data enabled self driven precision agriculture mechanism. *Enterprise Information Systems*, 14(9–10), 1494–1515. https://doi.org/10.1080/17517575.2020.1713406

Keswani, B., Mohapatra, A. G., Mohanty, A., Khanna, A., Rodrigues, J. J. P. C., Gupta, D., & De Albuquerque, V. H. C. (2019). Adapting weather conditions based IoT enabled smart irrigation technique in precision agriculture mechanisms. *Neural Computing and Applications*, 31(S1), 277–292. https://doi.org/10.1007/s00521-018-3737-1

Khalifeh, A., Al-Qammaz, A., Darabkh, K. A., Abualigah, L., Khasawneh, A. M., & Zinonos, Z. (2021). An AI based irrigation and weather forecasting system utilizing LoRaWAN and cloud computing technologies. In: *2021 IEEE Conference of Russian Young Researchers in Electrical and Electronic Engineering (ElConRus)* (pp. 443–448). https://doi.org/10.1109/ElConRus51938.2021.9396431

Khriji, S., El Houssaini, D., Kammoun, I., &Kanoun, O. (2021). Precision irrigation: An IoT-enabled wireless sensor network for smart irrigation systems. In: T. K. Hamrita (Ed.), *Women in Precision Agriculture: Technological Breakthroughs, Challenges and Aspirations for a Prosperous and Sustainable Future* (pp. 107–129). Springer International Publishing. https://doi.org/10.1007/978-3-030-49244-1_6

Kim, T. H., &AlZubi, A. A. (2024). AI-enhanced precision irrigation in legume farming: Optimizing water use efficiency. *Legume Research: An International Journal*. https://doi.org/10.18805/LRF-791

Kisekka, I., Oker, T., Nguyen, G., Aguilar, J., & Rogers, D. (2017). Revisiting precision mobile drip irrigation under limited water. *Irrigation Science*, 35(6), 483–500. https://doi.org/10.1007/s00271-017-0555-7

Koné, B. A. T., Grati, R., Bouaziz, B., &Boukadi, K. (2023). A new long short-term memory based approach for soil moisture prediction. *Journal of Ambient Intelligence and Smart Environments*, 15(3), 255–268. https://doi.org/10.3233/AIS-230035

Korzekwa, K. (2024). Evaluating overhead irrigation sprinkler packages in Utah. *Crops & Soils*. https://doi.org/10.1002/crso.20379

Kumar, P., &Dr.Manjula, S. H. (2024). Unlocking water conservation: Synergizing IoT and machine learning for agricultural irrigation efficiency. In: *2024 International Conference on Knowledge Engineering and Communication Systems (ICKECS)* (pp. 1–7). https://doi.org/10.1109/ICKECS61492.2024.10617221

Lal, R. (2013). Climate-strategic agriculture and the water-soil-waste nexus. *Journal of Plant Nutrition and Soil Science*, 176. https://doi.org/10.1002/jpln.201300189

Lamm, F. R., Porter, D. O., Bordovsky, J. P., Evett, S. R., O'Shaughnessy, S. A., Stone, K. C., Kranz, W. L., Rogers, D. H., &Colaizzi, P. D. (2019). Targeted, precision irrigation for moving platforms: Selected papers from a Center Pivot technology transfer effort. *Transactions of the ASABE*, 62(5), 1409–1415. https://doi.org/10.13031/trans.13371

Lee, D.-H., & Park, J.-H. (2024). Development of a UAS-based multi-sensor deep learning model for predicting napa cabbage fresh weight and determining optimal harvest time. *Remote Sensing*, 16(18), 3455. https://doi.org/10.3390/rs16183455

Li, Z., Zhao, L., Zhao, Z., Cai, H., Xing, L., & Cui, N. (2024). Estimating the transpiration of kiwifruit using an optimized canopy resistance model based on the synthesis of sunlit and shaded leaves. *Agricultural Water Management*, 306, 109193. https://doi.org/10.1016/j.agwat.2024.109193

Liang, Z., Liu, X., Xiong, J., & Xiao, J. (2020). Water allocation and integrative management of precision irrigation: A systematic review. *Water*, 12(11), 3135. https://doi.org/10.3390/w12113135

López-Martínez, M. D. J., Díaz-Flórez, G., Villagrana-Barraza, S., Castañeda-Miranda, C. L., Solís-Sánchez, L. O., Ortíz-Esquivel, D. I., De La Rosa-Vargas, J. I., & Olvera-Olvera, C. A. (2024). Pattern classification of an onion crop (Allium Cepa) field using convolutional neural network models. *Agronomy*, 14(6), 1206. https://doi.org/10.3390/agronomy14061206

Mabasha, S. (2024). Evaluating the machine learning based efficacy of decision tree and support vector machines in smart irrigation systems for precise irrigation status classification for optimizing water management in agriculture. *Journal of Electrical Systems*, 20(2s), 867–875. https://doi.org/10.52783/jes.1682

Mekonnen, M. M., & Hoekstra, A. Y. (2011). The green, blue and grey water footprint of crops and derived crop products. *Hydrology and Earth System Sciences*, 15(5), 1577–1600. https://doi.org/10.5194/hess-15-1577-2011

Mgendi, G. (2024). Unlocking the potential of precision agriculture for sustainable farming. *Discover Agriculture*, 2(1), 87. https://doi.org/10.1007/s44279-024-00078-3

Mizik, T. (2023). How can precision farming work on a small scale? A systematic literature review. *Precision Agriculture*, 24(1), 384–406. https://doi.org/10.1007/s11119-022-09934-y

Nsoh, B., Katimbo, A., Guo, H., Heeren, D. M., Nakabuye, H. N., Qiao, X., Ge, Y., Rudnick, D. R., Wanyama, J., Bwambale, E., & Kiraga, S. (2024). Internet of things-based automated solutions utilizing machine learning for smart and real-time irrigation management: A review. *Sensors*, 24(23), 7480. https://doi.org/10.3390/s24237480

OswaltManoj, S., & Ananth, J. P. (2020). MapReduce and optimized deep network for rainfall prediction in agriculture. *The Computer Journal*, 63, 900–912. https://doi.org/10.1093/comjnl/bxz164

Pandya, E. A. (2019). Innovative and sustainable water management: Adapting to a variable and changing climate. *Irrigation and Drainage*, 68, 123–124. https://doi.org/10.1002/ird.2334

Patil, S. B., Kulkarni, R. B., Patil, S. S., &Kharade, P. A. (2024). Machine learning based precision agriculture model for farm irrigation to optimize water usage. *IOP Conference Series: Earth and Environmental Science*, 1285(1), 012026. https://doi.org/10.1088/1755-1315/1285/1/012026

Patrizi, G., Bartolini, A., Ciani, L., Gallo, V., Sommella, P., & Carratu, M. (2022). A virtual soil moisture sensor for smart farming using deep learning. *IEEE Transactions on Instrumentation and Measurement*, 71, 1–11. https://doi.org/10.1109/TIM.2022.3196446

Peddi, P., Dasgupta, A., & Gaidhane, V. H. (2022). Smart irrigation systems: Soil monitoring and disease detection for precision agriculture. In: *2022 IEEE International IOT, Electronics and Mechatronics Conference (IEMTRONICS)* (pp. 1–7). https://doi.org/10.1109/IEMTRONICS55184.2022.9795747

Pereira, L. S. (2017). Water, agriculture and food: Challenges and issues. *Water Resources Management*, 31(10), 2985–2999. https://doi.org/10.1007/s11269-017-1664-z

Peter, M. J., Kalaiyarasi, R., Vijayashanthi, V., A, M. T., Menaga, D., & Suresh, P. M. (2024). IoT based smart irrigation system for precision agriculture in greenhouse environment. In: *2024 5th International Conference on Electronics and Sustainable Communication Systems (ICESC)* (pp. 411–416). https://doi.org/10.1109/ICESC60852.2024.10689981

Pongiannan, R. K., Brindha, R., Akbar, S. A., Pravin, A. R., J, F., & M, P. (2023). AI-based autonomous paddy farm using smart and precise irrigation system. In: *2023 International Conference on System, Computation, Automation and Networking (ICSCAN)* (pp. 1–5). https://doi.org/10.1109/ICSCAN58655.2023.10395335

Priya, P. K., Blessy, M, P., Suguna, R. K., Narmada, C., Devi, G., & Kalyan, G. R. (2023). Intelligent irrigation management: An IoT-driven approach with real-time data integration. In: *2023 International Conference on Innovative Computing, Intelligent Communication and Smart Electrical Systems (ICSES)* (pp. 1–9). https://doi.org/10.1109/ICSES60034.2023.10465371

Sayari, S., Mahdavi-Meymand, A., & Zounemat-Kermani, M. (2021). Irrigation water infiltration modeling using machine learning. *Computers and Electronics in Agriculture*, 180, 105921. https://doi.org/10.1016/j.compag.2020.105921

Shahriar, S., Peyal, H. I., Nahiduzzaman, M., & Pramanik, M. A. H. (2021). An IoT-based real-time intelligent irrigation system using machine learning. In: *2021 13th International Conference on Information & Communication Technology and System (ICTS)* (pp. 277–281). https://doi.org/10.1109/ICTS52701.2021.9608813

Sharma, A., Jain, A., Gupta, P., & Chowdary, V. (2021). Machine learning applications for precision agriculture: A comprehensive review. *IEEE Access*, 9, 4843–4873. https://doi.org/10.1109/ACCESS.2020.3048415

Singh, T., Kundroo, M., & Kim, T. (2024). WSN-driven advances in soil moisture estimation: A machine learning approach. *Electronics*, 13(8), 1590. https://doi.org/10.3390/electronics13081590

Sivakumar, V. G., Baskar, V. V., Vadivel, M., Vimal, S. P., & Murugan, S. (2023). IoT and GIS integration for real-time monitoring of soil health and nutrient status. In: *2023 International Conference on Self Sustainable Artificial Intelligence Systems (ICSSAS)* (pp. 1265–1270). https://doi.org/10.1109/ICSSAS57918.2023.10331694

Suhag, S., N. Singh, S. Jadaun, P. Johri, A. Shukla, & N. Parashar. (2021). IoT based soil nutrition and plant disease detection system for smart agriculture. In: *2021 10th IEEE International Conference on Communication Systems and Network Technologies (CSNT)* (pp. 478–483). https://doi.org/10.1109/CSNT51715.2021.9509719

Trabelsi, F., & Bel Hadj Ali, S. (2022). Exploring machine learning models in predicting irrigation groundwater quality indices for effective decision making in Medjerda River Basin, Tunisia. *Sustainability*, 14(4), 2341. https://doi.org/10.3390/su14042341

Traore, A., Ata-Ul-Karim, S. T., Duan, A., Soothar, M. K., Traore, S., & Zhao, B. (2021). Predicting equivalent water thickness in wheat using UAV mounted multispectral sensor through deep learning techniques. *Remote Sensing*, 13(21), 4476. https://doi.org/10.3390/rs13214476

Tzanakakis, V. A., Paranychianakis, N. V., & Angelakis, A. N. (2020). Water supply and water scarcity. *Water*, 12(9), 2347. https://doi.org/10.3390/w12092347

U.S. Government Accountability Office. (2024). *Precision Agriculture: Benefits and Challenges for Technology Adoption and Use*. U.S. GAO Reports. https://www.gao.gov/products/gao-24-105962

United Nations, Department of Economic and Social Affairs, Population Division. (2017). *World Population Prospects 2017 – Data Booklet*. New York: United Nations.

Wang, Y., Li, S., Qin, S., Guo, H., Yang, D., & Lam, H.-M. (2020). How can drip irrigation save water and reduce evapotranspiration compared to border irrigation in arid regions in northwest China. *Agricultural Water Management*, 239, 106256. https://doi.org/10.1016/j.agwat.2020.106256

Wei, H., Xu, W., Kang, B., Eisner, R., Muleke, A., Rodriguez, D., deVoil, P., Sadras, V., Monjardino, M., & Harrison, M. T. (2024). Irrigation with artificial intelligence: Problems, premises, promises. *Human-Centric Intelligent Systems*, 4(2), 187–205. https://doi.org/10.1007/s44230-024-00072-4

Yao, L., Xu, Z., & Chen, X. (2019). Sustainable water allocation strategies under various climate scenarios: A case study in China. *Journal of Hydrology*, 574, 529–543. https://doi.org/10.1016/j.jhydrol.2019.04.055

Zhang, J., Guan, K., Peng, B., Jiang, C., Zhou, W., Yang, Y., Pan, M., Franz, T. E., Heeren, D. M., Rudnick, D. R., Abimbola, O., Kimm, H., Caylor, K., Good, S., Khanna, M., Gates, J., & Cai, Y. (2021). Challenges and opportunities in precision irrigation decision-support systems for center pivots. *Environmental Research Letters*, 16(5), 053003. https://doi.org/10.1088/1748-9326/abe436

Zhao, H., Di, L., Guo, L., Zhang, C., & Lin, L. (2023). An automated data-driven irrigation scheduling approach using model simulated soil moisture and evapotranspiration. *Sustainability*, 15(17), 12908. https://doi.org/10.3390/su151712908

11 Precision Agriculture and Water Efficiency

Predictive Analytics, Irrigation Scheduling, and Weather Forecasting Models

Suresh Babu C., Guda Bhargavi, Banoth Bixapathi Nayak, and Chunchu Suchith Kumar

11.1 INTRODUCTION

Agriculture is the largest consumer of freshwater globally, accounting for nearly 70% of total freshwater withdrawals (FAO, 2020). As populations grow and food demand increases, the agricultural sector faces a mounting challenge to sustain productivity while managing water resources more efficiently. Climate change exacerbates this issue by altering weather patterns, making rainfall less predictable, and increasing the frequency and severity of droughts. These changes are particularly concerning in regions where water scarcity is already a significant issue, such as parts of Africa, South Asia, and the Middle East (Mekonnen & Hoekstra, 2016). Therefore, efficient water management is a key priority in efforts to ensure food security and protect the environment.

Water mismanagement in agriculture, often due to over-irrigation or inefficient irrigation methods, leads to significant water loss through evaporation, runoff, and leaching. This not only depletes vital water resources but also impacts soil health and reduces crop yields. Waterlogging, salinization, and nutrient leaching are common consequences of poor water management. In turn, these issues contribute to declining agricultural productivity, which threatens both local livelihoods and global food supply chains. As such, addressing the inefficiencies in agricultural water use is essential to improve food production while conserving water resources.

In response to these challenges, the concept of precision agriculture (PA) has gained prominence as a viable solution. PA refers to a farming management approach that leverages advanced technologies—such as sensors, geographic information systems (GIS), drones, and satellite imagery—to monitor crop needs and optimize input use at a granular level. The key idea behind PA is to apply water, fertilizers, pesticides, and other inputs only where and when they are needed, thereby reducing waste

DOI: 10.1201/9781003547099-13

and minimizing the environmental impact of farming activities (Zhang et al. 2019a). PA allows for site-specific management, which means that farmers can treat each portion of their field according to its unique conditions, such as soil moisture levels, topography, and crop growth stage. This is a significant improvement over traditional methods, which typically apply inputs uniformly across entire fields, often leading to resource overuse and inefficiencies.

An emerging trend within PA is the integration of artificial intelligence (AI) into these advanced farming systems. AI-driven solutions are now being used to analyze vast amounts of data collected from sensors, satellite images, and weather stations. By processing and interpreting this data, AI can provide real-time insights and predictive models that help farmers make more informed decisions about irrigation scheduling, nutrient management, and crop health monitoring. For instance, AI models can forecast soil moisture levels, predict future water needs based on weather patterns, and automate irrigation systems to deliver water precisely where and when it is needed (Zhang et al., 2019a).

The role of AI in PA extends beyond improving water efficiency. By optimizing resource use, AI technologies can also reduce the environmental footprint of farming. For example, precision irrigation powered by AI can minimize the energy required for water pumping, thereby lowering carbon emissions. Additionally, by reducing the need for excess water and fertilizer, AI can help mitigate the release of greenhouse gases like nitrous oxide, which is associated with nitrogen-based fertilizers (IPCC, 2019). As a result, the integration of AI into PA aligns with global efforts to achieve net-zero emissions, a target that aims to balance the amount of greenhouse gases emitted with the amount removed from the atmosphere.

11.2 PA AND WATER MANAGEMENT

11.2.1 PA: Definition and Technologies

PA represents a modern farming management concept that focuses on using technology to optimize field-level management concerning crop farming. It revolves around the premise that every field is not uniform; instead, there are variations in soil properties, water availability, crop health, and other environmental factors even within a single farm plot. PA aims to account for this variability by applying agricultural inputs—such as water, fertilizers, and pesticides—more efficiently and accurately, tailoring them to the specific conditions of different areas within a field (Mulla, 2013). By doing so, PA enhances crop productivity, resource use efficiency, and sustainability while minimizing environmental impacts.

The technologies that enable PA include GIS, satellite imagery, drones, soil sensors, and remote sensing technologies. These tools allow for the collection of vast amounts of real-time data from agricultural fields, which are then analyzed to inform decision-making in the farming process (Zhang et al., 2019a). For instance, GIS technology provides spatial data that help in mapping soil variability and crop performance across different zones. Similarly, satellite imagery and drones offer aerial views of the farm, enabling the detection of patterns in crop growth, nutrient deficiencies, or water stress. Soil sensors, on the other hand, provide direct measurements of soil properties like moisture, temperature, and nutrient levels, allowing for precise input application based on actual field conditions (Jones, 2004).

One of the most critical applications of PA is in water management. Agriculture accounts for the majority of global water withdrawals, and improper water use can lead to waste, soil degradation, and reduced crop productivity (FAO, 2020). By using soil moisture sensors and remote sensing technologies, farmers can continuously monitor moisture levels in the soil and determine exactly when and where irrigation is needed. This approach prevents over-irrigation, which can lead to waterlogging and nutrient leaching, and under-irrigation, which can stress plants and reduce yields (Jones, 2004).

A key technique under PA for improving water efficiency is precision irrigation. Precision irrigation refers to applying water in quantities and at times that match the needs of crops, which are often highly variable within a field. Technologies such as drip irrigation and variable rate irrigation (VRI) are central to this approach. Drip irrigation delivers water directly to the root zones of plants, minimizing water loss due to evaporation or runoff. VRI, on the other hand, adjusts the water application rates for different zones of the field based on factors like soil type and crop needs. This tailored approach can significantly enhance water use efficiency, reducing water consumption while ensuring optimal moisture conditions for plant growth (Zhang et al., 2019a).

PA also integrates AI and machine learning (ML) algorithms, which can analyze vast datasets from sensors and satellite images to predict irrigation requirements. These systems optimize water use by analyzing weather patterns, soil moisture, and crop health data to automate irrigation scheduling, delivering water only when and where it is most needed (Mulla, 2013). In this way, PA not only conserves water but also reduces the energy and costs associated with irrigation, contributing to overall agricultural sustainability and resilience to climate change.

11.2.2 Water Efficiency in Agriculture

Water efficiency in agriculture refers to the amount of crop yield obtained per unit of water used. With freshwater resources becoming increasingly limited due to climate change, population growth, and heightened competition for water, improving water efficiency in agriculture has become a critical priority. Efficient water management ensures that crops receive the right amount of water at the right time, reducing waste and optimizing yields. One of the most promising solutions for improving water efficiency in agriculture is the use of PA techniques, particularly when combined with AI.

Through real-time data collection, analysis, and automation, PA systems can optimize water use in several ways, resulting in more efficient irrigation practices and enhanced crop performance. These technologies not only help to reduce the overuse of water but also contribute to a more sustainable and climate-resilient agricultural system.

11.2.2.1 Irrigation Scheduling

Effective irrigation scheduling is key to improving water efficiency. PA systems, supported by sensors and AI-driven models, can help farmers determine the most suitable times for irrigation based on real-time data, such as soil moisture levels, weather conditions, and crop growth stages. By ensuring that water is applied only when needed, these systems prevent over-irrigation and under-irrigation, both of which can harm crop health and waste valuable water resources (Fereres & Soriano, 2007). Traditional irrigation methods often result in excessive water use, but with precise irrigation scheduling, water is applied in exact quantities to match the crop's needs. For example, farmers can avoid irrigating during periods when rainfall is expected or when soil moisture levels are already sufficient, thus reducing water loss due to runoff or deep percolation.

AI-enhanced irrigation scheduling takes this a step further by incorporating predictive models that anticipate future water requirements. These models can analyze weather forecasts, crop development stages, and soil conditions to predict when crops will need water and in what amounts, thus allowing for even more precise and efficient irrigation planning.

11.2.2.2 Evapotranspiration Monitoring

Evapotranspiration (ET) is the combined process of water loss from soil through evaporation and from plants through transpiration. Monitoring evapotranspiration is crucial for determining the water needs of crops, as it provides an estimate of how much water is being lost and must be replenished to maintain healthy growth. PA systems equipped with remote sensors and AI-driven models can accurately monitor evapotranspiration rates and calculate how much water should be applied to the crops at any given time (Irmak, 2015).

AI can enhance evapotranspiration models by integrating data from various sources—such as satellite images, on-ground sensors, and weather forecasts—to predict future water needs. These predictions are based on factors like temperature, humidity, wind speed, and solar radiation, all of which influence evapotranspiration rates. By accurately estimating water requirements, AI-powered systems can ensure that crops receive just the right amount of water, minimizing waste and conserving water resources.

11.2.2.3 Deficit Irrigation

Another technique to improve water efficiency is deficit irrigation, which involves applying less water than the full crop water requirement during certain non-critical growth stages. This technique conserves water while maintaining acceptable levels of crop yield, as plants are more tolerant to water stress during specific periods of their growth cycle (Vico & Porporato, 2011). While deficit irrigation can be risky if applied incorrectly, AI-driven models help farmers fine-tune the irrigation schedules to ensure that crops receive enough water during critical stages of development, such as flowering and fruiting, while reducing water use during less critical phases, such as vegetative growth.

By carefully managing water inputs through deficit irrigation, farmers can significantly reduce their overall water consumption without drastically affecting yields. This approach is particularly valuable in water-scarce regions, where it is crucial to conserve every drop of water while still maintaining productivity. AI systems, using real-time data and predictive analytics, can guide farmers in adopting deficit irrigation practices by identifying the optimal timing and volume of water application for different crops under varying environmental conditions.

11.3 THE ROLE OF AI IN WATER EFFICIENCY

AI has transformed various sectors, and its integration into agriculture—specifically in water management—has opened new possibilities for optimizing water use, improving crop yields, and reducing environmental impacts. AI-driven systems enhance the capabilities of PA by providing real-time data analysis, predictive modeling, and automation, which allow farmers to make data-driven decisions to achieve greater water efficiency. In water management, AI technologies contribute to water efficiency in several key areas:

11.3.1 Real-Time Monitoring and Data Analysis

AI enables the collection and analysis of real-time data from multiple sources, such as soil moisture sensors, weather stations, satellite imagery, and drones. This vast amount of data is processed and analyzed to provide farmers with up-to-date insights into soil moisture conditions, crop health, and weather forecasts, which are critical for effective water management. By analyzing these data points, AI can help determine the precise amount of water that crops need, reducing unnecessary water usage and preventing over-irrigation (Jones, 2004).

AI-based tools can also detect patterns and anomalies in the data, such as unexpected moisture levels or abnormal temperature changes, allowing farmers to adjust their irrigation strategies in real time. This capability reduces water waste, improves crop resilience, and enhances overall water use efficiency.

One of the most transformative aspects of AI in PA is its ability to enable real-time monitoring and data analysis. AI processes and analyzes data from a variety of sources such as soil moisture sensors, weather stations, satellite imagery, and drones to provide farmers with actionable insights regarding soil conditions, crop health, and weather patterns. These insights are critical for making informed decisions on water management, ensuring that the crops receive the exact amount of water they need when they need it. This precision significantly reduces water wastage and mitigates the risk of over-irrigation, which can harm crops and degrade soil quality (Jones, 2004).

11.3.1.1 AI in Soil Moisture Monitoring

Soil moisture sensors placed throughout a field continuously monitor moisture levels at various depths. The data gathered is transmitted to a central system, where AI algorithms analyze it in real time to predict the precise amount of water required for different areas of the field. These predictions are based on the moisture content, the specific crop type, and local climatic conditions. AI models also incorporate historical data to better forecast the future water requirements, optimizing irrigation schedules (Mulla, 2013). This technology allows for tailored irrigation, meaning water is applied only where it is needed, significantly improving water use efficiency.

For instance, AI systems use ML models that can adapt to the field's dynamic conditions. When paired with remote sensing technologies, AI can further refine water application by identifying spatial variability in soil moisture across the field. This combination of real-time data and predictive analysis is a powerful tool for avoiding both over-irrigation and under-irrigation, maximizing the crop yield per unit of water used.

11.3.1.2 Weather Forecast Integration for Irrigation Decisions

Another critical input to real-time data analysis is the integration of weather forecast data into AI-based water management systems. These systems are capable of collecting real-time weather data from local weather stations and forecasting models, analyzing this data to predict rainfall and evapotranspiration rates—the combination of water evaporation from the soil and transpiration from plants (Irmak, 2015a). With these insights, AI systems can predict when irrigation is unnecessary due to impending rainfall, ensuring water conservation by preventing irrigation before natural precipitation events.

Furthermore, weather data can be analyzed alongside crop-specific growth models to determine the best times for irrigation, ensuring crops receive adequate water during critical growth stages while conserving water during non-essential periods. For instance, in regions prone to drought or irregular rainfall, farmers can rely on AI-driven weather forecasts to schedule irrigation more effectively, preventing water shortages or wastage.

11.3.1.3 Anomaly Detection and Adjustments

Another crucial feature of AI-driven real-time monitoring is its ability to detect anomalies in the field's environmental conditions. For example, an unexpected drop in soil moisture or a sudden rise in ambient temperature can indicate potential threats to crop health, such as water stress or drought conditions. AI systems, equipped with pattern recognition algorithms, can identify these anomalies and alert farmers, enabling them to make timely adjustments to their water management practices (Zhang et al., 2019a). For example, AI can adjust irrigation schedules in response to unforeseen moisture deficits or sudden increases in water evaporation rates due to rising temperatures.

These systems go beyond simply responding to current conditions by incorporating predictive analytics to anticipate future irrigation needs. By examining historical data in combination with real-time inputs, AI models can predict upcoming changes in soil moisture levels and evapotranspiration rates, enabling farmers to adjust irrigation schedules proactively rather than reactively. This capability is particularly important in regions where water resources are scarce, as it allows for water conservation while maintaining high levels of crop productivity.

11.3.1.4 Enhancing Crop Resilience and Reducing Water Waste

Real-time data monitoring and analysis not only optimize water use but also enhance crop resilience. AI systems help farmers avoid both over-watering and under-watering, which can both negatively impact crop health. Over-irrigation can lead to waterlogging, root rot, and nutrient leaching, while under-irrigation can cause water stress, reducing crop yields. AI-based irrigation systems prevent these outcomes by ensuring that crops consistently receive the optimal amount of water for their growth stage, soil type, and local climate (Fereres & Soriano, 2007).

Additionally, by reducing water waste, AI-based tools contribute to broader goals of sustainability and water conservation. In water-scarce regions, such as parts of California's Central Valley, AI-driven irrigation management has helped farmers achieve significant reductions in water use without sacrificing crop yield. For example, vineyards that have implemented these systems have reported up to a 25% reduction in water use while maintaining or improving grape yield and quality (González, 2017).

11.3.1.5 Future Potential and Challenges

While real-time data analysis through AI offers immense potential, challenges remain in ensuring widespread adoption. For smallholder farmers in developing regions, the cost of AI technologies and limited access to high-quality data can be prohibitive. However, advancements in mobile technology and the increasing affordability of sensors and drones are making it easier for these farmers to adopt PA techniques (González, 2017). As these technologies become more accessible, real-time data analysis will play an increasingly critical role in global water management efforts.

11.3.2 Predictive Modeling and Forecasting

AI systems excel in predictive modeling, which is essential for forecasting water requirements in agriculture. By integrating historical data on weather, soil moisture, and crop development, AI can predict future irrigation needs with high accuracy. These models use ML algorithms to learn from past data and improve their predictions over time, ensuring that irrigation schedules are optimized based on current and future conditions (Shahzad et al., 2020).

For example, AI can predict drought conditions and recommend the best irrigation practices to conserve water without compromising crop health. This predictive ability is crucial for farmers in arid and semi-arid regions, where water scarcity is a pressing issue. AI models can also anticipate the impact of different irrigation practices on crop yields, helping farmers choose the most water-efficient strategies.

One of the most significant contributions of AI to PA is its ability to enhance predictive modeling and forecasting capabilities. These models are particularly essential for optimizing water requirements in agriculture, enabling farmers to make informed irrigation decisions based on accurate, data-driven forecasts. AI-driven predictive models integrate historical data related to weather patterns, soil moisture levels, and crop development to forecast future irrigation needs with high precision. This helps ensure that water is used efficiently, reducing waste while maintaining optimal crop health (Shahzad et al., 2020).

11.3.2.1 ML and Historical Data Integration

The backbone of AI's predictive modeling capabilities is ML, which allows systems to improve their predictions over time by learning from historical data. These models incorporate various datasets, such as past weather data, soil moisture trends, and crop-specific growth patterns, to predict future irrigation requirements. As new data is collected in real time, AI algorithms can refine their predictions, ensuring that irrigation schedules are continuously optimized based on both current conditions and anticipated future changes (Mulla, 2013).

For instance, AI systems can integrate real-time inputs with historical climate data to predict the likelihood of drought conditions or periods of heavy rainfall. These insights allow farmers to take proactive measures, such as adjusting irrigation practices or adopting deficit irrigation strategies to conserve water during droughts (Vico & Porporato, 2011). Moreover, by analyzing crop-specific water needs alongside weather predictions, AI systems can recommend the ideal water application rates to maximize yields without over-irrigation.

11.3.2.2 Drought Prediction and Water Conservation Strategies

AI's ability to predict drought conditions is particularly valuable in arid and semi-arid regions, where water scarcity poses a constant threat to agricultural productivity. In these areas, AI models use weather forecasts combined with soil moisture data to predict the onset of droughts, enabling farmers to adjust their irrigation strategies in advance. This helps avoid water waste during non-essential growth periods while ensuring that crops receive adequate water during critical growth stages.

For example, in areas where water resources are limited, AI-driven models can suggest alternative irrigation techniques such as deficit irrigation—a method that applies water only during key developmental phases of the crop, conserving water during less critical stages. These models are designed to maintain crop yields while significantly reducing the overall amount of water used (Fereres & Soriano, 2007). By predicting water stress and recommending appropriate irrigation strategies, AI helps farmers in water-scarce regions optimize their use of available resources while protecting their crops from drought-related damage.

11.3.2.3 Impact on Crop Yields and Water-Efficient Strategies

AI's predictive modeling capabilities are not limited to weather forecasting and drought prediction; they also extend to forecasting the impact of irrigation practices on crop yields. By analyzing the effects of different irrigation strategies, AI models can help farmers identify the most water-efficient practices that still yield optimal crop output. For instance, by simulating how various levels of water application affect crop growth, AI systems can recommend strategies that conserve water while maintaining or improving yields.

These models also incorporate evapotranspiration rates, which are critical for determining crop water needs. Evapotranspiration is the process by which water is transferred from the land to the atmosphere through evaporation from the soil and transpiration by plants. AI-driven models can predict changes in evapotranspiration rates based on current and forecasted weather conditions, ensuring that farmers apply the correct amount of water to compensate for any losses. This precision helps prevent over-irrigation, which can lead to nutrient leaching and soil degradation, as well as under-irrigation, which can cause water stress and reduced yields (Irmak, 2015).

11.3.2.4 Example: AI Predictive Models in Action

A practical example of AI-driven predictive modeling can be seen in California's vineyards, where water conservation is a high priority due to frequent droughts. These vineyards use AI-based models that integrate real-time data from soil moisture sensors and satellite imagery to predict water requirements. By forecasting periods of water stress and anticipating rainfall, these AI systems help vineyards implement deficit irrigation strategies, reducing water use by 25% while maintaining high grape yield and quality (González, 2017).

Similarly, in smallholder farms in India, AI-powered mobile applications are helping farmers manage water more efficiently. By incorporating local weather forecasts and real-time soil moisture data, these applications have enabled farmers to reduce water consumption by up to 30% while simultaneously improving crop yields, particularly in rice paddies where water management is crucial (González, 2017).

11.3.2.5 Continuous Learning and Adaptation

AI's ability to continuously learn from new data sets it apart from traditional irrigation management systems. As weather patterns and soil conditions evolve, AI models update themselves, ensuring that predictions and irrigation recommendations remain accurate and relevant. This dynamic learning process is especially important in the context of climate change, where unpredictable weather patterns and extreme events can challenge traditional water management practices. AI's capacity for continuous learning ensures that farmers can adapt to these changes, protecting their crops from water stress and ensuring sustainable water use in the long term.

11.3.2.6 Future Prospects and Challenges

The potential for AI-driven predictive modeling to revolutionize water management in agriculture is vast. However, several challenges remain, particularly in terms of data availability and access to technology. Smallholder farmers in developing regions may struggle with limited access to reliable weather data, real-time soil sensors, and advanced AI platforms. Additionally, the cost of implementing these systems can be prohibitive for many farmers, particularly in regions where infrastructure for data collection is underdeveloped (Shahzad et al., 2020).

Despite these challenges, the future of AI in agriculture looks promising. As sensor technology becomes more affordable and mobile applications continue to expand, more farmers will be able to benefit from AI's predictive modeling capabilities. This will be crucial for improving water use efficiency globally, particularly in the face of growing climate variability and water scarcity.

11.3.3 Automated Irrigation Systems

One of the most powerful applications of AI in water management is the automation of irrigation systems. AI-powered irrigation systems use real-time data to automatically control the timing, frequency, and amount of water applied to crops. These systems can adjust irrigation based on soil moisture levels, weather forecasts, and evapotranspiration rates, ensuring that crops receive just the right amount of water at the right time (Mulla, 2013).

Automated irrigation systems, such as smart drip irrigation and VRI, leverage AI to make real-time decisions about water distribution. This not only saves time and labor but also reduces water waste by ensuring that no water is applied when it is not needed. Moreover, automated systems can be integrated with mobile apps, allowing farmers to monitor and control irrigation remotely.

11.3.4 Water Stress Detection and Mitigation

AI plays a vital role in detecting water stress in crops before visible symptoms appear. Using computer vision and ML algorithms, AI systems can analyze images captured by drones or satellites to detect subtle changes in plant color, leaf temperature, and canopy structure, which are indicative of water stress (Bargoti & Underwood, 2017). By identifying water stress early, AI enables farmers to take corrective action—such as adjusting irrigation schedules or applying water-saving measures—before crop damage occurs.

This early detection is especially important in regions where water resources are limited, as it helps prevent crop loss and reduces the need for emergency irrigation, further enhancing water efficiency.

11.3.5 Optimizing Deficit Irrigation

AI also assists in the deficit irrigation strategy, where less water is applied to crops during non-critical growth stages to conserve water. AI models can identify the specific periods in a crop's growth cycle where water requirements are lower, allowing farmers to reduce irrigation without compromising yields (Vico & Porporato, 2011). By optimizing deficit irrigation schedules, AI helps farmers maintain productivity while using less water, contributing to long-term water sustainability.

In summary, AI's role in water efficiency is multifaceted, ranging from real-time data analysis and predictive modeling to the automation of irrigation systems and early detection of water stress. These AI-driven innovations allow farmers to optimize water use, conserve resources, and adapt to the challenges posed by climate change and water scarcity.

11.4 CASE STUDIES: AI AND WATER EFFICIENCY IN PA

11.4.1 California Vineyards

In the arid regions of California, where water scarcity is a persistent challenge, vineyards have increasingly turned to AI-driven irrigation systems to manage their water resources more efficiently. These systems leverage real-time data from soil moisture sensors, weather forecasts, and satellite imagery to optimize irrigation schedules based on the specific water needs of the vines. By automating the irrigation process, AI ensures that water is applied only when and where it is most needed, preventing over-irrigation and reducing water waste.

A case study from California's wine industry shows the significant impact of these technologies. Vineyards that implemented AI-driven irrigation systems reported a

25% reduction in water use without compromising the yield or quality of the grapes (González, 2017). The vineyards were able to maintain their high standards for grape production, with AI enabling them to monitor soil moisture and grape growth in real time, ensuring that water was applied only during critical growth stages.

This achievement highlights the potential of AI to enhance water efficiency in water-intensive crops such as grapes, offering a model for other agricultural sectors facing similar water scarcity issues. The success of AI in optimizing water use in California vineyards demonstrates that precision irrigation, powered by AI, can support sustainable agriculture practices while helping regions cope with limited water resources.

11.4.2 Smallholder Farmers in India

In India, where agriculture is predominantly dependent on monsoons and smallholder farmers manage a significant portion of the country's food production, AI-based mobile applications have been pivotal in transforming water management practices. Rice paddies, which are typically water-intensive, have been at the forefront of this innovation, particularly in regions with scarce water resources. By integrating real-time weather forecasts, soil moisture data, and other localized inputs, AI-driven mobile applications offer actionable insights to farmers, allowing them to make informed decisions about irrigation.

Smallholder farmers have reported a 30% reduction in water consumption through the use of these AI-enabled tools, which guide them on the best times to irrigate based on soil conditions and anticipated rainfall (González, 2017). These applications also help optimize the water usage across different stages of rice growth, ensuring that farmers do not over-irrigate during non-critical periods, thus conserving water.

In addition to reducing water use, these technologies have contributed to increased crop yields, further highlighting the benefits of precision water management. For farmers with limited access to irrigation infrastructure, the AI-driven insights are proving invaluable, helping them manage scarce water resources more effectively while boosting productivity. The adoption of such technologies is particularly important in rural India, where smallholder farmers face significant challenges due to fluctuating water availability and increasing climate variability.

11.5 CHALLENGES AND FUTURE DIRECTIONS

Despite the clear benefits of AI-driven PA in improving water efficiency, several challenges hinder the widespread adoption of these technologies, particularly in developing regions. Addressing these challenges and identifying future directions are essential for realizing the full potential of AI in sustainable water management.

11.5.1 High Initial Costs and Infrastructure Requirements

One of the primary challenges facing the adoption of AI and PA technologies is the high initial investment required. Installing soil sensors, satellite imagery systems, and AI-driven irrigation tools can be prohibitively expensive, particularly for smallholder farmers in developing countries. Additionally, these technologies often require reliable infrastructure such as high-speed internet, electricity, and data storage, which may not be readily available in rural and remote areas (Van Evert et al., 2017).

To overcome these barriers, governments and international organizations need to provide subsidies, low-interest loans, and infrastructure support to farmers, especially in water-scarce regions. Public-private partnerships can also play a key role in reducing costs and facilitating access to these technologies.

11.5.2 Lack of Technical Knowledge and Training

Many farmers, particularly those in developing regions, lack the technical knowledge required to operate AI-based systems and interpret the data they generate. This knowledge gap can prevent effective use of PA technologies, even when they are available. For instance, understanding how to adjust irrigation schedules based on soil moisture data or predicting crop water needs using AI algorithms may be unfamiliar to farmers who have traditionally relied on conventional methods of farming (Jones, 2004).

To bridge this gap, there is a need for capacity-building programs that provide training to farmers on the use of AI-based tools. Extension services, agricultural cooperatives, and non-governmental organizations (NGOs) can help deliver these training programs, ensuring that farmers are well-equipped to use AI tools effectively.

11.5.3 Data Privacy and Security Concerns

As PA relies heavily on data collection and analysis, concerns over data privacy and security are increasingly relevant. Farmers may be reluctant to adopt AI technologies if they are unsure of how their data will be used or who will have access to it. Additionally, the risk of cyberattacks or data breaches poses a significant threat to the reliability and security of AI-driven agricultural systems (Bronson & Knezevic, 2016).

To address these concerns, clear policies and regulations on data ownership, privacy, and security need to be established. These policies should define who owns the data collected by AI systems and outline the mechanisms in place to protect this data from misuse or unauthorized access.

11.5.4 Climate Change and Unpredictable Weather Patterns

While AI can enhance water efficiency by predicting irrigation needs and optimizing water use, the unpredictability of weather patterns due to climate change poses a significant challenge. In regions where extreme weather events such as droughts and floods are becoming more frequent, AI models based on historical data may be less effective in forecasting future conditions (Mekonnen & Hoekstra, 2016). This makes it difficult for farmers to rely solely on AI-driven models for water management, especially during periods of climate uncertainty.

To improve the resilience of AI systems to climate variability, future models should integrate climate change projections and real-time monitoring to adapt to rapidly changing environmental conditions. Incorporating ML algorithms that can adjust and evolve in response to new data will also be crucial in addressing the impacts of climate change on agricultural water management.

11.5.5 Policy and Regulatory Barriers

The widespread implementation of AI-driven PA requires supportive policies and regulatory frameworks that encourage innovation while protecting smallholder farmers. In many regions, agricultural policies may not fully accommodate the use of advanced technologies such as AI and IoT (Internet of Things). This can limit the availability of resources and incentives needed for farmers to adopt these tools (Gebbers & Adamchuk, 2010).

Governments and policymakers need to develop frameworks that promote the adoption of AI in agriculture, including tax incentives, funding for research and development, and support for infrastructure development. International collaboration on data sharing and AI research in agriculture will also be key to addressing global challenges such as water scarcity and food security.

11.5.5.1 Future Directions

To ensure that AI-driven PA continues to contribute to water efficiency and sustainability, several future directions should be pursued:

11.5.6 Integration of AI with Other Emerging Technologies

The integration of AI with other emerging technologies, such as blockchain and big data analytics, could enhance the transparency, traceability, and efficiency of water use in agriculture. Blockchain, for instance, can provide a secure and transparent system for tracking water usage, helping to verify that sustainable practices are being followed (Tan et al., 2021). Big data analytics can further improve the predictive accuracy of AI models by incorporating larger datasets from diverse sources, including historical climate data and real-time environmental sensors.

11.5.7 Focus on Smallholder Farmers and Marginalized Communities

While large-scale farms have been quick to adopt AI-driven systems, there needs to be a stronger focus on smallholder farmers and marginalized communities, who are most vulnerable to water scarcity and climate change. Developing low-cost AI solutions and making them accessible to these communities is critical for achieving global water efficiency and sustainability goals. Customized solutions that address the specific needs of small-scale farmers, such as mobile apps and portable soil sensors, will play a key role in this effort.

11.5.8 Enhanced Public-Private Collaboration

Public and private sector collaboration is essential to scaling up the implementation of AI in water management. Public funding for research and private investment in technology development can accelerate the creation of affordable, effective solutions. Partnerships between tech companies, agricultural organizations, and governments can also help disseminate AI technologies to farmers, particularly in developing countries.

11.5.9 Emphasis on Environmental Sustainability and Net-Zero Emissions

Finally, as the agricultural sector strives to achieve net-zero emissions, the role of AI in optimizing water use and reducing greenhouse gas emissions will become increasingly important. AI technologies can help minimize the environmental footprint of agriculture by reducing water waste, promoting regenerative agricultural practices, and optimizing the use of fertilizers and pesticides.

11.6 CONCLUSION

The integration of AI into PA is revolutionizing the way water resources are managed in agriculture. By optimizing water use, enhancing crop productivity, and contributing to net-zero emission targets, AI-driven technologies are paving the way for more sustainable and efficient agricultural practices. These technologies, such as predictive analytics, automated irrigation systems, and climate forecasting models, are enabling farmers to conserve water while maintaining or even increasing yields. For instance, AI applications in irrigation scheduling, evapotranspiration monitoring, and deficit irrigation are helping farmers apply water more precisely, avoiding waste and maximizing the efficiency of water use (Fereres & Soriano, 2007; Vico & Porporato, 2011).

Despite the clear benefits, challenges remain in the widespread adoption of AI technologies in agriculture. Issues such as data access, high initial costs, lack of technical knowledge, and climate variability continue to hinder the full-scale implementation of these solutions, particularly for smallholder farmers and in regions with limited infrastructure (Van Evert et al., 2017). Nevertheless, with ongoing advancements in AI and increased focus on capacity-building programs and public-private partnerships, the future of AI in agriculture is promising.

AI's role in improving water efficiency directly addresses the pressing issue of water scarcity, which is exacerbated by population growth, climate change, and increasing food demand (Mekonnen & Hoekstra, 2016). Additionally, by reducing water waste and optimizing resource use, AI technologies support broader goals of sustainable agriculture and environmental protection. As agriculture continues to evolve, AI-driven solutions will undoubtedly play a key role in securing global food systems while promoting sustainable water management practices.

11.7 DECLARATION OF GENERATIVE AI IN SCIENTIFIC WRITING

This book chapter, *Precision Agriculture and Water Efficiency*, has been written and edited by the authors with significant intellectual contributions. Generative AI tools, including ChatGPT and other AI-based language models, were utilized in a limited capacity to enhance the clarity, coherence, and grammatical accuracy of the content.

However, all scientific interpretations, conceptual frameworks, data analysis, and conclusions are the original work of the authors.

The authors take full responsibility for the integrity and authenticity of the content presented in this chapter. Any AI-generated text has been critically reviewed and revised to ensure compliance with academic and ethical standards in scientific writing.

REFERENCES

Bargoti, S., & Underwood, J. P. (2017). Image segmentation for fruit detection and yield estimation in apple orchards. *Journal of Field Robotics*, 34(6), 1039–1060.

Bronson, K., & Knezevic, I. (2016). Big data in food and agriculture. *Big Data & Society*, 3(1), 1–5.

FAO. (2020). *Water Use in Agriculture*. Retrieved from FAO website. https://openknowledge.fao.org/server/api/core/bitstreams/6e2d2772-5976-4671-9e2a-0b2ad87cb646/content

Fereres, E., & Soriano, M. A. (2007). Deficit irrigation for reducing agricultural water use. *Journal of Experimental Botany*, 58(2), 147–159.

Gebbers, R., & Adamchuk, V. I. (2010). Precision agriculture and food security. *Science*, 327(5967), 828–831.

González, A. (2017). AI-driven irrigation management in vineyards. *California Water Journal*, 34(3), 102–112.

IPCC. (2019). *Climate Change and Land*. Intergovernmental Panel on Climate Change, Geneva, Switzerland.

Irmak, S. (2015a). *Evapotranspiration and Crop Water Use*. Nebraska Extension Publications, Lincoln, Nebraska.

Irmak, S. (2015b). Evapotranspiration: Potential or reference. *Irrigation Science*, 33(1), 1–18.

Jones, H. G. (2004). Irrigation scheduling: Advantages and pitfalls of plant-based methods. *Journal of Experimental Botany*, 55(407), 2427–2436.

Mekonnen, M. M., & Hoekstra, A. Y. (2016). Four billion people facing severe water scarcity. *Science Advances*, 2(2), e1500323.

Mulla, D. J. (2013). Twenty-five years of remote sensing in precision agriculture: Key advances and remaining knowledge gaps. *Biosystems Engineering*, 114(4), 358–371.

Shahzad, B., Tanveer, M., Rehman, A., et al. (2020). Drought-tolerant crop varieties and yield under drought stress. *Agricultural Water Management*, 236, 106143.

Tan, B., Sundarakani, B., & Mohamed, N. (2021). Sustainable food supply chain management: Blockchain enhances transparency and traceability. *Sustainability*, 13(2), 635.

Van Evert, F. K., Fountas, S., Jakovetic, D., Crnojevic, V., Travlos, I., & Kempenaar, C. (2017). Big data for weed control and crop protection. *Weed Research*, 57(5), 218–233.

Vico, G., & Porporato, A. (2011). From rainfed agriculture to stress-avoidance irrigation: II. Sustainability, crop yield, and profitability. *Agricultural Water Management*, 98(5), 779–789.

Zhang, N., Wang, M., & Bai, Y. (2019a). Precision agriculture: A worldwide overview. *Computers and Electronics in Agriculture*, 36(2), 113–132.

Zhang, Q., Wang, C., & Bai, G. (2019b). Artificial intelligence in precision agriculture: A review. Journal of Agricultural and Food Chemistry, 67(17), 4791–4804.

Part III

Climate Action and Conservation

12 Aquatic Ecosystem Preservation Using Artificial Intelligence

Doaa A. El-Emam and Menna A. Shata

12.1 INTRODUCTION

The preservation of aquatic ecosystems represents a significant challenge that necessitates immediate action, particularly in light of the growing threats from anthropogenic activities and climate change (Jenny et al., 2020; Wang and Gu, 2021). Aquatic ecosystems, which include both marine environments (like oceans and coral reefs) and freshwater systems (like rivers, lakes, and wetlands), are very important for keeping the environment in balance, protecting biodiversity, and giving people the resources they need to stay alive (Marimuthu et al., 2020). A diverse array of species, many of which remain undocumented, inhabit these ecosystems and play a crucial role in global processes like carbon cycling, water purification, and climate regulation. However, various factors increasingly jeopardize these ecosystems, including pollution stemming from industrial and agricultural runoff, climate change-induced alterations in temperature and weather patterns, and overfishing, which results in the depletion of fish populations and disruption of food webs. These issues must be addressed to preserve aquatic ecosystems and the wildlife and humans that depend on them (Prakash, 2021; Ahmed et al., 2022).

In light of the challenges facing aquatic environments, the integration of advanced technologies, particularly artificial intelligence (AI), into conservation efforts represents a promising strategy for safeguarding these ecosystems (Bakker, 2022). AI encompasses a variety of technologies, including machine learning, neural networks, and data analytics, which enable the processing of extensive datasets with speed and accuracy. By harnessing these technologies, researchers and conservationists can enhance their understanding of aquatic ecosystems, more effectively identify potential threats, and implement targeted conservation strategies. AI is significant in this context because it has the potential to transform ecosystem management from manual data collection and analysis to more effective, data-driven methods that provide real-time information about ecosystem health and offer promising strategies for the preservation of aquatic ecosystems through enhanced data collection, improved resource management, and the creation of early warning systems (Bibri et al., 2024).

The rapid advancement of AI presents significant opportunities for transforming various fields, particularly in the realm of scientific research. As global challenges become increasingly intricate, integrating AI technologies can enhance our

DOI: 10.1201/9781003547099-15

understanding, analysis, and management of diverse scientific domains. From an ecological perspective, AI has the potential to revolutionize the processing and interpretation of extensive datasets commonly encountered in scientific research. This feature helps people make better choices about data-driven hypotheses, experimental designs, and conservation methods. This is especially important in fields like ecology where protecting biodiversity and the environment is very important. Economically, the effectiveness and cost-efficiency of AI applications in research are imperative factors influencing their acceptance among researchers, funding bodies, and policymakers. By elucidating the economic advantages of AI in scientific inquiry, we can encourage broader integration of these technologies, thereby optimizing resource allocation and enhancing research productivity. Socially, AI can empower research communities by providing accessible tools that improve local engagement in scientific endeavors. By

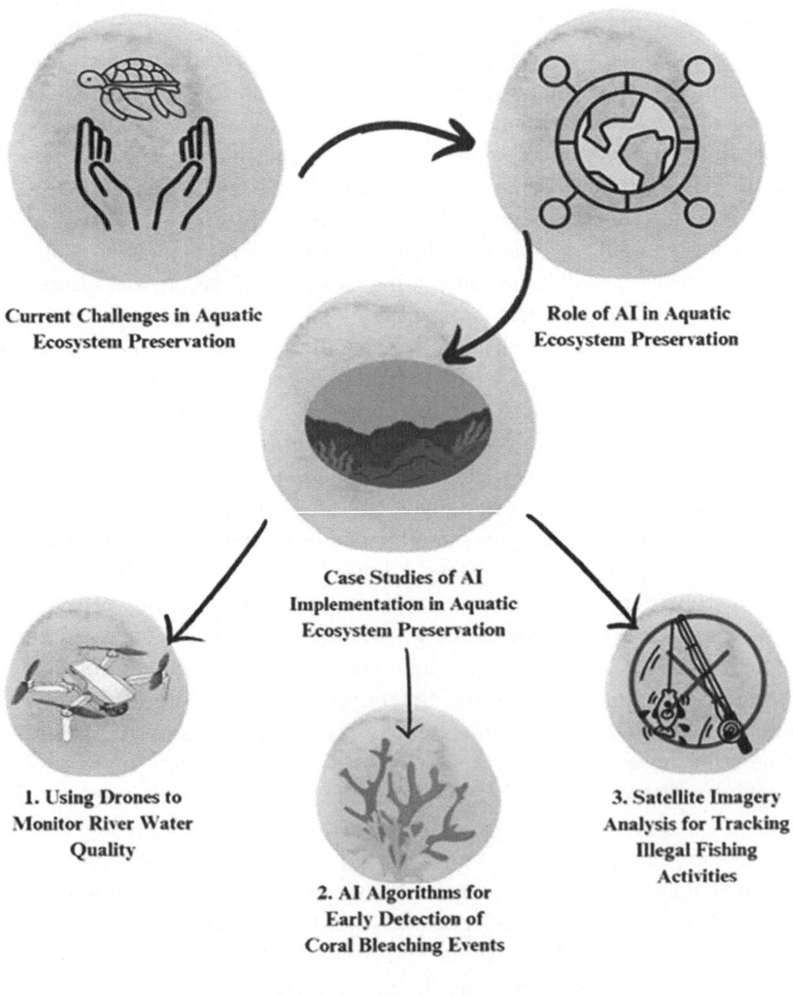

FIGURE 12.1 Aquatic ecosystem preservation using artificial intelligence.

democratizing sophisticated analytical techniques, AI enables researchers and communities to collaboratively advance knowledge and address pressing environmental challenges, fostering a greater sense of shared responsibility toward scientific stewardship and innovation (Vishnoi and Goel, 2024).

The integration of AI into environmental science represents a significant advancement in our efforts to understand and conserve biodiversity (Adanma and Ogunbiyi, 2024). This chapter aims to explore the multifaceted nature of AI, elucidating its mechanisms and applications in the context of ecological systems (Figure 12.1). We will examine various viewpoints concerning the impact of AI on biodiversity and sustainability, supported by case studies that illustrate its effectiveness. By highlighting these examples, we seek to underscore AI's transformative potential in creating a more sustainable and resilient future for our aquatic ecosystems. As we navigate the complexities of contemporary environmental challenges, understanding the role of AI becomes essential for fostering innovative solutions that promote ecological health.

12.2 CURRENT CHALLENGES IN AQUATIC ECOSYSTEM PRESERVATION

Overfishing, pollution, habitat destruction, and climate change are some of the major challenges facing aquatic ecosystems today. These threats not only endanger marine life but also have far-reaching consequences for human health and livelihoods. The loss of aquatic biodiversity can disrupt food chains and ecosystems, leading to decreased fish populations and economic instability for coastal communities. It is crucial to address these challenges in order to protect the delicate balance of aquatic ecosystems (Ogidi and Akpan, 2022). Overfishing leads to depletion of fish stocks, which can have a ripple effect on the entire ecosystem. Pollution, whether from plastic waste or chemical runoff, can damage marine habitats and poison marine life. Habitat destruction, such as the destruction of coral reefs or mangrove forests, can result in the loss of crucial breeding grounds for many species. Climate change exacerbates all of these issues, causing ocean acidification, rising sea levels, and extreme weather events that further threaten aquatic ecosystems (Prakash, 2021). By implementing sustainable fishing practices, reducing pollution, protecting habitats, and mitigating climate change, we can work toward preserving these vital ecosystems for future generations.

12.3 POLLUTION FROM VARIOUS SOURCES

Pollution of aquatic ecosystems significantly threatens marine life and habitats, originating from various sources, including agricultural runoff, oil spills, plastic waste, heavy metals, and wastewater discharge (Figure 12.2). Agricultural runoff introduces nutrients and chemicals from fertilizers and pesticides into water bodies, leading to nutrient pollution that can cause harmful algal blooms, deplete oxygen, and disrupt marine food webs, ultimately harming fish and other marine organisms (Häder et al., 2020).

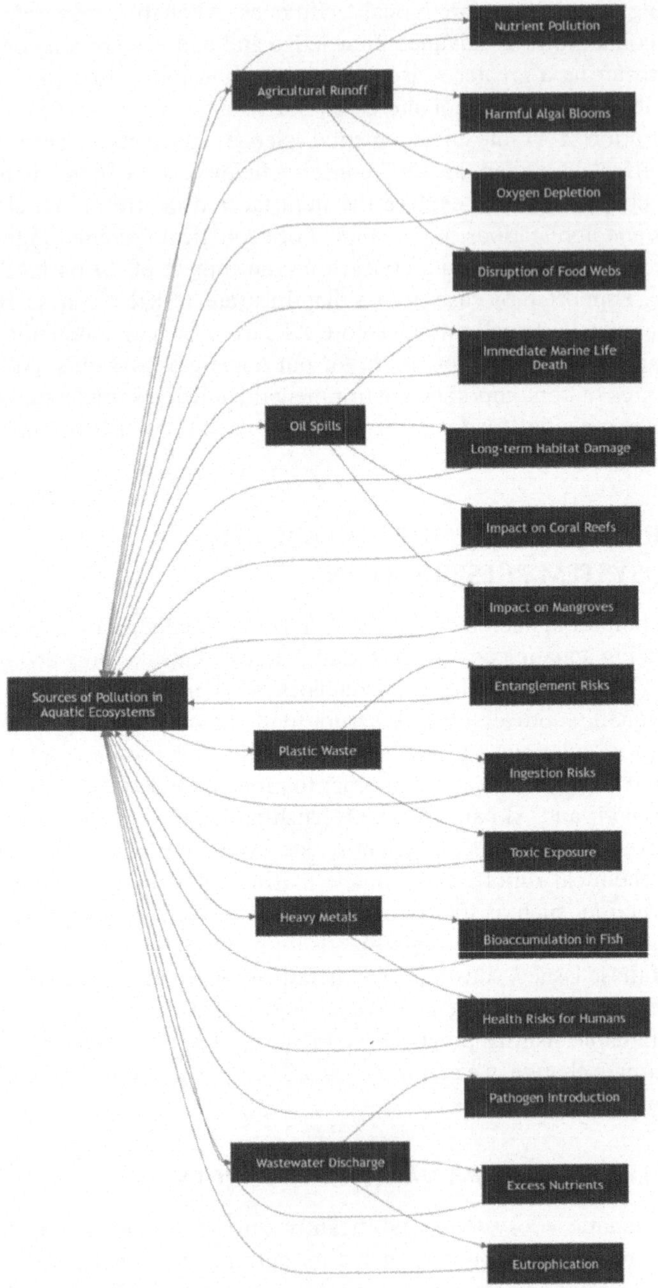

FIGURE 12.2 Sources of pollution in aquatic ecosystems.

Oil spills can have immediate catastrophic effects, killing marine life such as fish, seabirds, and mammals while also causing long-term habitat damage to critical areas like coral reefs and mangroves, which serve as nurseries for many species (Ogidi and Akpan, 2022; Thakur and Koul, 2022).

Plastic waste poses a severe risk through entanglement and ingestion, with marine animals suffering from injuries, suffocation, and toxic exposure from chemicals leached from plastics. Additionally, industrial activities contribute heavy metals like mercury and lead into aquatic systems, leading to bioaccumulation in fish and subsequent health risks for marine life and humans alike. Also, wastewater that hasn't been treated can bring pathogens and too many nutrients into waterways, which makes eutrophication worse and creates conditions that are bad for marine ecosystems. To combat these issues, it is essential to implement effective regulations, promote sustainable agricultural practices, increase public awareness about plastic pollution, and engage in cleanup efforts to restore the health of our oceans. Addressing pollution in aquatic ecosystems through a collaborative approach is crucial for safeguarding marine biodiversity and ensuring cleaner, healthier oceans and rivers for future generations (Isangedighi et al., 2020; Abdein and El-Emam, 2024).

12.4 CLIMATE CHANGE AND ITS IMPACTS ON AQUATIC ECOSYSTEMS

Climate change presents substantial threats to aquatic ecosystems through a variety of interconnected mechanisms (Figure 12.3). The increase in water temperatures results in altered species distributions, compelling many organisms to migrate toward cooler waters, thereby disrupting existing ecosystems and food webs (Viitasalo and Bonsdorff, 2022; Elemam and Eldeeb, 2023). Elevated metabolic rates in aquatic species can have detrimental effects on their growth, reproduction, and survival, particularly among organisms sensitive to temperature changes. Additionally, the oceans' uptake of excess carbon dioxide results in ocean acidification, negatively impacting calcifying organisms such as corals and mollusks (Gomez-Zavaglia et al., 2020). This phenomenon poses a significant threat to coral reefs by compromising their calcium carbonate structures, rendering them more susceptible to bleaching and physical damage, which ultimately undermines the biodiversity dependent on these critical habitats. The continued rise in sea levels further exacerbates the challenges faced by coastal ecosystems, inundating areas such as mangroves and salt marshes, leading to habitat loss for numerous fish species. Increasing salinity levels in freshwater systems can also significantly affect the viability of native species (Letcher, 2021). Furthermore, alterations in rainfall patterns can influence runoff, contributing to increased pollutants, sediments, and nutrients in aquatic environments, which exacerbate issues such as algal blooms. Conversely, reduced rainfall can diminish water levels, impacting the availability of freshwater and altering the habitats of various aquatic organisms. These environmental changes disrupt food webs by shifting primary productivity and modifying predator-prey interactions, often resulting in imbalances that threaten the overall health of ecosystems (Gomez-Zavaglia et al., 2020; Viitasalo and Bonsdorff, 2022). Additionally, the growing frequency of extreme weather events, including intense storms and flooding, can devastate aquatic habitats, introducing pollutants and altering sedimentation and nutrient distribution patterns. To effectively address the impacts of climate change on aquatic ecosystems, comprehensive strategies for both mitigation and adaptation are required. These strategies should prioritize habitat restoration, pollution reduction, and the implementation of sustainable long-term management practices. Moreover, enhancing research

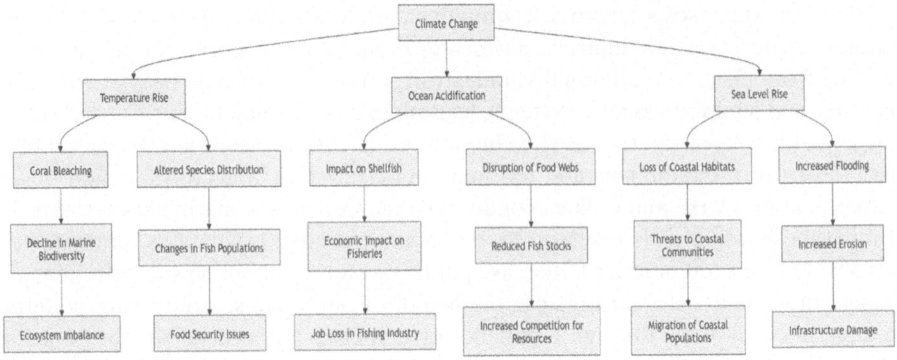

FIGURE 12.3 Interconnected mechanisms of climate change threatening aquatic ecosystems.

and monitoring efforts is crucial for developing effective conservation strategies that promote the resilience of aquatic ecosystems in the face of ongoing climatic changes.

12.4.1 Overfishing and Habitat Destruction

Overfishing and habitat destruction present critical challenges that threaten the health and sustainability of aquatic ecosystems. The sustainability of aquatic ecosystems is increasingly endangered by overfishing and habitat destruction. Overfishing results in the rapid depletion of fish populations, which disrupts food webs and threatens species that rely on fish as their primary source of nutrition (Letcher, 2021; Ahmed et al., 2022). This phenomenon not only affects marine biodiversity but also has significant economic implications for communities dependent on fishing for their livelihoods, potentially leading to food insecurity for vulnerable populations (Marques and Marques, 2020; Letcher, 2021).

Overfishing has profound ecological consequences, primarily through the phenomenon known as bycatch, which refers to the incidental capture of non-target species during fishing operations. Bycatch contributes to further population declines and exacerbates the risk of extinction for numerous marine species. Consequently, the depletion of target fish species can precipitate a cascade of detrimental effects throughout the marine food web, leading to altered species compositions and reduced biodiversity.

Simultaneously, habitat destruction driven by coastal development, pollution, climate change, and the introduction of invasive species significantly undermines the integrity of critical marine environments, such as coral reefs, mangroves, and wetlands (Santojanni et al., 2023). These habitats serve as vital shelter and breeding grounds for a multitude of marine organisms. Pollution resulting from agricultural runoff and plastic waste further degrades these ecosystems, impairing their ecological functions.

The interplay between overfishing and habitat destruction creates a self-reinforcing cycle of ecological degradation. Declining fish populations not only threaten the ecological balance but can also lead to the deterioration of their habitats, which further exacerbates the decline of fish stocks. This cycle underscores the complexity of addressing these issues in isolation.

Mitigating the effects of overfishing and habitat destruction necessitates a comprehensive approach. Key strategies include the implementation of sustainable fishing practices, habitat restoration initiatives, and the establishment of marine protected areas (MPAs) (Grorud-Colvert et al., 2021).

Addressing the intertwined challenges of overfishing and habitat destruction requires coordinated action from governments, non-governmental organizations (NGOs), scientists, and local communities. By advancing interdisciplinary approaches and collaborative efforts, it is possible to promote the long-term health and resilience of aquatic ecosystems.

12.5 ROLE OF AI IN AQUATIC ECOSYSTEM PRESERVATION

AI is crucial in preserving aquatic ecosystems through various applications (Figure 12.4). It enhances monitoring and data collection with remote sensing and Internet of Things (IoT) sensor networks, allowing real-time assessments of water quality and aquatic health. AI algorithms analyze satellite images to track changes like algal blooms and habitat degradation. In biodiversity conservation, AI helps identify aquatic species and simulates ecosystem dynamics to predict the impacts of climate change. Additionally, AI aids in pollution management by forecasting water quality trends and identifying pollution sources, while in fisheries management, it forecasts catches to prevent overfishing and detects illegal fishing. AI also engages the public through citizen science initiatives and supports habitat restoration efforts. Overall, AI significantly improves the management and sustainability of aquatic ecosystems (Wang, 2024)

12.5.1 Monitoring Water Quality

AI plays a critical and transformative role in the preservation of aquatic ecosystems, particularly in the monitoring of water quality. The integration of AI with sensors and IoT technology enables real-time data collection on key water quality parameters such as temperature, pH, turbidity, dissolved oxygen, and pollutant levels. Drones and autonomous underwater vehicles outfitted with advanced sensors are capable of accessing remote and difficult-to-reach areas, thereby providing comprehensive insights across extensive aquatic environments. AI algorithms, especially those based on machine learning, facilitate the rapid analysis of large datasets, allowing for the identification of patterns and trends in water quality over time (Divine et al., 2024). These algorithms also aid in predictive modeling, which forecasts future conditions based on historical data and environmental factors. Such capabilities are vital for the early detection of pollution incidents, as AI can recognize anomalies in water quality data and trigger automated alerts for environmental agencies when levels fall below established thresholds. Moreover, AI contributes to the assessment of ecosystem health by monitoring biodiversity through the analysis of imagery and data gathered from underwater cameras (Yang et al., 2022). This information is essential

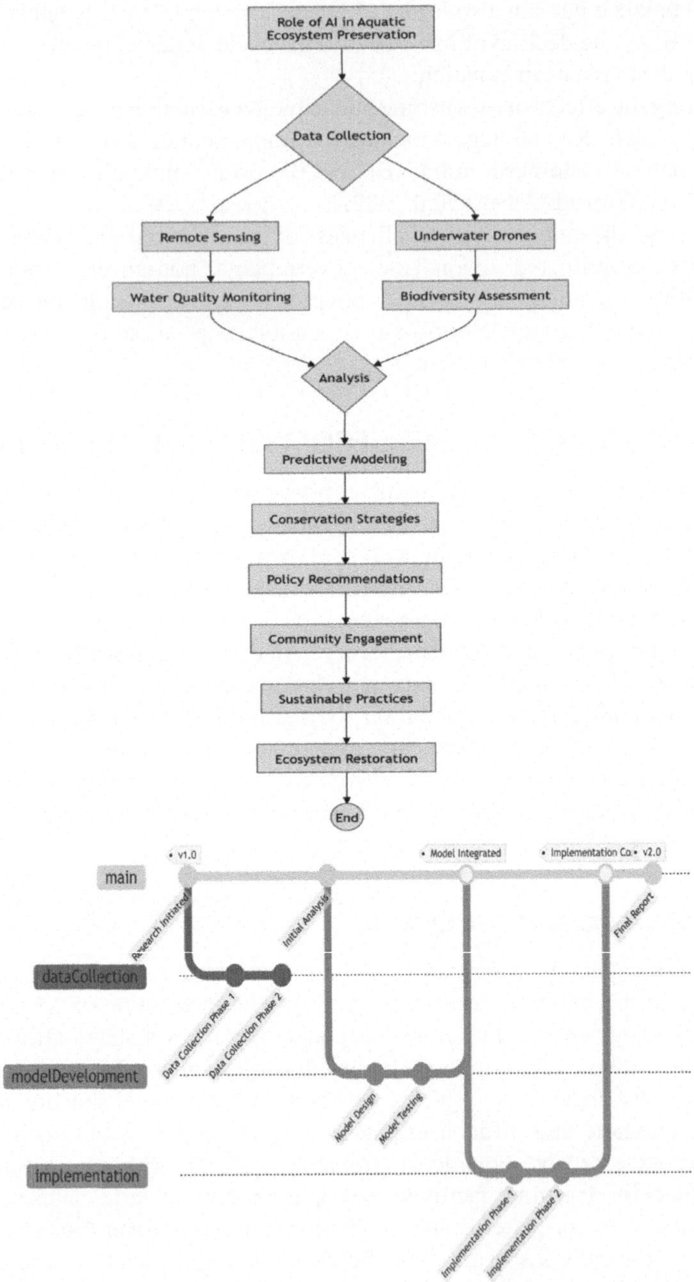

FIGURE 12.4 Application of AI in aquatic ecosystem preservation.

for guiding informed conservation decisions. AI can also model species distribution concerning water quality and environmental conditions, thereby enhancing targeted preservation efforts. The ability to integrate diverse data sources such as satellite imagery, meteorological data, and ground-based sensor readings provides a comprehensive view of aquatic system health (Divine et al., 2024).

AI-powered visualization tools create interactive maps and dashboards that enhance stakeholder comprehension of changes in water quality over time. Additionally, mobile applications promote community engagement by enabling citizens to report water quality issues, thus encouraging local participation in conservation initiatives and raising public awareness regarding environmental challenges. Furthermore, AI supports policymakers and environmental managers by furnishing data-driven insights and recommendations for effective water quality management. The automation of data analysis and reporting significantly reduces the costs typically associated with traditional monitoring methodologies, thereby facilitating the involvement of smaller organizations and communities in preservation activities (Rao et al., 2024). So, the integration of AI in the monitoring of water quality is revolutionizing the approach to aquatic ecosystem preservation by enabling timely responses to ecological changes and supporting sustainable management practices.

12.6 PREDICTING AND MITIGATING THREATS TO ECOSYSTEMS

The use of AI and advanced data analytics has greatly enhanced the creation of early warning systems aimed at detecting potential ecological crises (Figure 12.5). These systems rely on real-time data collection and analysis to monitor various environmental indicators, enabling stakeholders, such as environmental agencies, policymakers, and conservation organizations, to be alerted to emerging threats. This

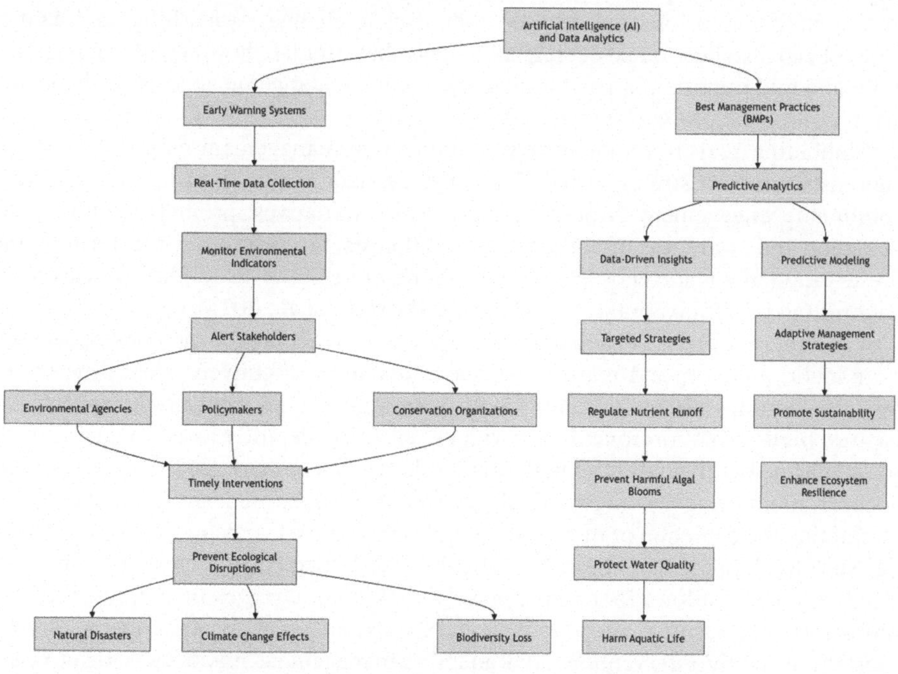

FIGURE 12.5 Leveraging AI to anticipate and address ecosystem threats.

proactive approach allows for timely interventions, which can be crucial in preventing or mitigating the impact of ecological disruptions, such as natural disasters, the effects of climate change, or biodiversity loss (Ncube and Ngulube, 2024).

To effectively reduce risks to ecosystems, it is essential to implement best management practices (BMPs) grounded in predictive analytics. By utilizing data-driven insights, stakeholders can deploy targeted strategies that tackle specific environmental challenges. For instance, regulating nutrient runoff from agricultural activities is vital in preventing harmful algal blooms, which can degrade water quality and harm aquatic life. Through the integration of predictive modeling and adaptive management strategies, we can better protect ecosystems, ultimately promoting sustainability and resilience in the face of human-induced pressures (Mumtaz et al., 2022; Mutsuddi, 2025).

12.6.1 Enhancing Conservation Efforts through Data Analysis and Modeling

Enhancing conservation efforts in aquatic ecosystems through advanced data analysis and modeling represents a significant advancement in environmental science. AI automates the collection and management of environmental data by using drones and underwater robots to monitor critical variables such as water quality, temperature, and biodiversity (Figure 12.6). It also employs satellite imagery to track ecosystem changes, identifying urgent issues like algal blooms and habitat degradation (Ditria et al., 2022). AI's ability to quickly process large datasets allows machine learning algorithms to detect patterns that may elude human analysts. This facilitates predictive analytics to forecast impacts from climate change, overfishing, and habitat loss. AI can also help make ecological models that simulate how aquatic ecosystems work and find important areas for intervention and endangered species (Isabelle and Westerlund, 2022; Silvestro et al., 2022)

Combining real-time monitoring with adaptive management empowers conservationists to adjust strategies based on ongoing data analysis. AI further enhances community engagement through citizen science initiatives, promoting public participation and awareness of conservation challenges. Overall, AI is transforming the preservation of aquatic ecosystems by enabling more effective and informed conservation strategies (Shivaprakash et al., 2022; Dwivedi et al., 2024).

The integration of AI in aquatic conservation is advancing the methodologies used to monitor, analyze, and manage marine ecosystems effectively. Real-time monitoring is significantly enhanced through AI-powered underwater drones, which can capture high-resolution images and videos, enabling detailed assessments of coral reefs, marine biodiversity, and pollution levels (Hemanth et al., 2024). Concurrently, acoustic monitoring leverages AI algorithms to analyze underwater soundscapes, facilitating the detection of marine mammals, fish species, and anthropogenic activities such as shipping and commercial fishing. Additionally, AI's capacity to process satellite imagery allows for comprehensive tracking of changes in water quality, the proliferation of harmful algal blooms, and the dynamics of coastal erosion, thereby supporting timely interventions and adaptive management practices (Dogan et al., 2024; Hemanth et al., 2024).

AI also significantly contributes to species identification and tracking, with image and video analysis automating the classification of marine species and streamlining biodiversity assessments. In the realm of bioacoustic analysis, AI technologies enhance the tracking and identification of marine organisms based on their acoustic signatures, providing insights into species distribution and behavioral patterns critical for conservation strategies. The capability for individual identification of marine species through unique markings, patterns, or behaviors further supports long-term monitoring initiatives and population dynamics research (Ditria et al., 2022; Dogan et al., 2024).

In terms of habitat mapping and restoration, AI models predict habitat distribution and quality based on various environmental parameters, thereby informing restoration efforts and conservation planning. AI-powered sonar technologies facilitate the creation of detailed seafloor maps that identify important habitats and potential anthropogenic threats, which is crucial for guiding conservation efforts in vulnerable marine environments. Moreover, AI plays a pivotal role in assessing the impacts of climate change on marine ecosystems. Climate modeling through AI simulations allows researchers to project the effects of rising temperatures, ocean acidification, and changing salinity, enabling the design of adaptive management strategies. Predictive modeling further empowers conservationists to anticipate future changes within these ecosystems, facilitating proactive conservation measures. AI also enhances the detection and prevention of illegal fishing activities. By analyzing vessel movements through advanced tracking algorithms, AI can identify suspicious behaviors indicative of illegal fishing practices. Additionally, AI-based image analysis scrutinizes satellite and aerial imagery to detect unauthorized fishing gear and activities, thus enhancing enforcement capabilities in MPAs (Ayoola et al., 2024; Ullah et al., 2024)

It also plays a significant role in environmental conservation through habitat modeling and seafloor mapping. By analyzing environmental variables, AI can predict

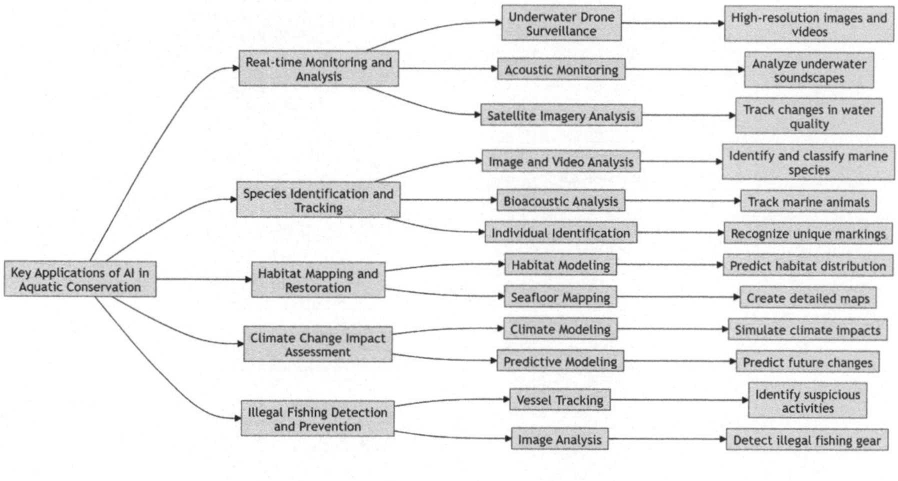

FIGURE 12.6 Key applications of AI in aquatic conservation.

habitat distribution and quality, which is essential for effective habitat restoration and conservation planning. Researchers have developed machine learning models that consider factors like water quality, substrate type, and historical data to identify areas suitable for restoration. These detailed maps aid in designing MPAs and conducting environmental impact assessments (Yang et al., 2023; Dwivedi et al., 2024).

These AI applications contribute to more informed and effective strategies for preserving and restoring marine environments. Continued research and development in this area holds promise for further advancements in the management of marine resources and biodiversity preservation.

12.6.2 Challenges and Ethical Considerations in AI Applications for Ecosystem Preservation

The application of AI in aquatic ecosystem preservation presents numerous opportunities, but it also raises several significant challenges and ethical considerations (Figure 12.7). Addressing these issues is critical to ensuring the responsible and equitable implementation of AI technologies in conservation efforts.

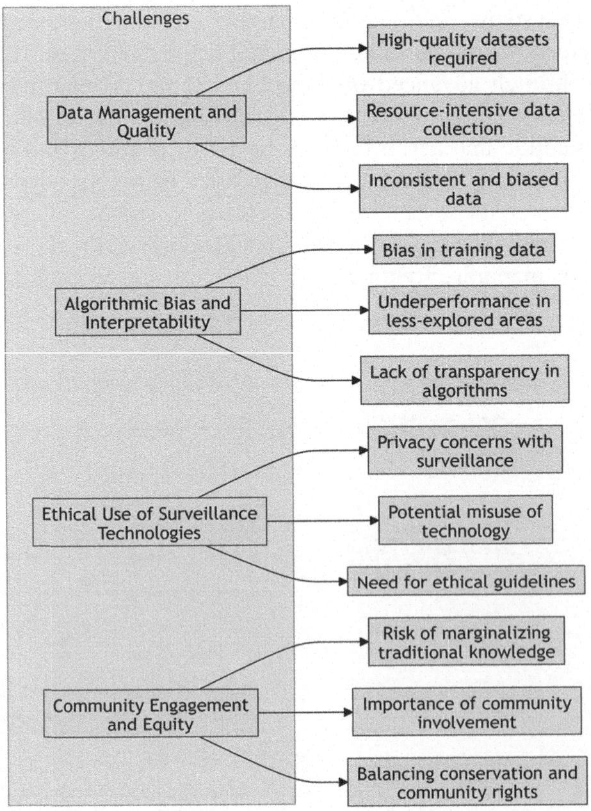

FIGURE 12.7 AI challenges in aquatic ecosystem preservation.

12.6.2.1 Data Management and Quality

One of the primary challenges in utilizing AI for aquatic ecosystem preservation is the management of large volumes of environmental data. Effective AI systems rely on high-quality, accurately labeled datasets to train algorithms (Nishant et al., 2020). However, collecting and curating such datasets can be resource-intensive and logistically complex. In many cases, data may be incomplete, biased, or inconsistent, which can lead to inaccurate predictions and misinformed conservation strategies. Ensuring rigorous data management practices, including standardized data collection methods and validation processes, is essential to improve the reliability of AI applications in this field (Alloghani, 2023).

12.6.2.2 Algorithmic Bias and Interpretability

AI algorithms may inherit biases contained in their training data, resulting in skewed outcomes that can disproportionately impact certain species or habitats. For instance, training an AI model primarily on data from well-studied regions may lead to underperformance in less-explored areas, potentially misrepresenting biodiversity or habitat conditions. Additionally, many AI models, especially deep learning algorithms, function as "black boxes," making it difficult for researchers and practitioners to understand the basis for their decisions. This lack of transparency can undermine trust in AI-generated recommendations and create challenges for regulatory compliance and accountability (Nishant et al., 2020; Gambín et al., 2021; Rane et al., 2024).

12.6.2.3 Ethical Use of Surveillance Technologies

The deployment of surveillance technologies powered by AI, such as drones and acoustic monitoring systems, raises ethical questions regarding privacy and the potential for misuse. Although these tools are intended for conservation purposes, they may inadvertently infringe on the rights of local communities and stakeholders. For example, excessive surveillance of fishing activities could lead to tensions between regulators and fishers, particularly if not accompanied by strong community engagement and transparent communication. It is essential to establish ethical guidelines and frameworks that respect community rights while balancing conservation objectives (Mutsuddi, 2024; Mutsuddi, 2025).

12.6.2.4 Community Engagement and Equity

The successful implementation of AI in aquatic conservation relies heavily on the engagement of local communities and stakeholders. However, there is a risk that the introduction of high-tech solutions could marginalize traditional knowledge and practices. Communities that depend on aquatic resources for their livelihoods must be actively involved in the design and execution of AI-driven projects to ensure that their perspectives and needs are respected. Moreover, equitable access to AI technologies must be considered to prevent the technological divide from exacerbating existing social inequalities (Gundeti et al., 2024; Mutsuddi, 2024).

12.6.2.5 Integration into Existing Conservation Frameworks

Integrating AI technologies into existing conservation frameworks and practices presents logistical and operational challenges. Conservation managers may lack the necessary training and expertise to effectively utilize AI tools, leading to underutilization or misuse of these technologies. Furthermore, the incorporation of AI into decision-making processes may require changes to organizational structures and workflows, necessitating a cultural shift within conservation institutions. Addressing these challenges will require dedicated training, capacity building, and collaboration among ecologists, technology developers, and policymakers (Mutsuddi, 2025; Palakurti, 2025).

12.6.2.6 Long-term Sustainability and Maintenance

AI systems require ongoing maintenance, updates, and retraining to remain effective as environmental conditions and ecological dynamics evolve. Ensuring the long-term sustainability of AI initiatives in aquatic conservation involves significant logistical and financial commitments. Funding for these systems often depends on grants or short-term projects, leading to potential gaps in support once initial funding runs out. Establishing stable funding mechanisms and partnerships can help secure the continued operation of AI applications in conservation (Dwivedi et al., 2021; Chauhan et al., 2024).

12.6.2.7 Regulatory and Legal Frameworks

The rapid development and deployment of AI technologies in aquatic conservation may outpace existing regulatory and legal frameworks. There is a pressing need for policies that address the ethical implications of AI use, including data governance, accountability for algorithmic decisions, and mechanisms for appeal in the event of errors or adverse outcomes. Policymakers must engage with scientists, ethicists, and local communities to develop comprehensive frameworks that ensure responsible AI deployment in environmental contexts (Javaid et al., 2022; Alam et al., 2024).

12.7 FUTURE DIRECTIONS: IMPLICATIONS AND RECOMMENDATIONS

The integration of AI into aquatic ecosystem preservation is a rapidly evolving field with significant potential for advancing conservation efforts. However, as outlined in previous sections, various challenges must be addressed to ensure the effective and ethical use of AI technologies. In this section, we outline future directions for research and practice, along with implications and recommendations for overcoming these challenges in aquatic ecosystem conservation (Nishant et al., 2020; Ditria et al., 2022; Isabelle and Westerlund, 2022; Alshami et al., 2024; Bibri et al., 2024; Gundeti et al., 2024).

12.7.1 Enhancing Data Collection and Quality

- **Standardizing Data Protocols**: To improve the quality and comparability of datasets, it is crucial to establish standardized data collection protocols for AI applications in aquatic ecosystems. This includes developing guidelines for data types, formats, and metadata to facilitate interoperability across different systems and research initiatives.

- **Investing in Citizen Science**: Engaging citizen scientists in data collection can enhance the volume and diversity of data available for AI training. By harnessing mobile technologies and applications, communities can contribute valuable observations of local aquatic ecosystems, leading to richer datasets and increased public involvement in conservation efforts.
- **Leveraging Multimodal Data**: Future research should explore the integration of multimodal data sources such as combining acoustic, visual, and satellite data into AI analysis frameworks. This holistic approach can enhance the robustness of AI models and improve their predictive accuracy, providing a more comprehensive understanding of aquatic ecosystems.

12.7.2 Addressing Algorithmic Bias and Transparency

- **Developing Bias Mitigation Strategies**: Researchers and practitioners should prioritize the development of methodologies for detecting and mitigating algorithmic bias. This includes diversifying training datasets and employing algorithmic fairness techniques to ensure that AI applications do not inadvertently favor certain species or habitats over others.
- **Promoting Explainable AI (XAI)**: Investing in the development of explainable AI algorithms will enhance trust and accountability in AI-driven conservation decisions. Researchers should focus on creating AI systems that provide transparent insights into how decisions are made, enabling stakeholders to understand and verify AI outputs.

12.7.3 Strengthening Ethical Frameworks and Community Engagement

- **Implementing Ethical Guidelines**: Establishing clear ethical guidelines for the deployment of AI in aquatic conservation is essential. These guidelines should address issues of privacy, data governance, and the ethical use of surveillance technologies. Engaging ethicists in the development of these guidelines can ensure they reflect a broad spectrum of perspectives.
- **Empowering Local Communities**: Future initiatives should prioritize community engagement by involving local stakeholders in the design, implementation, and monitoring of AI projects. Workshops, public consultations, and co-management approaches can foster trust and ensure that conservation efforts align with community needs and values.
- **Education and Capacity Building**: Training programs that enhance the technological literacy of conservation practitioners and local communities will promote effective use and management of AI tools. Investment in educational initiatives will empower stakeholders to leverage AI for their conservation needs while ensuring equitable access to technology.

12.7.4 Integrating AI into Existing Conservation Frameworks

- **Cross-Disciplinary Collaboration**: Successful integration of AI in conservation efforts requires collaboration among ecologists, data scientists, technologists, and policymakers. Interdisciplinary teams can ensure that AI applications are relevant to specific ecological contexts while addressing technical complexities.

- **Developing Adaptive Management Frameworks**: Future research should focus on creating adaptive management systems that incorporate AI insights into decision-making processes. These frameworks should allow for iterative learning and responsiveness to changing ecological conditions, enabling more effective management strategies over time.

12.7.5 Ensuring Long-term Sustainability and Funding

- **Establishing Sustainable Funding Mechanisms**: To support long-term AI initiatives in aquatic conservation, stakeholders should explore diverse funding sources, including public-private partnerships, governmental grants, and philanthropic contributions. Establishing endowments or trust funds specifically for AI conservation initiatives can also provide financial stability.
- **Monitoring and Evaluation Frameworks**: Implementing robust monitoring and evaluation frameworks will assess the effectiveness of AI applications in conservation. These frameworks should track both ecological outcomes and the socio-economic impacts of AI technologies, ensuring that conservation objectives are met.

12.7.6 Building Comprehensive Regulatory Frameworks

- **Engaging Policymakers and Legal Experts**: Involving policymakers and legal experts in the development of regulations concerning AI in conservation is crucial. Collaborative efforts can lead to the establishment of frameworks that govern data use, algorithmic accountability, and biodiversity protection.
- **Promoting International Cooperation**: Given the transboundary nature of many aquatic ecosystems, fostering international cooperation on AI deployment and data sharing will be necessary to bolster global conservation efforts. Multinational agreements can facilitate collaboration among countries to address shared challenges in aquatic ecosystem preservation.

12.8 CONCLUSION

The preservation of aquatic ecosystems is increasingly critical due to escalating threats from anthropogenic activities and climate change. These ecosystems, encompassing both marine and freshwater environments, are vital for maintaining ecological balance, supporting biodiversity, and providing essential resources for human survival. However, they face significant challenges, including pollution, overfishing, habitat destruction, and climate change, which jeopardize their health and sustainability. The integration of AI into conservation strategies offers a transformative approach to address these challenges. AI technologies, such as machine learning and data analytics, enhance the monitoring and management of aquatic ecosystems by enabling real-time data collection, predictive modeling, and effective resource management. These capabilities facilitate the identification of threats, the assessment of ecosystem health, and the implementation of targeted conservation efforts. Furthermore, AI can democratize scientific inquiry by engaging local communities in conservation initiatives, thereby fostering a shared responsibility for environmental stewardship. Despite the promising potential of AI in aquatic ecosystem preservation, several challenges and ethical considerations must be addressed. These include data management issues, algorithmic bias, ethical implications of surveillance technologies, and the need for community engagement. Ensuring the responsible and equitable application of AI requires the establishment of robust ethical frameworks, transparent methodologies, and inclusive practices that respect local knowledge and rights. Future directions for research and practice should focus on enhancing data quality, addressing algorithmic biases, strengthening community engagement, integrating AI into existing conservation frameworks, ensuring long-term sustainability, and developing comprehensive regulatory frameworks. By adopting these strategies, stakeholders can optimize the use of AI in preserving aquatic ecosystems, ultimately contributing to a more sustainable and resilient future for both biodiversity and human communities.

12.9 DECLARATION OF GENERATIVE AI IN SCIENTIFIC WRITING

During the preparation of this chapter, the authors used generative AI and AI-assisted technologies in the writing process (rephrasing, grammar checking, and chart production). After using this tool, the authors reviewed and edited the content as needed and took full responsibility for the content of the published chapter.

REFERENCES

Abdein, R. H., & El-Emam, D. A. (2024). The removal of microplastics as a source for toxic chemicals in aquatic environments: characteristics, occurrence and removal mechanisms. In: Haddout, S., Priya, K., Hoguane, A. M. (eds) *Proceedings of The 2nd International Conference on Climate Change and Ocean Renewable Energy. CCORE 2022. Springer Proceedings in Earth and Environmental Sciences*. Springer, Cham.

Adanma, U. M., & Ogunbiyi, E. O. (2024). Artificial intelligence in environmental conservation: evaluating cyber risks and opportunities for sustainable practices. *Computer Science & IT Research Journal*, 5(5), 1178–1209.

Ahmed, S. F., Kumar, P. S., Kabir, M., Zuhara, F. T., Mehjabin, A., Tasannum, N., … & Mofijur, M. (2022). Threats, challenges and sustainable conservation strategies for freshwater biodiversity. *Environmental Research*, 214, 113808.

Alam, M., Khan, I. R., Siddiqui, F., & Alam, M. A. (2024). Artificial intelligence as key enabler for safeguarding the marine resources. In: De, D., Sengupta, D., Tran, T.A. (eds) *Artificial Intelligence and Edge Computing for Sustainable Ocean Health* (pp. 409–451). Cham: Springer Nature Switzerland.

Alloghani, M. A. (2024). Using AI to monitor marine environmental pollution: Systematic review. In: *Artificial Intelligence and Sustainability. Signals and Communication Technology*. Springer, Cham. (pp. 87–97).

Alshami, A., Ali, E., Elsayed, M., Eltoukhy, A. E., & Zayed, T. (2024). IoT innovations in sustainable water and wastewater management and water quality monitoring: a comprehensive review of advancements, implications, and future directions. *IEEE Access*, 12, 58427–58453.

Ayoola, V. B., Idoko, P. I., Eromonsei, S. O., Afolabi, O., Apampa, A. R., & Oyebanji, O. S. (2024). The role of big data and AI in enhancing biodiversity conservation and resource management in the USA. *World Journal of Advanced Research and Reviews*, 23(2), 1851–1873.

Bakker, K. (2022). Smart oceans: artificial intelligence and marine protected area governance. *Earth System Governance*, 13, 100141.

Bibri, S. E., Krogstie, J., Kaboli, A., & Alahi, A. (2024). Smarter eco-cities and their leading-edge artificial intelligence of things solutions for environmental sustainability: a comprehensive systematic review. *Environmental Science and Ecotechnology*, 19, 100330.

Chauhan, D., Bahad, P., & Jain, J. K. (2024). Sustainable AI: environmental implications, challenges, and opportunities. In: Lakshmi, D., Ravi, S. T., Rajesh, K. D., Seifedine K. (eds) Explainable AI (XAI) for Sustainable Development (pp. 1–15). Chapman and Hall/CRC. New York.

Ditria, E. M., Buelow, C. A., Gonzalez-Rivero, M., & Connolly, R. M. (2022). Artificial intelligence and automated monitoring for assisting conservation of marine ecosystems: a perspective. *Frontiers in Marine Science*, 9, 918104.

Divine, I. L. O. H., Aguma, C. P., & Olagunju, A. O. (2024). Integrating AI enhanced remote sensing technologies with IOT networks for precision environmental monitoring and predicative ecosystem management. *World Journal of Advanced Research and Reviews*, 23(2), 2156–2166.

Dogan, G., Vaidya, D., Bromhal, M., & Banday, N. (2024). Artificial intelligence in marine biology. In: Ambreen, H., Nazir, A. G., Henna, H., Bashir, J.(eds) *A Biologist s Guide to Artificial Intelligence* (pp. 241–254). Academic Press, Elsevier

Dwivedi, D. N., Mahanty, G., & nath Dwivedi, V. (2024). Intelligent conservation: a comprehensive study on ai-enhanced environmental monitoring and preservation. In: Roopa, C. R., Gowri Ganesh, N. S., Daniel, J. A., Awais, A., Santhosh, R. (eds) *The Convergence of Self-Sustaining Systems with AI and IoT* (pp. 215–226). IGI Global.

Dwivedi, Y. K., Hughes, L., Ismagilova, E., Aarts, G., Coombs, C., Crick, T., … & Williams, M. D. (2021). Artificial Intelligence (AI): multidisciplinary perspectives on emerging challenges, opportunities, and agenda for research, practice and policy. *International Journal of Information Management*, 57, 101994.

Elemam, D. A., & Eldeeb, A. R. (2023). Climate change in the coastal areas: consequences, adaptations, and projections for the Northern Coastal Area, Egypt. *Scientific Journal for Damietta Faculty of Science*, 12(2), 19–29.

Gambín, Á. F., Angelats, E., González, J. S., Miozzo, M., & Dini, P. (2021). Sustainable marine ecosystems: deep learning for water quality assessment and forecasting. *IEEE Access*, 9, 121344–121365.

Gomez-Zavaglia, A., Mejuto, J. C., & Simal-Gandara, J. (2020). Mitigation of emerging implications of climate change on food production systems. *Food Research International*, 134, 109256.

Grorud-Colvert, K., Sullivan-Stack, J., Roberts, C., Constant, V., Horta e Costa, B., Pike, E. P., … & Lubchenco, J. (2021). The MPA Guide: a framework to achieve global goals for the ocean. *Science*, 373(6560), eabf0861.

Gundeti, R., Vuppala, K., & Kasireddy, V. (2024). The future of AI and environmental sustainability: challenges and opportunities. In: Hemachandran, K., Raul, V. R., Zita, Z. P., Abejide, A. (eds) *Exploring Ethical Dimensions of Environmental Sustainability and Use of AI* (pp. 346–371.). IGI Global.

Häder, D. P., Banaszak, A. T., Villafañe, V. E., Narvarte, M. A., González, R. A., & Helbling, E. W. (2020). Anthropogenic pollution of aquatic ecosystems: emerging problems with global implications. *Science of the Total Environment*, 713, 136586.

Hemanth, B., Prabhu, P., Shetti, K. M., Shetty, D. D., & Zhenkar, T. M. (2024). AI applications for clean energy and sustainability: utilizing artificial intelligence for monitoring marine ecosystems. In: Budi, A. R., Bhupinder, S., Christian, K., Komal, V. (eds) *AI Applications for Clean Energy and Sustainability* (pp. 16–31). IGI Global.

Isabelle, D. A., & Westerlund, M. (2022). A review and categorization of artificial intelligence-based opportunities in wildlife, ocean and land conservation. *Sustainability*, 14(4), 1979.

Isangedighi, I. A., David, G. S., & Obot, O. I. (2020). Plastic waste in the aquatic environment: impacts and management. In: Nollet, L. M. L., Khwaja, S. S. (eds) *Analysis of Nanoplastics and Microplastics in Food* (pp. 15–43). CRC Press.

Javaid, M., Haleem, A., Singh, R. P., Suman, R., & Gonzalez, E. S. (2022). Understanding the adoption of Industry 4.0 technologies in improving environmental sustainability. *Sustainable Operations and Computers*, 3, 203–217.

Jenny, J. P., Anneville, O., Arnaud, F., Baulaz, Y., Bouffard, D., Domaizon, I., … & Weyhenmeyer, G. A. (2020). Scientists' warning to humanity: rapid degradation of the world's large lakes. *Journal of Great Lakes Research*, 46(4), 686–702.

Letcher, T. (Ed.). (2021). *The Impacts of Climate Change: A Comprehensive Study of Physical, Biophysical, Social, and Political Issues*. Elsevier Science Publishing Co Inc, United States.

Marimuthu, K. A. S. I., Umasankar, A. K. U. L. A., & Baskaran, X. R. (2020). Sustainable aquatic biodiversity-prospects and conservation. In: *Pavithra S. Degeneration to Regeneration-Alternative Approaches* (p. 49). Royal Book Publishing – International, India.

Marques, L., & Marques, L. (2020). Collapse of biodiversity in the aquatic environment. In: *Marques L. Capitalism and Environmental Collapse* (pp. 275–301). Springer, Cham.

Mumtaz, N., Izhar, T., Pandey, G., & Labhasetwar, P. K. (2022). Utilizing artificial intelligence for environmental sustainability. In: Ashutosh, K. D., Sushil, K. N., García-Díaz, V.(eds) *Artificial Intelligence for Renewable Energy Systems* (pp. 259–279). Woodhead Publishing.

Mutsuddi, I. (2024). Ethical considerations in the intersection of AI and environmental science. In: Bhupinder, S., Christian, K., Komal, V., Soumi, D. *Maintaining a Sustainable World in the Nexus of Environmental Science and AI* (pp. 283–310). IGI Global.

Mutsuddi, I. (2025). Data analytics, machine learning, and IoT for environmental governance. In: Imran. H., Mahmudul Haque, A. K. M, Akram, Ullah S. M. (eds) *Intersecting Environmental Governance with Technological Advancements* (pp. 265–286). IGI Global Scientific Publishing.

Ncube, M. M., & Ngulube, P. (2024). Enhancing environmental decision-making: a systematic review of data analytics applications in monitoring and management. *Discover Sustainability*, 5(1), 290.

Nishant, R., Kennedy, M., & Corbett, J. (2020). Artificial intelligence for sustainability: challenges, opportunities, and a research agenda. *International Journal of Information Management*, 53, 102104.

Ogidi, O. I., & Akpan, U. M. (2022). Aquatic biodiversity loss: impacts of pollution and anthropogenic activities and strategies for conservation. In: Chibueze Izah, S. (eds) *Biodiversity in Africa: Potentials, Threats and Conservation* (pp. 421–448). Singapore: Springer Nature Singapore.

Palakurti, N. R. (2025). Obstacles and resolutions: integrating artificial intelligence in sustainable development. In: Pawan, W., Pavika, S., Nagender, A., Elngar, A. A., Silva, N.(eds) *Artificial Intelligence and Machine Learning for Sustainable Development* (pp. 1–18). CRC Press, Boca Raton

Prakash, S. (2021). Impact of climate change on aquatic ecosystem and its biodiversity: an overview. *International Journal of Biological Innovations*, 3(2), 312–317.

Rane, J., Mallick, S. K., Kaya, O., & Rane, N. L. (2024). Enhancing black-box models: advances in explainable artificial intelligence for ethical decision-making. In: *Future Research Opportunities for Artificial Intelligence in Industry*, 4, 136–180.

Rao, C. N., Rao, A. V., Shanmugasundar, G. et al. (2024). Marine ecosystem monitoring based on remote sensing using underwater image analysis for biodiversity conservation model. *Remote Sensing in Earth Systems Sciences* 7, 309–318.

Santojanni, F. B., Miner, H., Hain, H., & Sutton, G. (2023). The impact of climate change on biodiversity in coastal ecosystems. *Jurnal Ilmu Pendidikan dan Humaniora*, 12(3), 167–182.

Shivaprakash, K. N., Swami, N., Mysorekar, S., Arora, R., Gangadharan, A., Vohra, K., … & Kiesecker, J. M. (2022). Potential for artificial intelligence (AI) and machine learning (ML) applications in biodiversity conservation, managing forests, and related services in India. *Sustainability*, 14(12), 7154.

Silvestro, D., Goria, S., Sterner, T., & Antonelli, A. (2022). Improving biodiversity protection through artificial intelligence. *Nature Sustainability*, 5(5), 415–424.

Thakur, A., & Koul, B. (2022). Impact of oil exploration and spillage on marine environments. In: Papita, D., Suvendu, M. Jitendra, K. P. (eds) *Advances in Oil-Water Separation* (pp. 115–135). Elsevier.

Ullah, F., Saqib, S. & Xiong, YC.(2025). Integrating artificial intelligence in biodiversity conservation: bridging classical and modern approaches. *Biodiversity Conservation,* 34, 45–65.

Viitasalo, M., & Bonsdorff, E. (2022). Global climate change and the Baltic Sea ecosystem: direct and indirect effects on species, communities and ecosystem functioning. *Earth System Dynamics*, 13(2), 711–747.

Vishnoi, S., & Goel, R. K. (2024). Climate smart agriculture for sustainable productivity and healthy landscapes. *Environmental Science & Policy*, 151, 103600.

Wang, L. (2024). Advances in monitoring and managing aquatic ecosystem health: integrating technology and policy. *International Journal of Aquaculture*, 14(2), 0012.

Wang, Y. S., & Gu, J. D. (2021). Ecological responses, adaptation and mechanisms of mangrove wetland ecosystem to global climate change and anthropogenic activities. *International Biodeterioration & Biodegradation*, 162, 105248.

Yang, L., Driscol, J., Sarigai, S., Wu, Q., Lippitt, C. D., & Morgan, M. (2022). Towards synoptic water monitoring systems: a review of AI methods for automating water body detection and water quality monitoring using remote sensing. *Sensors*, 22(6), 2416.

Yang, X., Zhang, X., Zhang, P., Bidegain, G., Dong, J., Hu, C., … & Guo, H. (2023). Ensemble habitat suitability modeling for predicting optimal sites for eelgrass (*Zostera marina*) in the tidal lagoon ecosystem: implications for restoration and conservation. *Journal of Environmental Management*, 330, 117108.

13 Artificial Intelligence and Water Conservation

Thulisekari Prasanna, Guda Bhargavi, and Burra Shyamsunder

13.1 INTRODUCTION

Fresh water is a scarce and vital commodity that directly shapes economic and ecological activities concerning the earth. But water resources around the globe remain threatened by a number of factors, for instance, population growth, urbanization, industrialization, and climate change, as pointed out by UNESCO in 2021. In this context, water is one of the most needed resources in efficient management to ensure that it will not be depleted soon, as people may require it. Conventional approaches to water management are normally inadequate when it comes to standard and large problems, hence the need to integrate new approaches such as AI (Wagle et al., 2022; FAO, 2020).

AI has become an innovation to perform in water resource management to predict, monitor, and decide. On the basis of data analytics, machine learning (ML) technologies incorporated with IoT in agriculture, urban planning, and industrial usage can bring out the best of AI in water usage. For instance, smart irrigation systems that harness AI can detect soil moisture, weather conditions, or crop demand, in order to reduce usage of large amounts of water while promoting increased production of crops. Likewise, AI can identify some leaks and losses in urban water supply systems and then prevent resource wastage and costs (Pahl-Wostl, 2019).

Customization for the environment, as well as other associated environmental issues, is also within the domain of the development of AI. ML capabilities can be used to forecast droughts, floods, and water quality problems that can be addressed before they arise (Ahmed et al., 2020). Satellite and airborne data with AI applied offer accurate information about the quantity and distribution of groundwater, rivers' flows, and precipitation required for proper decision-making by authorities and investors.

However, there are some issues, which are data unavailability, integration difficulty, and the disparity in the availability of technological infrastructure between the developed and the developing nations or regions (Wagle et al., 2022). Addressing these barriers is critical to fully harnessing AI's benefits for water resource management and achieving global water security in alignment with Sustainable Development Goal 6: Achieving Sustainable Water and Sanitation for All.

DOI: 10.1201/9781003547099-16

13.2 OVERVIEW OF THE KEY CONCEPTS AND DEFINITION OF AI

AI can be defined as a technique through which computers replicate human intelligence in executing tasks including learning and problem solving (Russell & Norvig, 2020). Some of the critical AI approaches consist of ML, deep learning, and neural networks to deliver adaptability and proficiency in predictions.

13.2.1 Trends of AI Usage in Environmental and Water Science

Artificial intelligence (AI) started being used in environmental science in climate modeling and moved to water quality, resource computing, and smart watering (Kourgialas & Karatzas, 2018). Advanced AI applies with IoT and Remote Sensing for real-time water management integrated with IoT and RS.

A revolutionary step toward the sustainable management of natural resources is the incorporation of AI into environmental and water science. Predictive modeling for floods, droughts, and groundwater dynamics, as well as real-time monitoring for water quality and ecosystem changes using AI-enhanced sensors, IoT, and satellite images, are important themes. AI guides climate adaption measures including carbon sequestration and ecosystem restoration while optimizing resource allocation in urban and agricultural systems, ensuring effective irrigation and minimizing water waste. Additionally, AI facilitates policymaking through scenario analysis and public participation; it is frequently combined with blockchain and drone technologies for accurate and transparent management. These developments highlight AI's capacity to successfully tackle ecological issues and help achieve global sustainability objectives in water management and environmental preservation.

13.2.2 Advantages of AI in Water Conservation

AI helps in improving the chances of making the right decision and minimizing wastage of resources, specifically water. For example AI solution that includes smart irrigation scheduling, estimating the water demand, and enabling leakage detection in the watering distribution system makes it sustainable (Ahmed et al., 2021).

13.3 THE AI APPLICATIONS FOR WATER CONSERVATION

13.3.1 Forecast of Water Demand and Supply

AI models help to predict the necessary quantity of water, leveraging the historical and actual data to facilitate advanced management. For instance, an electrical neural network is used in the estimation of agricultural water demand depending on climatic and crop factors (Zhang et al., 2019a).

13.3.1.1 Case Studies

In smart cities such as Singapore, demand forecasting for water using artificial neural networks has chosen the smart supply to produce water with 15% wastage of water (Public Utilities Board, 2020).

13.3.2 Real-Time Monitoring and Control Systems

Smart devices such as IoT sensors with AI modules track water quality parameters (for instance, pH and turbidity) as well as quantity. This makes it easy to detect any contamination or scarcity very early, as evident with the Ganges River Basin Monitoring Project (Kumar et al., 2020).

13.3.3 Smart Irrigation Systems

Through weather data, soil moisture, and crop information, AI-irrigation controls a watering plan. Data on soil quality from satellite imagery alongside ML models enhance water use efficiency and crop yields up to 30% in dry areas (Jiang et al., 2022).

13.3.4 Water Loss and Leaks Detection

AI identifies faults in the flow rates and pressure in pipelines, ensuring that there are no losses. AI has been effective in use for detection of leaks with the use having been deemed economical in cities like Tokyo where leakages were reduced by 18% (Suzuki et al., 2021).

13.3.5 Water Recycling and Reuse

AI enhances the operation of wastewater treatment by indicating the chemical requirements and energy required. For example, in membrane bioreactor systems, AI solutions have improved the rate of water recovery (Singh et al., 2021).

13.4 INTEGRATION OF AI AND REMOTE SENSING IN WATER RESOURCE MANAGEMENT

Applying AI with remote sensing technology has brought improved methods of analyzing data, monitoring, and modeling of water resources management. Remote sensing extends spatial and temporal data for water resources at a broader scale, (Ghosh & Patel, 2022) and AI optimizes the examination and understanding of that data for decision-making.

13.4.1 AI and Remote Sensing in Water Resource Management

Cores of surface water bodies are acquired through satellite imagery and aerial photographs, especially through the use of drones. It is noted that AI algorithms, especially in deep learning, can be applied for water body detection, classification, and tracking its changes, including reservoirs, lakes, rivers, etc.

Example: CNj, CNN has also been used successfully in mapping surface water bodies using data from Landsat and Sentinel 2 satellite data with high accuracy detecting a seasonal fluctuation in water extent (Feng et al. 2020).

13.4.2 Monitoring Droughts

Therefore, AI evaluates remote sensing data to determine regions vulnerable to drought by using vegetation indices, such as the NDVI (Normalized Difference Vegetation Index) and the soil's moisture content. Drought intensity and duration indices are predicted by the ML algorithms that use climate variables, including rainfall and temperature as proxies.

Example: (Mishra et al., 2021) in their study that embraced the use of AI as well as MODIS satellite data have developed the drought vulnerability maps for the Indian subcontinent, which help in the effective preparation of drought management.

13.4.3 Challenges of Flood Observation and Forecast

AI models work on remote sensing information to infer about the risks of flood and future flood occurrences. These models employ variables including rainfall, river discharge, and altitude.

Example: The combination of AI and synthetic aperture radar (SAR) data from Sentinel-1 satellites has improved flood extent mapping, especially in areas covered by clouds (Rahman et al., 2020).

13.4.4 Groundwater Monitoring

Geophysics integration with remote sensing data predicts groundwater levels through changes in surface water, vegetation, and soil. AI also estimates the rates of parameter recharging from groundwater based on hydro-meteorological data.

Example: SVMs were applied with the GRACE satellite data to quantify changes in groundwater storage in California's Central Valley (Ahmed et al., 2020).

13.4.5 Watershed Management

With the help of AI and remote sensing, watershed data can be collected through detecting and mapping of the land use, for erosion and resulting runoff. This is useful because it informs wise planning of future use of the watersheds while informing conservation of the watersheds.

Example: The principles of AI have been used in multispectral remote sensing data analysis for proper management of the Basin in the Blue Nile of Ethiopia (Mengistu et al., 2021).

13.4.6 Water Quality Monitoring

Remote sensing measures the physical parameters of water including turbidity, chlorophyll, and algal blooms. These data are needed for AI algorithms to monitor large regions' water quality.

Example: Application of AI with hyperspectral remote sensing data has successfully detected algal blooms in coastal waters to enhance the control of water quality in fishery (Zhang et al., 2019b).

13.4.7 Opportunities of the Application of AI and Remote Sensing in Water Resource Management

Large-Scale Monitoring: Affords evaluation of vast territories with hardly any onsite inspections.

Cost-Effective: Saves time for having to gather data manually, process, and analyze it.

Real-Time Insights: Provides current information for action, for example, in case of flooding or drought.

Precision: AI strongly improves the aspect of resource estimation and generates more accurate predictive models.

Data Quality: Remote sensing data often requires preprocessing to eliminate noise and ensure reliability.

Computational Resources: Processing large datasets demands significant computational power.

Skill Gap: Expertise in both AI and remote sensing is required for effective implementation.

Accessibility: Limited access to high-resolution satellite data in developing countries.

13.4.8 Future Trends

1. **AI-Driven Drone Technology**: Further integration of drones with AI that can be used on a limited, local basis to monitor water resources.
2. **Integration with IoT**: Integrating remote sensing data with sensors for Internet of Things for a water management system.
3. **Advanced AI Models**: ML and more specifically generative AI as well as reinforcement learning for dynamic resource management, and more specifically for water resources.

13.5 FACETS OF AI IN SUSTAINABLE WATER MANAGEMENT

13.5.1 Opportunities of Applying AI in Sustainable Water Management

The reinforcement of AI has become a revolutionary tool for implementing sustainable water management solutions for scarcity of water, vivid pollution, and wasteful use of water resources. AI, thus, improves decision-making in agricultural practices, water usage planning—especially in urban areas—as well as in the industrial sector.

13.5.1.1 Higher Accuracy of Water Distribution and Conservation

Real-time data is employed by AI algorithms to develop better techniques in irrigation, industrial processes, and household consumption.

Agriculture: Remote sensing data in conjunction with weather forecasts and soil moisture sensors help irrigated systems to employ efficient irrigation regimes to minimize wastage of water.

Example: Mohan and colleagues also observed that the implementation of AI in irrigation systems leads to only 30% water usage in vineyards.

Industry: By using AI for predictive analytics, the wastage of water in industrial applications can be identified and reduced.

13.5.1.2 Enhanced Monitoring and Management

AI analyzes data collected by devices and remote sensing systems, and therefore, it can monitor water resources in real-time

Water Quality: AI models work on data regarding pH level, turbidity, and extent of pollutants for early detection of menace.

Example: Park et al. (2021) provided methods of ML to mitigate the impact of the hazardous algal blooms on the aquatic environment.

Water Distribution: AI enhances the delivery systems of water by analyzing leakages and also the rate of flow in water systems.

13.5.1.3 Cost-Effectiveness

Automated solutions decrease the financial demands of overseeing the structures, labor expenses for monitoring, and resource utilization.

Example: Advanced Metering Infrastructure and IoT systems that form smart water grids reduced operating costs by up to 20% for European municipal water providers (Smith et al., 2020a, p. 68). Automation reduces the effect of human-prone issues like data collection and detection of any leak, enabling continuous monitoring of water resources.

Water Quality: AI models analyze data on pH levels, turbidity, and pollutants, ensuring early detection of contamination.

Example: ML models developed by Park et al. (2021) accurately predicted harmful algal blooms, mitigating their impact on aquatic ecosystems.

Water Distribution: AI improves the efficiency of water distribution networks by detecting leaks and optimizing flow rates.

13.5.1.4 Disaster Preparedness and Response

Through analysis and risk evaluation, AI provides protection against water-related calamities including floods and droughts.

Flood Forecasting: Flood event identification is done by combining satellite images, data from river gauges, and weather predictions.

Example: ML-based flood prediction in Bangladesh led to better evacuation performance, the number of fatalities and losses reduced (Haque et al., 2021).

Drought Management: AI models predict where drought is likely to occur and help policymaking on rationing of water.

13.6 RELATION TO THE GOALS OF SUSTAINABLE DEVELOPMENT (SDGS)

AI helps contribute to the United Nations Sustainable Development Goal 6, "Clean Water and Sanitation" as it supports early identification needed to optimize water consumption, its distribution and quality. AI also supports the evaluation of the water deficit, prediction of water shortage, and usage of water in irrigation and industries.

Water Efficiency and Equity: AI can help balance water use within sectors and regions, especially those that are water-stressed. It also finds application in the enhancement of irrigation systems and water distribution system. AI helps in managing the groundwater levels, the rainfall patterns and the likelihood of water availability in the particular season.

Example: AI models in Indian agriculture in irrigation scheduling, this will prove to be beneficial for the Indian farmer and help in water use efficiency to conserve water resources (Kumar et al., 2022).

Water Quality Monitoring: AI-based models constantly monitor the water quality data and assist the authorities in early detection of any pollution so that no one will suffer from waterborne diseases.

Example: In the UK, there are examples of using AI-driven sensors to monitor water quality in real-time, to detect pollution events and prevent contamination (Roberts et al., 2021).

13.7 WATER CONTROL AUTHORITY USES ITS WASTEWATER TREATMENT PLANT TO PROVIDE TREATMENT AND REUSE SERVICES

Harnessing the capability of AI, wastewater treatment can be enhanced, water quality can be improved, and water reuse can be enabled. The AI models can contribute to the reduction of energy consumed in treatment processes, improvement of efficiency in treatment plants and usage of the treated water for irrigation, industrial use, and counting toward the urban demand.

Optimization of Treatment Processes: It is notable that using ML, treatment results may be forecasted, and such factors as chemical feed and aeration may be optimized in real-time to increase the efficiency of the procedure and decrease expenditures.

Example: The application of AI models in Singapore has been applied to enhance energy efficiency at the country's largest WAN, which owns the largest wastewater treatment plant, and enhance treated water quality in the facility (Tan et al., 2021).

Water Reuse: AI is used in developing strategies that enhance water recycling and reuse for industries and in metropolitan regions.

Example: AI in California is employed in monitoring water use, water reuse opportunities, as well as recycling water in urban areas (Zhao et al., 2020).

13.8 REDUCTION OF WATER WASTAGE

Facilitating real-time monitoring of water use across sectors, AI can help greatly prevent wastage by pointing out and detecting borrower issues in water systems. Using predictive analytics and real-time analysis, AI aids in controlling wastage, leakage, or misuse and uneconomical consumption.

Leak Detection and Control: AI systems are capable of diagnosing the conditions of the damage of sensors for leakage detection in water pipes and distribution networks, so certain municipalities and industries might save water.

Example: In the United States, for example, AI models are employed to identify leakage in the water supply system in the urban setting, hence improving water supply system effectiveness by 20% in some cities according to Tanaka et al. (2021).

Optimized Water Distribution: AI models predict water demand with the help of data about previous weeks and months, weather conditions, and population increase, and, as a result, the distribution of water is quite accurate.

Example: ML demand forecasting of water requirements in the water supply system in Dubai led to the minimization of overstocking and wastage through efficient storage of water and its distribution (Al-Dosary et al., 2020).

13.9 INTEGRATED WATER RESOURCE MANAGEMENT

Integrated Water Resource Management (IWRM) is an integration process of using water resources for development within the framework of sustainable development. AI helps IWRM in the sense that it offers valuable tools for decision-making, helping to improve the use of available water resources both in terms of quantity and quality, depending on the sectors and regions involved.

Data Integration: AI systems combine data sets from hydrological, climatic, and socioeconomic domains to supply an extensive picture to water availability and demand.

Example: AI models are employed in the Nile Basin to enhance hydroclimatic data to fairly allocate water resources among nations (El-Masri et al., 2021).

Optimized Resource Allocation: AI is helpful when determining water usage by agricultural, industrial, or domestic consumers using real numbers and forecasts.

Example: In Australia, AI-based systems have been deployed to assist in water allocation decisions during periods of drought, reducing water scarcity impacts (Smith et al., 2020b).

13.10 TWO TYPES OF TRENDS: FUTURE TRENDS AND OPPORTUNITIES

ML, deep learning, and sensors offer a bright future for AI technology in the water management system. It is therefore expected that as the sophistication of the AI systems advances, there will be great opportunities for improving water use, the distribution and management of water, as well as advanced water treatment technologies.

AI in Water Recycling: Future AI systems will be important in enhancing the functionality of water recycling plants by ensuring the use of ML in real-time plants.

Smart Water Grids: Smart water grids integrated with AI will help to achieve online monitoring, leakage detection, and demand forecasting, (Jayaraman & Ganesan, 2022) and the development of robust water distribution systems.

13.11 CASE STUDIES

Various international cases prove that AI is successfully applied to water management and preservation.

Case Study 1: In Barcelona, Spain, AI systems are increasingly being deployed in the management of water supply and demand, supply and forecast of water demand and leakage. This has been achieved, and this has contributed to a 15% reduction in water loss at the distribution network of the city (Roberts et al., 2020).

Case Study 2: Current advanced applications of AI include the use of precision irrigation in agriculture to conserve water in Israel. The efficiency in water usage in farming has been enhanced by 30% through the use of AI through the integration of weather data, sensors for the determination of the moisture content of the soil, and the water requirements by general crops (Zhang et al., 2019c)

13.12 POLICY IMPLICATIONS AND RECOMMENDATIONS

The application of AI in water conservation is poised to revolutionize the way we manage this critical resource. However, its successful implementation requires thoughtful policy measures to address various technological, social, and economic challenges (Ghosh & Patel, 2022). Below are key policy implications and recommendations derived from the research:

13.12.1 Establishing Robust Regulatory Frameworks

AI applications in water management may raise concerns about privacy, data security, and ethical use of technology. Governments should create clear and comprehensive guidelines that aligns sustainability goals (Singh & Sharma, 2021):

- Ensure transparency in data collection and usage.
- Protect the privacy of users and stakeholders involved in AI-powered systems.
- Promote ethical AI applications that align with sustainability goals.

These frameworks should evolve with advancements in AI to remain relevant and effective.

13.12.2 Investment in AI and Water Infrastructure

To maximize the benefits of AI, governments and stakeholders must invest in both technological and physical infrastructure, especially in water-scarce regions. This includes:

- Funding research and development for innovative AI solutions tailored to local water challenges.
- Enhancing existing water infrastructure to integrate AI technologies such as smart sensors and meters.
- Providing financial incentives for adopting AI-driven water conservation systems.

13.12.3 Promoting Public-Private Partnerships

Collaboration between public institutions, private companies, and research organizations is essential for the successful deployment of AI solutions, Public–Private Partnerships can:

- Accelerate the development of advanced AI technologies for water conservation.
- Facilitate the sharing of resources, expertise, and data.
- Drive large-scale implementation of AI tools in both urban and rural settings.

13.12.4 Capacity Building and Training

AI adoption requires skilled personnel who can operate, manage, and innovate these technologies. Governments and organizations should:

- Implement training programs for water managers, policymakers, and engineers.
- Encourage educational institutions to integrate AI and water conservation topics into their curricula.
- Support knowledge exchange programs to learn from global best practices.

13.12.5 Encouraging Community Engagement

Community participation is crucial for the success of AI in water conservation. Policies should focus on:

- Raising awareness about the benefits of AI tools for water management.
- Empowering local communities with user-friendly AI applications that promote sustainable water practices.
- Ensuring equitable access to AI technologies for marginalized groups.

13.13 CONCLUSION

ML, especially AI, is transforming water conservation by providing unique and effective ways of managing the resource sustainably. By way of predictive analytics tools, real-time monitoring, smart irrigation systems, and other AI-enabled tools, the water is now managed, minimizing wastage and maximizing utilization across the sectoral health. AI is valuable to the stakeholders as it provides information extracted from large data sets on the best ways to manage water resources sustainably.

AI working in conjunction with IoT, remote sensing, and ML algorithms has thus been helpful in solving problems such as water scarcity, poor usage, and effects resulting from climate change. Ranging from farming as a system to integrated urban water systems, these technologies offer sustainable solutions for water use efficiency at large. Moreover, the systems enable the demands of water as well as leakage prediction, and even expecting climate change situations, which aid in efficient water management.

First and foremost, AI can unlock the way to approach the challenge of rising demands by applying such a limited and valuable resource as water. But to achieve these benefits, AI needs more investment; this is in terms of research, infrastructure, and human capacity development, particularly in the developing world where water scarcity is rife. Specifically, improvements in data sharing and collaboration, along with practical guidelines for the ethical application of AI technologies, should be incorporated into policies to promote equal availability of such solutions among different populations.

In conclusion, one emerges with a belief that AI has a great potential to positively spur sustainable management of water resources. In welcoming and advancing these technologies and encouraging international collaboration, the existing and future water problems can be solved by implementing existing solutions.

REFERENCES

Ahmed, S., Khan, S., & Singh, R. (2021). Artificial intelligence applications in water resource management. *Journal of Environmental Science and Technology*, 15(4), 567–589.

Ahmed, S., Singh, R., & Khan, M. (2020). Applications of machine learning in drought prediction and groundwater monitoring. *Journal of Hydrology*, *588*, 125–137.

Al-Dosary, A., Abed, A., & Nasser, F. (2020). Optimizing water distribution systems in arid cities using AI and predictive analytics. *Urban Water Journal*, 17(5), 541–552.

El-Masri, B., Ibrahim, M., & El-Sharif, H. (2021). AI for integrated water resource management in the Nile Basin: A case study. *Water Resources Management*, 35(10), 3357–3371.

FAO. (2020). *Artificial Intelligence and Water Conservation: Opportunities and Challenges.* Food and Agriculture Organization of the United Nations [Rome].

Feng, M., Ouyang, Y., & Liu, W. (2020). Mapping water bodies using Sentinel-2 imagery and convolutional neural networks. *Remote Sensing of Environment*, 247, 111–129.

Ghosh, S., & Patel, S. (2022). Policy frameworks for AI adoption in water management. *Environmental Policy and Governance*, 32(4), 333–344.

Haque, A., Ahmed, S., & Rahman, M. (2021). AI-based flood forecasting and disaster management: A case study from Bangladesh. *Natural Hazards*, 106(3), 1456–1472.

Jayaraman, V., & Ganesan, K. (2022). Future of AI in smart water grids: Opportunities and challenges. *Smart Cities*, 5(2), 142–155.

Jiang, Y., Zhang, H., & Wang, L. (2022). Artificial intelligence applications for smart irrigation and water efficiency in arid regions. *Agricultural Water Management, 264*, 107522.

Kourgialas, N. N., & Karatzas, G. P. (2018). The role of AI in environmental monitoring. *Water Resources Management*, 32(3), 783–797.

Kumar, R., Singh, A., & Tiwari, P. (2020). AI and IoT for water management: Case studies from India. *Journal of Sustainable Technology*, 9(2), 210–225.

Mengistu, T., Awulachew, S. B., & Melaku, T. (2021). Watershed management planning using AI and remote sensing techniques. *International Journal of Environmental Science and Technology*, 18(3), 765–780.

Mishra, A. K., Singh, R., & Pandey, R. (2021). Drought vulnerability mapping using AI and MODIS satellite data: A case study. *Natural Hazards*, 110(4), 1535–1554.

Pahl-Wostl, C. (2019). The role of governance modes and meta-governance in the transformation towards sustainable water governance. *Environmental Science & Policy, 91*, 6–16.

Park, J., Kim, H., & Lee, S. (2021). Machine learning models for predicting harmful algal blooms using water quality data. *Journal of Environmental Management*, 289, 112–124.

Public Utilities Board. (2020). *Smart Water Management in Singapore*. Singapore: Singapore Government Press.

Rahman, M. T., Sultana, S., & Chowdhury, R. (2020). Flood mapping using Sentinel-1 SAR data and machine learning algorithms. *Journal of Flood Risk Management*, 13(4), e12642.

Roberts, S., Smith, J., & Turner, M. (2021). AI-powered water quality monitoring in the UK: Enhancing water safety. *Environmental Monitoring and Assessment*, 193(2), 78–89.

Roberts, S., Turner, M., & Baker, J. (2020). AI applications in urban water management: A case study from Barcelona. *Environmental Monitoring and Assessment*, 192(3), 107–119.

Russell, S., & Norvig, P. (2020). *Artificial Intelligence: A Modern Approach* (4th ed.). Pearson.

Singh, M., & Sharma, P. (2021). Strategies for implementing AI in water resource management: Policy recommendations. *Water Resources Policy Review*, 43(5), 667–678.

Singh, V., Sharma, R., & Gupta, P. (2021). AI applications in water recycling processes. *Sustainable Water Resources*, 18(5), 345–360.

Smith, D., Baker, T., & Garcia, M. (2020a). Cost savings through AI-driven smart water grids in urban management. *Urban Water Journal*, 17(5), 457–469.

Smith, D., Baker, T., & Garcia, M. (2020b). Optimizing water allocation in times of drought using AI in Australia. *Journal of Hydrology*, 579, 121–132.

Suzuki, T., Tanaka, Y., & Yamamoto, K. (2021). AI-based leak detection in urban water supply systems: A case study from Tokyo. *Water Supply*, *21*(5), 2498–2510.

Tan, J., Ong, W., & Lee, H. (2021). Optimization of wastewater treatment in Singapore using AI: A case study. *Environmental Technology & Innovation*, 22, 100–110.

Tanaka, Y., Suzuki, K., & Yamamoto, N. (2021). Leak detection in urban water networks using artificial intelligence. *Water Science and Technology*, 84(6), 1345–1356.

Wagle, N., Sharma, P., & Patel, R. (2022). Artificial intelligence in sustainable water management: Opportunities and challenges. *Journal of Water Resources and Sustainability, 14*(3), 201–215.

Zhang, W., Zhou, Y., & Chen, L. (2019a). Precision irrigation using AI systems in Israel: A success story. *Agricultural Water Management*, 209, 115–124.
Zhang, Y., Li, H., & Sun, J. (2019b). Monitoring algal blooms using hyperspectral remote sensing and artificial intelligence. *Environmental Monitoring and Assessment*, 191(12), 789.
Zhao, M., Wang, Q., & Huang, J. (2020). Optimization of wastewater treatment processes using neural networks. *Water Research*, 177, 115–128.

14 Floodplain Width Prediction in Flood Disaster Management

Mohammad Uzair Anwar Qureshi, Isa Ebtehaj, and Hossein Bonakdari

14.1 INTRODUCTION

Accurately predicting floodplain width is critical for sustainable urban development and effective flood risk mitigation, both essential in pursuing net-zero emissions. Urbanization increasingly exacerbates flood risks, as expanding infrastructure and rising populations in flood-prone areas alter natural hydrological systems and escalate exposure to flood hazards (Fitriyati et al., 2024). This trend is especially significant as urban areas grow within floodplain regions, leading to greater impervious surfaces, reduced infiltration, and heightened surface runoff. Climate change compounds these risks by intensifying rainfall patterns, raising river levels, and potentially significantly increasing the population exposed to flood hazards if proactive measures are not implemented (Noori & Bonakdari, 2023). The challenge of reliable floodplain prediction is intensified by a range of hydrological uncertainties due to climate variability, river morphology changes, and human interventions that alter watercourses over time (Shah et al., 2015). The complex, interwoven effects of these factors mean that traditional flood prediction models may struggle to account for the dynamic interactions driving flood behaviors, particularly in urbanizing landscapes (Cardi et al., 2023). However, integrating urban growth prediction models with climate-sensitive flood risk scenarios enhances planners' ability to anticipate and prepare for worst-case flood events (Kim & Newman, 2020). This integrated approach enables cities to assess the impacts of future land-use changes on flood risks and to implement adaptive planning measures that protect vulnerable communities while fostering sustainable growth.

Sustainable flood risk management aligns with several UN Sustainable Development Goals (SDGs), including SDG 11 (Sustainable Cities and Communities) and SDG 13 (Climate Action), underscoring the need for proactive, data-driven approaches to mitigate the devastating impacts of climate-induced flooding. Addressing the dual challenges of climate change and flood-related disasters demands resilient urban planning strategies beyond traditional reactive measures. By integrating advanced floodplain prediction models into urban and environmental policy, planners can make informed decisions that reduce flood risks and contribute to a lower-carbon, resilient urban future (Binns, 2022). Recent studies underscore the critical role of accurate

 DOI: 10.1201/9781003547099-17

floodplain mapping in fostering urban resilience and adapting to climate change. Brandt et al. (2021) propose a tool that generates "flood uncertainty zones," offering urban planners a framework to incorporate spatial uncertainties into their planning processes. Zahmatkesh et al. (2021) emphasize the influence of climate change on flood probability, developing probabilistic floodplain maps that incorporate climate impacts and highlight the substantial role of hydrologic model input as a critical source of uncertainty in floodplain delineation. Their results indicate increased flood probability and an expansion of flood-prone areas under climate change scenarios. These studies illustrate the need for sophisticated flood modeling tools that address the significant uncertainties in floodplain mapping and adapt to the heightened flood risks expected with future climate change (Zahmatkesh et al., 2021).

Despite its critical importance, accurately predicting floodplain width presents considerable challenges, especially in the context of climate change and rapidly evolving urban landscapes (Mishra et al., 2022). Prediction efforts face significant limitations due to various causes and hydrological uncertainties stemming from climate-related factors (Shah et al., 2015). Variability in precipitation, including both temporal and spatial shifts in rainfall patterns, can drastically alter the magnitude, extent, and frequency of flood events, complicating floodplain delineation (Amiri et al., 2024). These changes are further influenced by shifts in storm tracks and seasonal precipitation, which are critical for assessing future flood risks under climate change (Gomez et al., 2024). Additionally, river morphology—shaped by ongoing processes like sedimentation and erosion—plays a vital role in determining discharge patterns and floodplain extent (Karamouz et al., 2010). As sediment deposits and erosion reshape river channels, they create unpredictable changes in flow paths and floodplain boundaries, adding a layer of complexity to accurate floodplain width prediction. This dynamic interaction between climate variability, river morphology, and urban development demands advanced, adaptable models that can incorporate these fluctuating factors to improve floodplain predictions and support resilient, future-focused flood management strategies. Quantifying floodplain width enables decision-makers to evaluate inundation potential, delineate high-risk zones, and assess implications for adjacent land uses. These hydrodynamic insights form the basis for targeted mitigation efforts and risk-informed urban planning, making assessing and accurately measuring floodplain/river width a vital step in building a thorough understanding of the magnitude of a specific flood event. It helps assess the potential inundation depth, pinpoint flood risk zones (Benito et al., 2001; Duží et al., 2017), and plan flood mitigation and management strategies (Diehl et al., 2021). Recent studies have highlighted the potential of using two-dimensional (2D) hydraulic modeling with the Hydrological Engineering Center-River Analysis System (HEC-RAS) to generate detailed flood discharge scenarios and improve flood hazard mapping accuracy (Priya et al., 2020). When integrated with Geographic Information Systems (GIS), HEC-RAS gains enhanced capabilities for visualizing and assessing flood risk, offering a powerful tool for urban planning and disaster preparedness (Azouagh et al., 2018). However, its application in real-time flood forecasting introduces both opportunities and challenges. In its 2D form, HEC-RAS is computationally demanding, which can be a barrier to real-time predictions that require rapid data processing (Lago et al., 2023). Additionally, the time-consuming nature of HEC-RAS simulations presents a limitation for real-time flood forecasting, where timely decision-making is essential (Idfi et al., 2019). These limitations have sparked interest in exploring alternative and complementary technologies like Artificial Intelligence (AI) techniques that can enhance flood prediction and floodplain management by addressing the need for quicker and more adaptable solutions. While AI techniques offer remarkable potential in flood prediction and floodplain management – demonstrating significant improvements in both accuracy and computational efficiency over traditional methods – numerous challenges persist (Ebtehaj & Bonakdari, 2022). Ensuring data availability and quality remains a critical challenge, as AI models require large and accurate datasets for effective training (Algarni, 2023). Accurate floodplain predictions rely on integrating various data sources, including climate records, hydrological data, and urban development patterns; however, these sources are often fragmented or incomplete. Floodplain characteristics differ significantly due to local topography, climate, river morphology, and land-use practices, making it challenging to apply models universally along the river. Predicting floodplain width also requires models to account for substantial uncertainties in hydrological processes and future climate scenarios. Variations in precipitation patterns, river flows, and storm frequency, driven by climate change, introduce complex variability that is challenging to model accurately (Soltani et al., 2021). AI models may struggle to generalize to unprecedented climate conditions or extreme weather events that have not been observed previously. Particularly in recent years, floods surpassing the 100-year flood event threshold have become increasingly common, serving as stark reminders of climate change's impact on flood dynamics. Such extreme events, which lack historical precedent, can undermine AI model accuracy since they fall outside the typical datasets used to train these models (Cardi et al., 2023).

To overcome these limitations, integrating AI with numerical tools that solve fundamental hydrodynamic equations and simulate flow behavior across multiple river cross-sections offers a promising approach for generating comprehensive datasets (Muñoz-Carpena et al., 2023). The software offers a plethora of capabilities for assessing and simulating river hydraulics, encompassing aspects such as floodplain inundation scenarios (Abdessamed & Abderrazak, 2019), sediment conveyance (Sathya et al., 2023), waterway profiles (Dysarz et al., 2017), and flow rates (Al-Hussein et al., 2022). By examining river hydrodynamics in a detailed and physics-based manner, these tools can produce data that captures the interactions between river morphology and flood behavior, providing a more robust foundation for AI models. Additionally, this approach facilitates the incorporation of crucial local variables, such as topography, climate, and land-use practices, addressing the issue that AI models trained on data from one region often fail to perform effectively in another. Enhancing AI models with numerical simulations rooted in fundamental hydrodynamic principles can thus provide more adaptable, reliable floodplain predictions, especially for regions facing new, climate-driven flood risks.

This study aims to advance floodplain width prediction by integrating numerical modeling with an AI algorithm based on the Flexible Group Method of Data Handling (FGMDH). The FGMDH technique, particularly well-suited to floodplain prediction, offers distinct advantages in handling complex, nonlinear relationships inherent in hydrological data (Basri et al., 2021). FGMDH can automatically identify and select the most influential variables for floodplain width prediction by iteratively constructing and optimizing model layers (Ebtehaj & Bonakdari, 2024). This self-organizing approach enhances model accuracy and reduces computational load by focusing only on relevant inputs, which is crucial for generating reliable predictions in dynamic environments like floodplains (Elkurdy et al., 2022). Moreover, FGMDH produces explicit, interpretable equations, enabling decision-makers to calculate floodplain characteristics easily and precisely, even in scenarios that deviate significantly from historical norms (Ebtehaj & Bonakdari, 2024). Within the context of water management, this AI algorithm offers advanced capabilities for analyzing complex, multivariate datasets that characterize hydrological processes, especially in regions like Ottawa, where seasonal variability and climate-driven events significantly influence river dynamics. A comprehensive dataset of river flow discharge scenarios will be generated using a 2D numerical model (HEC-RAS), simulating the full range of hydrodynamic behaviors, including extreme floods expected under various climate scenarios. These simulations produce data essential for accurately capturing the Ottawa River's floodplain response to both typical and atypical flood events, considering dynamic factors like river topography, slope, channel bed and floodplain roughness, upstream discharge, and cross-section locations. By incorporating these variables, the FGMDH model can automatically identify and select the most influential factors by iteratively constructing and optimizing model layers. This self-organizing approach enhances model accuracy and reduces computational load by focusing on relevant inputs, which is crucial for generating reliable predictions in dynamic environments like floodplains. This adaptability and interpretability make FGMDH a powerful tool in predicting floodplain responses under changing climate and river morphology, offering a practical, data-driven solution to the complex challenge of floodplain management.

14.2 MATERIALS AND METHODS

14.2.1 Study Area

Historically, the Ottawa River has been the site of significant flooding events that underscore its complex hydrological behavior and susceptibility to extreme hydrometeorological events. Notably, in the spring of 2017, intense rainfall coupled with rapid snowmelt contributed to a major flood that devastated communities along the river, including Ottawa and Gatineau (Teufel et al., 2019). The 2017 flood event was remarkable not only for its magnitude but also for highlighting the vulnerabilities within the region's flood management systems. Just two years later, in 2019, the Ottawa River once again experienced unprecedented flooding, with discharge reaching 5,980 m^3/s—exceeding the anticipated threshold for a 100-year flood event. This disaster led to widespread evacuations and prolonged states of emergency, surpassing the record set in 2017 (Kirchmeier-Young et al., 2021). These shifts in climate could lead to higher frequencies of flood-inducing rainfall events, posing heightened risks to communities and infrastructure along the river. Consequently, adaptive strategies and robust flood risk management frameworks are essential to mitigate these future risks (Noori & Bonakdari, 2023).

The study area is situated upstream of Ottawa, near Morris Island, along a tributary of the Ottawa River that runs parallel to the Mississippi Valley Watershed. This selected reach spans approximately 14 km and encompasses 22 surveyed cross-sections derived from high-resolution light detection and ranging (LiDAR) and river bathymetric datasets provided by the Mississippi Valley Conservation Authority (Figure 14.1). The study area is located 50 Km northwest of the City of Ottawa. The bed elevation of the river and the surrounding floodplain ranges from 50 m to 60 m above sea level. The undulating topography of the surrounding floodplains is mainly croplands and herbaceous grasslands. Within this reach, two key floodplain regions—located near Quyon and Constance Bay—have been identified as susceptible to inundation. These floodplains have experienced recurrent flood events in recent years, primarily driven by spring snowmelt and intensified precipitation patterns associated with climate change. In May 2019, Quyon experienced severe flooding, leading to the evacuation of over 100 residents due to concerns over the structural integrity of the dike protecting the area from the Ottawa and Quyon rivers (Canadian Broadcasting Corporation, 2019). Concurrently, Constance Bay underwent evacuations in April 2019 as the Ottawa River water levels rose significantly, attributed to accelerated snowmelt and excessive rainfall (Global News, 2019). These flood events underscore the importance of detailed hydrological and hydraulic analysis for accurate flood risk assessment and management.

14.2.2 Numerical Tool

The Hydraulic Engineering Centre's River Analysis System (HEC-RAS), developed by the U.S. Army Corps of Engineers, is a widely recognized tool for simulating

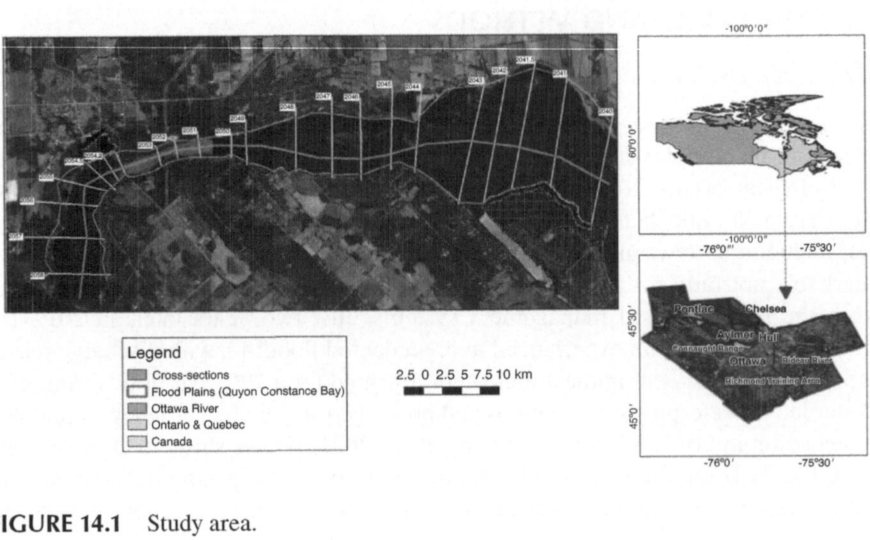

FIGURE 14.1 Study area.

river hydraulic processes. Known for its accuracy and versatility, HEC-RAS is often used as a benchmark to assess other hydrodynamic models. It supports both steady and unsteady flow simulations for subcritical, supercritical, and mixed flow regimes, offering modules for sediment transport, ice cover dynamics, dam breach analysis, and water quality assessments. The HEC-RAS modeling process combines detailed river geometry with hydrological data to simulate flood scenarios and predict inundation areas. By applying the Saint Venant equations for mass and momentum conservation, Manning's equation for frictional effects, and accounting for energy losses, HEC-RAS provides a robust framework for flood mapping.

In 1D mode, HEC-RAS integrates channel and overbank flows, improving flood level predictions. Flow routing can be defined within each branch for branching channels, though users must manually specify distribution across branches. Unsteady 1D simulations use the finite difference method to solve the St. Venant equations for continuity and momentum and provide a realistic perspective about possible flood threats by improving the accuracy of official flood maps. Its 2D mode enables fully two-dimensional modeling within specified areas and allows dynamic links to 1D segments for integrated hybrid modeling. Within 2D domains, it solves the shallow water equations, i.e., Saint Venant Equation or the Diffusion wave equation, to quantify longitudinal and lateral variability of the water depth and flow velocities in the floodplains (Zeiger & Hubbart, 2021). 2-D models are used to simulate the floodplain flow to demonstrate the extent of floods and to overcome the 1-D limitations to improve the uncertainty of results (Sharma & Regona, 2021). As highlighted by Yazdan et al. (2022), the coupled 1D/2D approach in HEC-RAS enables users to apply one-dimensional (1D) modeling within the main river channel while utilizing two-dimensional (2D) computational grids in the surrounding floodplain. This hybrid method optimizes computational efficiency by reducing the total number of calculation points, resulting in faster run times without compromising accuracy. In the current study, this approach was employed to balance precision and computational speed, allowing the generation of a comprehensive dataset that captures the hydrodynamics of flow across all 22 cross-sections. This rich dataset supports detailed floodplain mapping and flood risk assessment, providing essential inputs to feed the AI model for enhanced predictive capabilities. For hydraulic analysis within HEC-RAS, coupling geometric models of river systems with high-resolution elevation data is essential. Typically, geometric models are combined with digital elevation models (DEMs) derived from LiDAR technology, sourced from local or national agencies. These DEMs are provided as raster databases with grid resolutions that vary depending on site-specific data availability. Once obtained, the DEM can be imported into HEC-RAS as a terrain file to support modeling. Selecting an optimal DEM resolution is crucial for generating accurate hydrodynamic model outputs. Therefore, understanding the impact of DEM resolution on model precision is vital for selecting the most effective resolution. The model was executed and calibrated using data from Hydrometric Station 02KF005 on the Ottawa River, provided by the Environmental and Natural Resources Department of the Government of Canada. A wide range of river discharge representing a 2-year flood, 10-year flood, 100-year flood, 1,000-year flood, and 10,000-year flood were selected to simulate the hydrodynamic response to snowmelt and spring thaw events, capturing the variability in river flow conditions during the critical seasonal transitions and keeping in mind the unprecedented events that may occur due to climate change.

14.2.3 Flexible Group Method of Data Handling (FGMDH)

The Group Method of Data Handling (GMDH) is an inductive self-organizing algorithm used for modeling complex systems and identifying patterns in datasets (Letessier et al., 2023). It was first introduced by Alexey Ivakhnenko in the 1960s (Es-Haghi et al., 2021) and has since been applied in various fields such as engineering, economics, medicine, and environmental sciences (Scott & Hutchinson, 2020; Ebtehaj et al., 2021; Mehrankia et al., 2022; Piri et al., 2024). The method is useful for creating predictive models and identifying relationships in datasets where traditional modeling approaches may struggle due to complexity or non-linearity.

GMDH builds models as systems of polynomial equations, setting it apart from other machine learning algorithms by generating explicit mathematical expressions for predictions, which provide interpretable models that clarify the influence of each input variable on the output. Its key features include self-organization (Safari et al., 2019), where the algorithm autonomously determines the model structure, including input selection and relationship formulation; interpretability (Cardi et al., 2023), as it produces equations that clearly define the relationships between variables; scalability (Elkurdy et al., 2022), enabling its application across various problems such as regression, classification, and time-series forecasting; and flexible complexity (Ebtehaj & Bonakdari, 2024), as it iteratively refines its structure to match the problem's intricacy without requiring prior knowledge of the data's relationships.

The GMDH model is based on approximating the target function y as a polynomial expansion of the input variables $x_1, x_2,\ldots,x_N$ as shown in Equation 14.1:

$$\hat{y} = f^*(x_1,\ldots,x_N) = s_0 + \sum_{i=1}^{N} s_i x_i + \sum_{i=1}^{N}\sum_{j=1}^{N} s_{ij} x_i x_{ij} + \ldots \tag{14.1}$$

This general form can include terms of increasing complexity (e.g., cubic, quartic), but the second-order polynomial is most commonly used for simplicity and interpretability.

When considering two inputs, x_1 and x_2, as shown in Equation 14.2, the second-order polynomial expansion becomes:

$$\hat{y} = s_0 + s_1 x_1 + s_2 x_2 + s_{11} x_1^2 + s_{22} x_2^2 + s_{12} x_1 x_2 \tag{14.2}$$

Where s_0 is the intercept or bias term, s_1 and s_2 denote coefficients for the linear terms, s_{11} and s_{22} represent coefficients for the quadratic terms, and s_{12} is the coefficient of the interaction term.

A criterion function E is used to evaluate each candidate polynomial model's performance. As shown in Equation 14.3, the most common is **Root Mean Square Error (RMSE)**:

$$E = \sqrt{\frac{1}{M}\sum_{k=1}^{M}\left(y_k - \hat{y}_k\right)^2} \tag{14.3}$$

where M is the number of samples in the validation dataset, y_k shows the actual observed values for the kth samples, and $\hat{y}_k$ denotes the predicted values from the polynomial model for the kth sample.

The coefficients s_0, s_i, s_{ij} are determined by solving the least-squares problem. The goal is to minimize the error E, leading to Equation 14.4, which are normal equations:

$$\mathbf{H}^T\mathbf{H}\mathbf{S} = \mathbf{H}^T\mathbf{y} \tag{14.4}$$

where $\mathbf{S} = \left[s_0, s_1, s_2, s_{11}, s_{22}, s_{12}\right]^T$ is the vector of the unknown coefficients to be optimized, $\mathbf{y} = \left[y_1, y_2, ..., s_M\right]^T$ represents the vector of observed outputs, and $\mathbf{H}$ denotes the design matrix containing the polynomial terms. For two inputs (x_1, x_2), $\mathbf{H}$ is defined as follows in Equation 14.5:

$$\mathbf{H} = \begin{bmatrix} 1 & x_{1,1} & x_{2,1} & x_{1,1}^2 & x_{2,1}^2 & x_{1,1}x_{2,1} \\ 1 & x_{1,2} & x_{2,2} & x_{1,2}^2 & x_{2,2}^2 & x_{1,2}x_{2,2} \\ \vdots & \vdots & \vdots & \vdots & \vdots & \vdots \\ 1 & x_{1,M} & x_{2,M} & x_{1,M}^2 & x_{2,M}^2 & x_{1,M}x_{2,M} \end{bmatrix} \tag{14.5}$$

The GMDH algorithm proceeds iteratively by constructing models layer by layer. For a given layer, l, the output of the models from the previous layer ($\hat{y}^{(l-1)}$) is used as the input for the current layer. This can be written in Equation 14.6 as:

$$\hat{y}^{(l)} = f^{(l)}\left(\hat{y}_1^{(l-1)}, \hat{y}_2^{(l-1)}, ..., \hat{y}_k^{(l-1)}\right) \tag{14.6}$$

where $\hat{y}_i^{(l-1)}$ are outputs of the previous layer and $f^{(l)}$ is the polynomial model of the current layer.

The algorithm stops when the performance of the models no longer improves. Mathematically, this can be defined in Equation 14.7 as:

$$E^{(l)} - E^{(l-1)} \leq \varepsilon \tag{14.7}$$

where $E^{(l)}$ is the error of the models in the current layer and ε denotes the small threshold to determine convergence.

GMDH is a versatile and powerful tool for predictive modeling, particularly suited to systems with nonlinear and complex interactions among variables. Its ability to self-organize and generate interpretable models makes it an attractive choice for researchers and practitioners alike. However, GMDH has certain limitations, including its susceptibility to overfitting, reliance solely on second-order polynomials, restriction to only two input variables in each polynomial, and, notably, the generation of new neurons exclusively from the adjacent layer. The Flexible GMDH (FGMDH) is introduced in this study to address these limitations.

In the FGMDH, to address the issue of overfitting in GMDH, the Akaike Information Criterion (AIC) is employed as a selection criterion. The AIC balances the trade-off between the accuracy of the model and its complexity, thereby helping to prevent overfitting. Overfitting occurs when a model becomes too complex, fitting not only the underlying data patterns but also the noise, compromising its ability to generalize to unseen data.

The AIC is defined in Equation 14.8 as:

$$\mathrm{AIC} = k\ln(E^2) + 2C \tag{14.8}$$

where k represents the number of observations in the dataset, E is the root mean square error defined in Equation 14.3, and C is the number of tuned parameters in the final model (i.e., the vector of the unknown coefficients for all generated neurons, like Equation 14.2).

The AIC works by penalizing models with a high number of parameters (C), ensuring that simpler models are preferred unless the added complexity provides a significant improvement in data fit. For FGMDH, incorporating the AIC into the model selection process ensures that the algorithm does not favor overly complex polynomial structures or neurons, thereby improving generalization. This approach allows FGMDH to strike a balance between model accuracy and simplicity, making it more robust for predictive modeling tasks. The use of AIC enhances the overall performance and interpretability of the generated models while mitigating the risks associated with overfitting.

Moreover, the developed FGMDH has the capability to utilize all generated neurons from the previous layers, as well as the input variables, for neuron generation in every layer. Additionally, the inputs for each neuron can consist of either two or three variables in this method. Consequently, three new types of polynomials, in addition to the ones proposed in Equation 14.2, are introduced in the FGMDH as follows:

$$\hat{y} = s_0 + s_1x_1 + s_2x_2 + s_3x_3 + s_{11}x_1^2 + s_{22}x_2^2 + s_{33}x_3^2 + s_{12}x_1x_2 + s_{13}x_1x_3 + s_{23}x_2x_3 \tag{14.9}$$

$$\hat{y} = s_0 + s_1x_1 + s_2x_2 + s_{11}x_1^2 + s_{22}x_2^2 + s_{12}x_1x_2 + s_{122}x_1x_2^2 + s_{112}x_1^2x_2 + s_{111}x_1^3 + s_{222}x_2^3 \tag{14.10}$$

$$\begin{aligned}\hat{y} = {} & s_0 + s_1x_1 + s_2x_2 + s_3x_3 + s_{11}x_1^2 + s_{22}x_2^2 + s_{33}x_3^2 \\ & + s_{12}x_1x_2 + s_{13}x_1x_3 + s_{23}x_2x_3 + s_{111}x_1^3 + s_{222}x_2^3 + s_{333}x_3^3 + s_{112}x_1^2x_2 \\ & + s_{113}x_1^2x_3 + s_{221}x_2^2x_1 + s_{223}x_2^2x_3 + s_{331}x_3^2x_1 + s_{332}x_3^2x_2 + s_{123}x_1x_2x_3\end{aligned} \tag{14.11}$$

Due to the significant structural modifications introduced in FGMDH compared to the GMDH, the coefficients, denoted as $\mathbf{S} = [s_0, s_1, s_2, s_{11}, s_{22}, s_{12}]^T$ in Equation 14.2, along with the matrix in Equation 14.5, are updated accordingly. While the original GMDH employs only one polynomial type (Equation 14.2), FGMDH incorporates three additional types: a second-order polynomial with three inputs (Equation 14.9), a third-order polynomial with two inputs (Equation 14.10), and a third-order polynomial with three inputs (Equation 14.11). As a result, these equations are adjusted within FGMDH in three distinct ways.

14.2.3.1 2nd Order Polynomial with Three Inputs

$$\mathbf{S} = [s_0, s_1, s_2, s_3, s_{11}, s_{22}, s_{33}, s_{12}, s_{13}, s_{23}] \tag{14.12}$$

$$L = \begin{bmatrix} 1 & x_{1,1} & x_{2,1} & x_{3,1} & x_{1,1}^2 & x_{2,1}^2 & x_{3,1}^2 & x_{1,1}x_{2,1} & x_{1,1}x_{3,1} & x_{2,1}x_{3,1} \\ 1 & x_{1,2} & x_{2,2} & x_{3,2} & x_{1,2}^2 & x_{2,2}^2 & x_{3,2}^2 & x_{1,2}x_{2,2} & x_{1,1}x_{3,2} & x_{2,1}x_{3,2} \\ \vdots & \vdots & \vdots & \vdots & \vdots & \vdots & \vdots & \vdots & \vdots & \vdots \\ 1 & x_{1,M} & x_{2,M} & x_{3,M} & x_{1,M}^2 & x_{2,M}^2 & x_{3,M}^2 & x_{1,M}x_{2,M} & x_{1,M}x_{3,M} & x_{2,M}x_{3,M} \end{bmatrix} \tag{14.13}$$

14.2.3.2 Third-Order Polynomial with Two Inputs

$$\mathbf{S} = [s_0, s_1, s_2, s_{11}, s_{22}, s_{12}, s_{122}, s_{112}, s_{111}, s_{222}] \tag{14.14}$$

$$L = \begin{bmatrix} 1 & x_{1,1} & x_{2,1} & x_{1,1}^2 & x_{2,1}^2 & x_{1,1}x_{2,1} & x_{1,1}^2x_{2,1} & x_{1,1}x_{2,1}^2 & x_{1,1}^3 & x_{2,1}^3 \\ 1 & x_{1,2} & x_{2,2} & x_{1,2}^2 & x_{2,2}^2 & x_{1,2}x_{2,2} & x_{1,2}^2x_{2,2} & x_{1,2}x_{2,2}^2 & x_{1,2}^3 & x_{2,2}^3 \\ \vdots & \vdots & \vdots & \vdots & \vdots & \vdots & \vdots & \vdots & \vdots & \vdots \\ 1 & x_{1,M} & x_{2,M} & x_{1,M}^2 & x_{2,M}^2 & x_{1,M}x_{2,M} & x_{1,M}^2x_{2,M} & x_{1,M}x_{2,M}^2 & x_{1,M}^3 & x_{2,M}^3 \end{bmatrix} \tag{14.15}$$

14.2.3.3 Third-Order Polynomial with Three Inputs

$$\mathbf{S} = \begin{bmatrix} s_0, s_1, s_2, s_3, s_{11}, s_{22}, s_{33}, s_{12}, s_{13}, s_{23}, s_{111}, s_{222}, \\ s_{333}, s_{112}, s_{113}, s_{221}, s_{223}, s_{331}, s_{332}, s_{123}x_1x_2x_3 \end{bmatrix} \tag{14.16}$$

$$L = \begin{bmatrix} 1 & x_{1,1} & x_{2,1} & x_{3,1} & x_{1,1}^2 & x_{2,1}^2 & x_{3,1}^2 & x_{1,1}x_{2,1} & x_{1,1}x_{3,1} & x_{2,1}x_{3,1} \\ 1 & x_{1,2} & x_{2,2} & x_{3,2} & x_{1,2}^2 & x_{2,2}^2 & x_{3,2}^2 & x_{1,2}x_{2,2} & x_{1,2}x_{3,2} & x_{2,2}x_{3,2} \\ \vdots & \vdots & \vdots & \vdots & \vdots & \vdots & \vdots & \vdots & \vdots & \vdots \\ 1 & x_{1,M} & x_{2,M} & x_{3,M} & x_{1,M}^2 & x_{2,M}^2 & x_{3,M}^2 & x_{1,M}x_{2,M} & x_{1,M}x_{3,M} & x_{2,M}x_{3,M} \end{bmatrix} \cdots$$

$$\cdots \begin{matrix} x_{1,1}^3 & x_{2,1}^3 & x_{3,1}^3 & x_{1,1}^2x_{2,1} & x_{1,1}^2x_{3,1} & x_{2,1}^2x_{1,1} & x_{2,1}^2x_{3,1} & x_{3,1}^2x_{1,1} \\ x_{1,2}^3 & x_{2,2}^3 & x_{3,2}^3 & x_{1,2}^2x_{2,2} & x_{1,2}^2x_{3,2} & x_{2,2}^2x_{1,2} & x_{2,2}^2x_{3,2} & x_{3,2}^2x_{1,2} \\ \vdots & \vdots & \vdots & \vdots & \vdots & \vdots & \vdots & \vdots \\ x_{1,M}^3 & x_{2,M}^3 & x_{3,M}^3 & x_{1,M}^2x_{2,M} & x_{1,M}^2x_{3,M} & x_{2,M}^2x_{1,M} & x_{2,M}^2x_{3,M} & x_{3,M}^2x_{1,M} \end{matrix}$$

$$\begin{matrix} x_{3,1}^2x_{2,1} & x_{1,1}x_{2,1}x_{3,1} \\ x_{3,2}^2x_{2,2} & x_{1,2}x_{2,2}x_{3,2} \\ \vdots & \vdots \\ x_{3,M}^2x_{2,M} & x_{1,M}x_{2,M}x_{3,M} \end{matrix} \Bigg] \quad (14.17)$$

14.2.3.4 Developed Methodology

The developed methodology in the current study, illustrated in Figure 14.2, adopts a comprehensive approach to floodplain mapping and prediction, integrating hydraulic modeling with AI to support sustainable land-use planning aimed at net-zero emissions. This approach is designed to account for a wide range of flood conditions, including those that reflect the behavior of the Ottawa River under climate change impacts—even predicting floods beyond those recorded historically. Initially, essential data, including river geometry, streamflow characteristics, hydraulic properties, flood hydrographs, topographic features, and boundary conditions, are gathered to establish a robust numerical model. This model undergoes rigorous calibration and validation to represent the river's hydrodynamics accurately.

Once validated, the model is used to simulate river dynamics across diverse flow regimes, creating an extensive data bank that captures flood inundation characteristics under various conditions. This data bank is subsequently divided into training (70%) and testing (30%) datasets, preparing it for AI-driven flood mapping. The AI algorithms are then trained using these datasets to enable accurate floodplain prediction. By applying second- and third-order polynomial combinations, the AI model generates predictive equations that allow rapid floodplain mapping, essential for real-time applications.

The outputs of the AI model contribute to sustainable land-use planning by enabling precise floodplain zoning. This aligns with net-zero emissions objectives by supporting resilient, low-carbon flood management strategies. Additionally, this

AI-driven approach allows the integration of predictions into real-time flood monitoring systems, optimizing flood defense infrastructure through targeted insights and supporting community preparedness. By minimizing post-flood carbon impacts through precision damage assessment, the methodology enhances adaptive capacity while promoting climate resilience and sustainable urban planning.

14.2.4 Statistical Validation of Model Accuracy

This research utilized various statistical metrics to assess the effectiveness and precision of the flood prediction models. The metrics applied encompass correlation measures, absolute error metrics, and relative error measures.

Correlation-based metrics, such as Nash-Sutcliffe efficiency (NSE) and the coefficient of determination (R^2), evaluate the relationship between observed and predicted values, providing model fit and predictive performance measures. R^2 ranges from 0 to 1, with higher values denoting a more substantial alignment between observed and predicted data. NSE, on the other hand, ranges from negative infinity to 1, where 1 signifies a perfect match, 0 indicates predictions are as accurate as the observed mean, and negative values suggest the model performs worse than the observed mean.

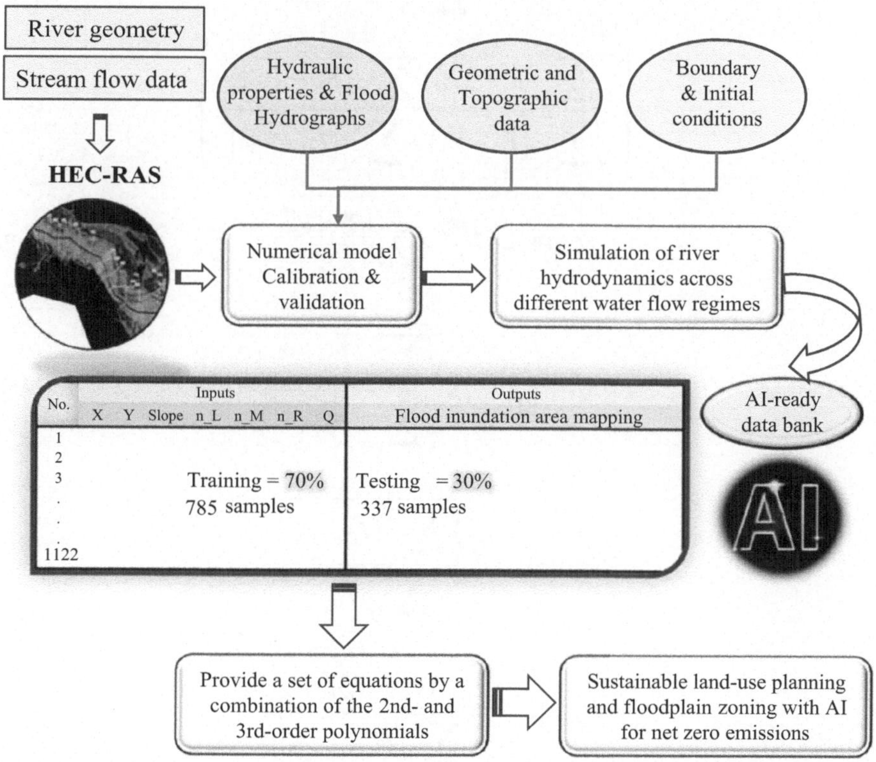

FIGURE 14.2 Integration of hydraulic modeling and AI for enhanced floodplain mapping and prediction.

Absolute metrics, including the percent bias (PBIAS), normalized root mean square error (NRMSE), and RMSE-to-observed standard deviation ratio (RSR), quantify the absolute difference between predicted and observed values, providing insights into model accuracy and systematic bias. Ideally, PBIAS should be near zero, indicating negligible bias. NRMSE values have a minimum of zero, representing optimal performance, while the maximum can extend to infinity, indicating poor accuracy. RSR values range from 0 to a large positive number, with lower values signifying better model performance.

Relative metrics, such as the mean absolute percentage error (MAPE), assess the proportional accuracy of model predictions by measuring the average error magnitude as a percentage of observed values. This enables effective comparisons across datasets with varying scales. Like RSR and NRMSE, smaller MAPE values indicate better model performance, with values nearing zero reflecting excellent predictive accuracy. The mathematical formulations for these metrics are provided in Equations 14.18–14.23 (Table 14.1).

$$\text{NSE} = 1 - \frac{\sum_{i=1}^{k}\left(P_M^i - P_O^i\right)^2}{\sum_{i=1}^{k}\left(P_M^i - \overline{P_O}\right)^2} \tag{14.18}$$

$$R^2 = \left(\frac{\sum_{i=1}^{k}\left(P_O^i - \overline{P_O}\right)\sum_{i=1}^{k}\left(P_M^i - \overline{P_M}\right)}{\sqrt{\sum_{i=1}^{k}\left(P_O^i - \overline{P_O}\right)^2 \sum_{i=1}^{k}\left(P_M^i - \overline{P_M}\right)^2}}\right)^2 \tag{14.19}$$

$$\text{MAPE} = \frac{100}{k} \times \sum_{i=1}^{k} \frac{\left|P_O^i - P_M^i\right|}{P_O^i} \tag{14.20}$$

$$\text{RSR} = \frac{\text{RMSE}}{\text{SD}_O} = \frac{\sqrt{\sum_{i=1}^{k}\left(P_O^i - P_M^i\right)^2}}{\sqrt{\sum_{i=1}^{k}\left(P_O^i - \overline{P_O}\right)^2}} \tag{14.21}$$

$$\text{NRMSE} = \frac{\sqrt{\frac{1}{k}\sum_{i=1}^{k}\left(P_O^i - P_M^i\right)^2}}{\sum_{i=1}^{k}\left(P_O^i\right)} \tag{14.22}$$

$$\text{PBIAS} = \frac{\sum_{i=1}^{k}\left(P_M^i - P_O^i\right)}{\sum_{i=1}^{k} P_O^i} \times 100 \tag{14.23}$$

where P_O is the observed flow discharge, P_M represents the modeled flow discharge, and $\overline{P_O}$ denotes the average of the observed flow discharge in this study. The

TABLE 14.1
Evaluation Metrics for Assessing Model Accuracy Using Various Statistical Indices

Descriptive Performance	R^2	NSE	NRMSE	RSR	PBIAS
Unsatisfactory	$R^2 < 0.5$	NSE < 0.4	30% < NRMSE	0.7 < RSR	PBIAS ≥ ±25%
Acceptable	-	0.4 < NSE < 0.5	-	-	-
Satisfactory	$0.5 < R^2 < 0.6$	0.5 < NSE < 0.65	20% < NRMSE < 30%	0.6 < RSR < 0.7	±15% ≥ PBIAS > ±25%
Good	$0.6 < R^2 < 0.7$	0.65 < NSE < 0.75	10% < NRMSE < 20%	0.5 < RSR < 0.6	±10% ≥ PBIAS > ±5%
Very good	$0.7 < R^2 < 1$	0.75 < NSE < 1	NRMSE < 10%	0 < RSR < 0.5	±5% ≥ PBIAS

descriptive performance of the R^2, NSE, NRMSE, RSR, and PBIAS (Moriasi et al., 2007; Ayele et al., 2017; Grégoire et al., 2023)

14.3 RESULTS

14.3.1 HEC-RAS Riverbed Modeling

The riverbed surface was modeled using a series of cross-sections derived from high-resolution LiDAR data combined with detailed bathymetric surveys. These cross-sections, referred to as river stations in HEC-RAS, were arranged sequentially along the channel, with station numbers increasing from downstream to upstream. The spatial spacing between consecutive cross-sections, quantified as Downstream Reach Lengths, was further divided into three distinct components: Left Overbank, Channel, and Right Overbank Reach Lengths. This segmentation method captures the effects of channel meandering and variations in flow pathways across the floodplain, ensuring an accurate representation of hydraulic behavior.

Each cross-section was laterally referenced using a coordinate system that starts from the left bank and spans across the channel to the right bank. These cross-sections were anchored along a one-dimensional centerline, with elevation data accurately reflecting the terrain profile. The Mississippi Valley Conservation Authority (MVCA) provided this high-resolution elevation information as part of their floodplain mapping initiative, ensuring the model's precision and consistency with regional hydraulic characteristics.

To illustrate, the river cross-section at station 2040 shown in Figure 14.3, which represents the final cross-section in the model, has Downstream Reach Lengths of zero, highlighting its terminal position in the channel profile. This configuration, shown in Figure 14.3, exemplifies how the HEC-RAS model incorporates cross-sectional data to simulate river hydrodynamics and floodplain characteristics accurately.

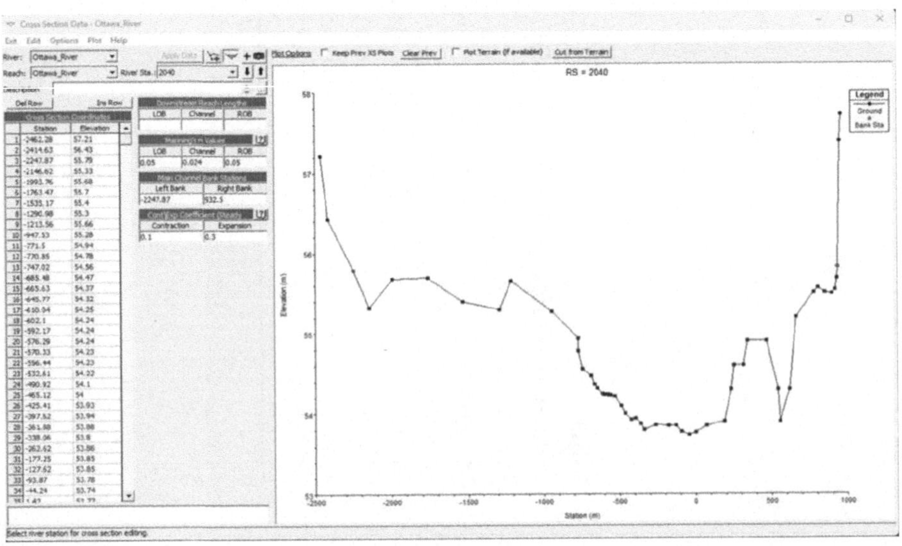

FIGURE 14.3 HEC-RAS cross-section data (XS 2040).

14.3.1.1 Geometric Representation in 2D HEC-RAS Modeling

A critical geometric component of the 2D HEC-RAS model used in this study is incorporating a High-Resolution Digital Elevation Model (HRDEM) to represent the floodplain and surrounding terrain accurately. The HRDEM, sourced from the CanElevation Series provided by the Government of Canada, ensures reliability and consistency in terrain data. Imported at a resolution of 1m x 1m grid cells, the HRDEM offers a detailed and precise representation of elevation, which is essential for capturing the topographic variability that significantly influences hydrodynamic flow behavior within floodplains and adjacent areas.

To seamlessly integrate river channel geometry with floodplain terrain, bathymetric cross-sectional data of the river channel were superimposed onto the HRDEM. This approach enabled a smooth transition between the high-resolution floodplain topography and the detailed bathymetric profiles of the riverbed. This integration was critical for accurately simulating hydraulic interactions both within the river channel and across the floodplain, ensuring realistic floodplain mapping and hydrodynamic flow predictions, as shown in Figure 14.4. The projected coordinate system for the model was **"NAD_1983_CSRS_UTM_Zone_18N"**, aligning with regional geospatial standards and ensuring accurate spatial referencing throughout the study.

14.3.2 Flow Data and Boundary Conditions

For our model, flow data provided by Environment and Climate Change Canada's Hydrometric Station 02KF005, located at Britannia (coordinates: 45°21'04″ N, 75°49'35″ W), served as the primary input. This station offers a continuous 110-year record of discharge and water surface elevation (WSEL) data, ensuring a robust dataset for calibrating and validating the hydraulic model. A flow hydrograph was developed using historical discharge data from the Britannia station to define the initial conditions. The simulation period was set for a timeframe marked by significant discharge variability due to snowmelt and spring thaw events. During this period, discharge values ranged from a peak of 3,000 m^3/s to 8,000 m^3/s corresponding to a 2-year, 5-year, 100-year, 1,000-year, and 10,000-year flood, representing the flow conditions at the upstream boundary (cross-section 2058). For comprehensive scenario testing, a discharge increment of 100 m^3/s was applied for each successive simulation, resulting in a total of 51 unique model runs. This range of simulations effectively captured the hydrodynamic response under varying flow conditions, reflecting the critical seasonal variability of the Ottawa River. The normal depth criterion for the downstream boundary conditions was employed to ensure a steady-state WSEL as flow exited the model domain. The normal depth was calculated using the average bed slope derived from cross-sectional data across the modeled reach, ensuring consistency with natural river hydraulics. This configuration facilitated a realistic downstream flow regime, aligning with the physical behavior of the river. Using these detailed configurations, the model achieved high fidelity in representing the real-time hydrodynamics of the Ottawa River. The robust setup allowed for reliable calibration and validation against observed datasets, ensuring confidence in the model's predictive floodplain analysis and management capabilities.

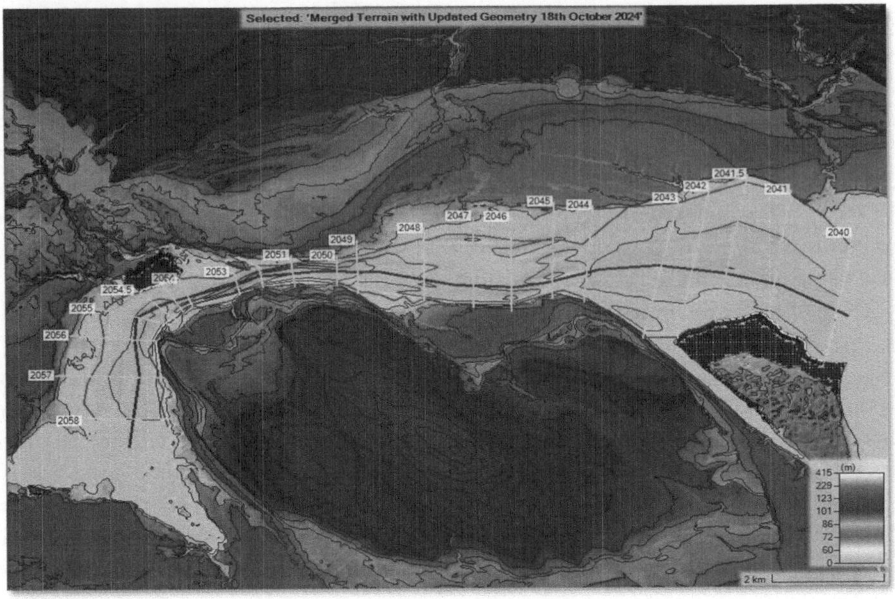

FIGURE 14.4 Contoured Terrain layer of the study area.

The comprehensive dataset generated by the HEC-RAS simulations forms the foundation for training the AI model presented in the next section. The model captures the hydrodynamic complexity of the Ottawa River and its floodplain by simulating a wide range of flow scenarios, including extreme discharge events beyond historical records. This rich dataset provides the AI model with high-resolution inputs necessary for accurate floodplain prediction and flood risk assessment. Employing AI to analyze and predict flood dynamics not only enhances the efficiency of floodplain mapping but also facilitates sustainable land-use planning and flood management strategies.

14.4 AI-BASED MODEL

Figure 14.5 illustrates a comparison between the traditional GMDH model (top) and the Flexible GMDH model (FGMDH) (bottom) in predicting the channel width using multiple input variables such as nR, nM, nL (Manning roughness at right, middle, and left), S (slope), Latitude, Longitude, and Q (Discharge). In the GMDH model, the structure progresses in a layer-wise manner, where each layer builds neurons representing polynomial relationships (e.g., x11, x15, x23), and connections are restricted to adjacent neurons from consecutive layers. This fixed architecture limits the flexibility of the model, as neurons in a new layer can only utilize outputs from the immediately preceding layer. In the GMDH structure, nR and nM in the input layer are "dead neurons" that are not used in any connections or computations, reflecting a lack of flexibility in selecting relevant inputs.

In contrast, the FGMDH model introduces greater flexibility, allowing neurons to connect directly across multiple layers. This architecture enables more complex and diverse feature interactions, improving the model's ability to capture intricate patterns within the data. In the FGMDH model, while some neurons, such as nM, are still "dead neurons," the selection of active neurons is more dynamic and adaptive compared to the GMDH structure. This highlights FGMDH's ability to better prioritize relevant features for modeling despite still having unused inputs. The FGMDH structure, as depicted, connects neurons (e.g., x10, x29, x36) in a less constrained manner, facilitating a more robust and adaptive prediction of channel width. The critical difference lies in the FGMDH model's ability to bypass the rigid layer-to-layer progression, leading to potentially more accurate and efficient results.

Figure 14.6 shows scatter plots that compare the predicted and actual channel widths for the FGMDH and the classical GMDH models. The results are presented for both training and testing datasets, providing an evaluation of the performance of the two models in predicting channel widths. Each subplot includes a diagonal 1:1 reference line, indicating perfect prediction and ±10% boundaries, representing acceptable error margins.

In the top row, the FGMDH model's performance is displayed. The left plot shows the training dataset results, where the predicted channel widths are tightly clustered

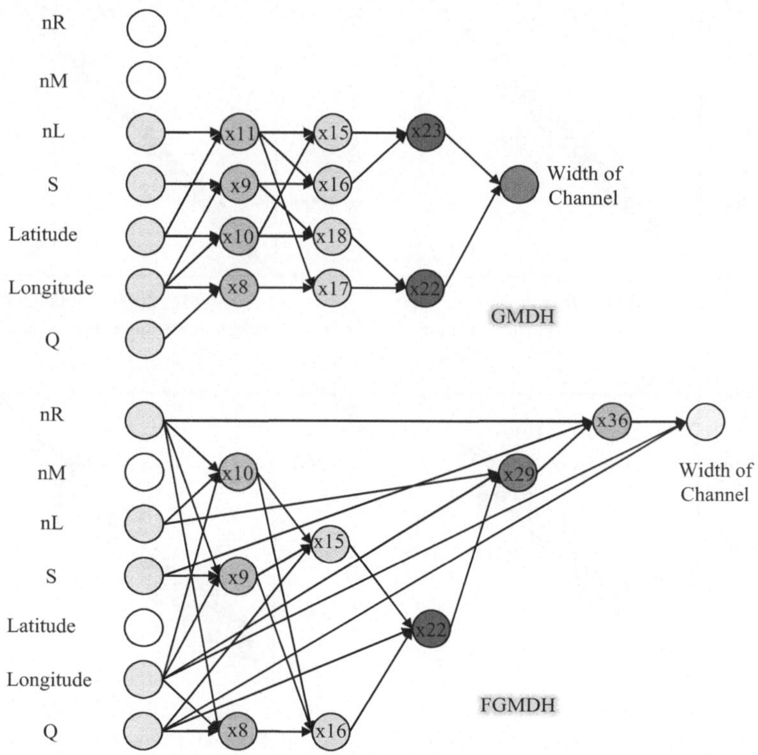

FIGURE 14.5 The structure of the classical GMDH and FGMDH for channel width prediction.

around the 1:1 line, with most points falling within the ±10% error bounds. This indicates a high level of accuracy during training. Similarly, the right plot presents the testing dataset results for FGMDH. The predicted values remain closely aligned with the actual values, and most points stay within the acceptable error range, suggesting that the FGMDH model generalizes well to unseen data. The compact clustering of data points and their proximity to the reference line highlight FGMDH's robustness and adaptability.

In the bottom row, the GMDH model's performance is shown for both training (left) and testing (right) datasets. For the training dataset, the scatter plot reveals a more scattered distribution of predicted values compared to FGMDH, with several points deviating from the ±10% boundaries. This indicates that the GMDH model has lower accuracy during training. The testing dataset results for GMDH (bottom-right plot) also demonstrate significant deviations from the 1:1 line, with many points falling outside the ±10% error range. This suggests that the GMDH model struggles to generalize effectively to new data, leading to less reliable predictions.

Overall, this figure highlights the superiority of the FGMDH model over the GMDH model in predicting channel widths. FGMDH consistently produces

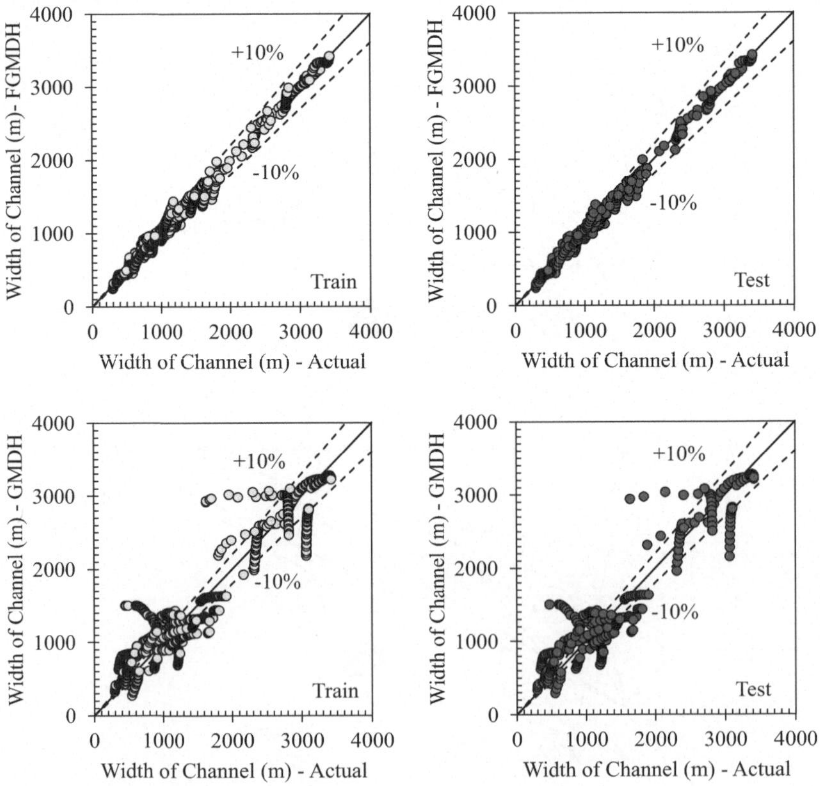

FIGURE 14.6 Scatter plots comparing predicted and actual channel widths for the Flexible GMDH (FGMDH) and classical GMDH models using training and testing datasets.

predictions that are closer to the actual values and within acceptable error margins for both training and testing datasets. In contrast, the GMDH model exhibits greater variability and larger prediction errors, particularly in the testing dataset. These results emphasize the advantages of FGMDH's flexible architecture, which enables it to capture complex relationships more effectively than the classical GMDH model.

Table 14.2 shows the performance evaluation metrics for the FGMDH and traditional GMDH models during the training and testing stages based on R^2, NSE, NRMSE, RSR, PBIAS, and MAPE. The metrics are categorized into descriptive performance levels, such as "Very Good," "Satisfactory," or "Good," as indicated in the table.

For the FGMDH model, the R^2 and NSE values are consistently 0.99 during both training and testing, indicating an almost perfect fit between the predicted and actual channel widths. The NRMSE values are extremely low (4.68% for training and 4.61% for testing), which further highlights the FGMDH model's high accuracy. Similarly, the RSR values of 0.07 during both stages and PBIAS values of 0.36% (training) and 0.13% (testing) confirm the model's robust performance with minimal bias. The MAPE values of 4.80% (training) and 4.72% (testing) emphasize the model's ability to maintain low prediction errors, resulting in a "Very Good" descriptive performance across all metrics.

In contrast, the GMDH model demonstrates inferior performance compared to FGMDH. The R^2 and NSE values are 0.88 during both training and testing, indicating a weaker relationship between predicted and actual values. The NRMSE values of 22.06% (training) and 21.93% (testing) are significantly higher than FGMDH, resulting in a "Satisfactory" descriptive performance for this metric. The RSR values of 0.35 during both stages are much higher than FGMDH, indicating poorer model precision. The PBIAS values of 8.89% (training) and 8.43% (testing) show more substantial bias compared to FGMDH. Furthermore, the MAPE values of 25.12% (training) and 25.00% (testing) indicate higher prediction errors, which classify the

TABLE 14.2
Performance Evaluation Metrics for the Flexible GMDH (FGMDH) and Traditional GMDH Models during Training and Testing

Method	Stage	R^2	NSE	NRMSE (%)	RSR	PBIAS (%)	MAPE (%)
FGMDH	Train	0.99	0.99	4.68	0.07	0.36	4.80
		Very good	Very good	Very good	very good	Very good	–
	Test	0.99	0.99	4.61	0.07	0.13	4.72
		Very Good	Very good	Very good	Very good	Very good	–
GMDH	Train	0.88	0.88	22.06	0.35	8.89	25.12
		Very good	Very good	Satisfactory	Very good	Good	–
	Test	0.88	0.88	21.93	0.35	8.43	25.00
		very good	Very good	Satisfactory	Very good	Good	–

model's performance as "Good" rather than "Very Good." As a result, it can be concluded that the FGMDH model outperforms the traditional GMDH model in all evaluated metrics, achieving "Very Good" descriptive performance for every indicator. The GMDH model, while still adequate, falls short in accuracy, precision, and bias compared to the more flexible FGMDH approach.

Figure 14.7 represents the cumulative frequency distributions of relative absolute errors (RAE) for the FGMDH and GMDH models. The x-axis denotes the cumulative frequency percentage, while the y-axis represents the threshold of relative absolute error (%). The plotted curves compare the error distributions of the two models, highlighting the proportion of predictions with RAE below a given threshold. The FGMDH curve (red) shows significantly better performance compared to the GMDH curve (blue). For FGMDH, approximately 20% of the cumulative predictions have RAE below 1%, and nearly 50% of the predictions fall within a 3% RAE. Furthermore, FGMDH reaches 87% cumulative predictions within a 10% RAE and achieves 100% cumulative frequency at 25% RAE, indicating that all predictions fall below this threshold. This demonstrates the FGMDH model's superior accuracy and error consistency, as nearly all predictions have minimal error. On the other hand, the GMDH model exhibits a slower cumulative frequency growth, indicating less precise predictions. Only about 4% of cumulative predictions have an RAE below 1%, and around 30% fall within a 10% RAE. The model reaches 100% cumulative frequency at a much higher 230% RAE, demonstrating significantly less accuracy and a wider error distribution compared to FGMDH. Overall, this figure highlights the superiority of the FGMDH model in achieving lower RAE across its predictions. Its rapid convergence to 100% cumulative frequency at a lower threshold demonstrates its robustness and reliability for channel width prediction compared to the traditional GMDH model.

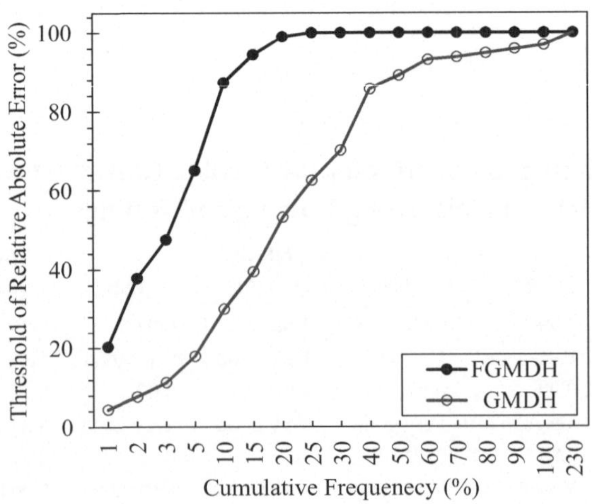

FIGURE 14.7 Cumulative frequency distributions of relative absolute errors for FGMDH and GMDH models.

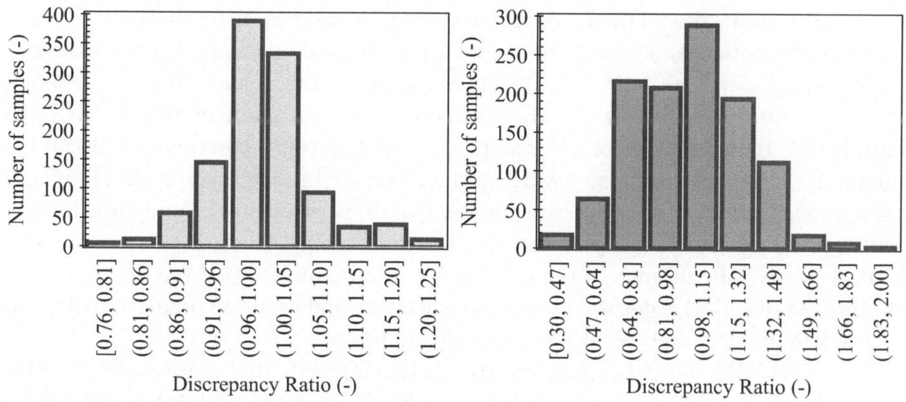

FIGURE 14.8 Histograms illustrating the distribution of Discrepancy Ratios (DR) for FGMDH and GMDH models.

Figure 14.8 illustrates histograms comparing the distribution of Discrepancy Ratios (DR) for the FGMDH and GMDH models. These ratios indicate the consistency of predicted results relative to actual values. The DR values ideally cluster around 1.0, representing perfect agreement between predictions and observations. In the FGMDH model, the distribution is tightly clustered around a DR of 1.0, with the majority of samples falling within the range 0.94–1.060.94 – 1.060.94–1.06. This indicates that most predictions are highly accurate, with minimal deviation from the actual values. The absence of significant outliers reflects the robustness and reliability of the FGMDH model, which maintains consistent performance across the dataset. In contrast, the GMDH model displays a broader and more dispersed distribution of DR values. While there is a concentration near 1.0, a considerable number of predictions deviate significantly, both under-predictions and over-predictions. The distribution extends beyond the range observed for FGMDH, with a notable spread into lower and higher DR values. This broader distribution highlights the GMDH model's reduced predictive stability and accuracy.

Overall, it can be concluded that the FGMDH's narrower DR distribution demonstrates its robust modeling capabilities and superior ability to capture complex relationships with greater reliability. In comparison, the GMDH model exhibits less consistent performance and greater variability in its predictions.

14.5 CONCLUSION

This study presents a novel approach to floodplain width prediction by integrating 2D numerical hydrodynamic modeling using HEC-RAS with the FGMDH algorithm. The results highlight the advantages of combining detailed hydrodynamic simulations with advanced AI techniques to overcome the challenges posed by urbanization and climate change in flood disaster management.

The HEC-RAS model provided a comprehensive dataset that accurately captured the hydrodynamic behavior of the Ottawa River across various discharge scenarios, ranging from 3,000 m^3/s to 8,000 m^3/s incorporating 2-year, 10-year, 100-year, 1,000-year, and 10,000-year floods. By incorporating high-resolution LiDAR and bathymetric data, the model offered precise spatial representations of floodplain inundation, flow velocities, and water surface elevations. The hybrid 1D-2D modeling approach effectively addressed the complexities of channel-floodplain interactions, providing critical insights into the river's response to extreme snowmelt and rainfall events. The outputs of the HEC-RAS simulations formed the foundation for training the FGMDH algorithm, enabling a detailed and accurate prediction of floodplain width under both typical and extreme conditions.

The FGMDH model significantly outperformed the traditional GMDH approach across all evaluation metrics, achieving an R^2 of 0.99 and NRMSE values of 4.68% and 4.61% for training and testing, respectively. It demonstrated exceptional generalization capabilities, with MAPE values of 4.80% for training and 4.72% for testing datasets. Unlike GMDH, the FGMDH model utilized its flexible architecture to dynamically select and prioritize influential variables such as Manning roughness coefficients, coordinates, slope, and discharge. This adaptability ensured accurate floodplain delineation, even under non-historical and extreme flood scenarios, making it a highly reliable tool for flood risk management.

This integrated methodology contributes to sustainable urban planning by enabling precise floodplain zoning, supporting real-time flood monitoring systems, and enhancing adaptive capacities to climate-induced flood risks. The outputs facilitate informed decision-making for infrastructure planning and disaster preparedness, reducing the reliance on carbon-intensive post-flood recovery activities and aligning with net-zero emissions objectives. Additionally, the interpretability of the FGMDH model's explicit equations empowers policymakers and urban planners to understand the underlying dynamics influencing floodplain behavior, fostering trust and transparency in flood risk assessments.

Despite these advancements, some limitations warrant further exploration. The computational intensity of HEC-RAS simulations, particularly in real-time applications, remains a challenge. Additionally, the FGMDH model's performance relies on the availability of high-quality input data, which may not be accessible in all regions. Future research could explore integrating additional climate variables and remote sensing technologies to enhance the robustness and scalability of the methodology. Moreover, extending the approach to different geographical regions with varying hydrological and morphological characteristics would provide further validation of its generalizability.

To sum up, integrating HEC-RAS with FGMDH offers a transformative approach to floodplain management, addressing both current and emerging challenges. The findings underscore the potential of combining advanced AI techniques with physics-based numerical modeling to develop resilient, data-driven flood mitigation strategies. This study provides a pathway for more sustainable and adaptive flood risk management in a changing climate by bridging the gap between technology and practical application.

REFERENCES

Abdessamed, D., & Abderrazak, B. (2019). Coupling HEC-RAS and HEC-HMS in rainfall–runoff modeling and evaluating floodplain inundation maps in arid environments: Case study of Ain Sefra city, Ksour Mountain. SW of Algeria. *Environmental Earth Sciences*, 78, 586. https://doi.org/10.1007/s12665-019-8604-6

Algarni, M. (2023). Deploying artificial intelligence for optimized flood forecasting and mitigation. In: *20th ACS/IEEE International Conference on Computer Systems and Applications (AICCSA)*, Giza, Egypt, 2023, 1–6. doi: 10.1109/AICCSA59173.2023.10479337

Al-Hussein, A.A., Khan, S., Ncibi, K., Hamdi, N., & Hamed, Y. (2022). Flood analysis using HEC-RAS and HEC-HMS: A case study of Khazir River (Middle East – Northern Iraq). *Water*, 14, 3779. https://doi.org/10.3390/w14223779

Amiri, A., Soltani, K., Ebtehaj, I., & Bonakdari, H. (2024). A novel machine learning tool for current and future flood susceptibility mapping by integrating remote sensing and geographic information systems. *Journal of Hydrology*, 632, 130936. https://doi.org/10.1016/j.jhydrol.2024.130936

Ayele, G.T., Teshale, E.Z., Yu, B., Rutherfurd, I.D., & Jeong, J. (2017). Streamflow and sediment yield prediction for watershed prioritization in the Upper Blue Nile River Basin, Ethiopia. *Water*, 9(10), 782. https://doi.org/10.3390/w9100782

Azouagh, A., Bardai, R., E., Hilal, I., & Messari, J.S.E. (2018). Integration of GIS and HEC-RAS in floods modeling of Martil River (Northern Morocco). *European Scientific Journal ESJ*, 14(12), 130. https://doi.org/10.19044/esj.2018.v14n12p130

Basri, B., Mokhtar, N.A., Jan, N.A.M., Marsani, M.F., Ramli, M.F., Majid, M., & Yaacob, F.F. (2021). Hydroclimatic data prediction using a new ensemble group method of data handling coupled with artificial bee colony algorithm. *Sains Malaysiana*, 51(8), 2655–2668. https://doi.org/10.17576/jsm-2022-5108-24

Benito, G., Benito-Calvo, A., Gallart, F., Martín-Vide, J.P., Regües, D., & Bladé, E. (2001). Hydrological and geomorphological criteria to evaluate the dispersion risk of waste sludge generated by the Aznalcollar mine spill (SW Spain). *Environmental Geology*, 40, 417–428. https://doi.org/10.1007/s002540000230

Binns, A.D. (2022). Sustainable development and flood risk management. *Journal of Flood Risk Management*, 15(2). https://doi.org/10.1111/jfr3.12807

Brandt, S.A., Lim, N.J., Colding, J., & Barthel, S. (2021). Mapping flood risk uncertainty zones in support of urban resilience planning. *Urban Planning*, 6(3), 258–271. https://doi.org/10.17645/up.v6i3.4073

Canadian Broadcasting Corporation. (2019). Quyon, Que., Residents Return Home after Forced Evacuation. https://www.cbc.ca/news/canada/ottawa/quyon-residents-return-home-after-mandatory-flood-evacuation-1.5123796

Cardi, J., Dussel, A., Letessier, C., Ebtehaj, I., Gumiere, S.J., & Bonakdari, H. (2023). Modeling hydrodynamic behavior of the Ottawa River: Harnessing the power of numerical simulation and machine learning for enhanced predictability. *Hydrology*, 10(9), 177. https://doi.org/10.3390/hydrology10090177

Diehl, R.M., Gourevitch, J.D., Drago, S., & Wemple, B.C. (2021). Improving flood hazard datasets using a low-complexity, probabilistic floodplain mapping approach. *PLoS One*, 16, e0248683. https://doi.org/10.1371/journal.pone.0248683

Duží, B., Vikhrov, D., Kelman, I., Stojanov, R., & Juricka, D. (2017). Household measures for river flood risk reduction in the Czech Republic. *Journal of Flood Risk Management*, 10, 253–266. https://doi.org/10.1111/jfr3.12132

Dysarz, T., Szałkiewicz, E., & Wicher-Dysarz, J. (2017). Long-term impact of sediment deposition and erosion on water surface profiles in the Ner River. *Water*, 9, 168. https://doi.org/10.3390/w9030168

Ebtehaj, I., & Bonakdari, H. (2022). A reliable hybrid outlier robust non-tuned rapid machine learning model for multi-step ahead flood forecasting in Quebec, Canada. *Journal of Hydrology*, 614, 128592. https://doi.org/10.1016/j.jhydrol.2022.128592

Ebtehaj, I., & Bonakdari, H. (2024). Generalized structure of group method of data handling: Novel technique for flash flood forecasting. *Water Resources Management*, 38(9), 3235–3253. https://doi.org/10.1007/s11269-024-03811-1

Ebtehaj, I., Sammen, S.S., Sidek, L.M., Malik, A., Sihag, P., Al-Janabi, A.M.S., ... & Bonakdari, H. (2021). Prediction of daily water level using new hybridized GS-GMDH and ANFIS-FCM models. *Engineering Applications of Computational Fluid Mechanics*, 15(1), 1343–1361. https://doi.org/10.1080/19942060.2021.1966837

Elkurdy, M., Binns, A.D., Bonakdari, H., Gharabaghi, B., & McBean, E. (2022). Early detection of riverine flooding events using the group method of data handling for the Bow River, Alberta, Canada. *International Journal of River Basin Management*, 20(4), 533–544. https://doi.org/10.1080/15715124.2021.1906261

Es-Haghi, M.S., Abbaspour, M., Abbasianjahromi, H., & Mariani, S. (2021). Machine learning-based prediction of the seismic bearing capacity of a shallow strip footing over a void in heterogeneous soils. *Algorithms*, 14(10), 288. https://doi.org/10.3390/a14100288

Fitriyati, N., Arifin, H.S., Kaswanto, R.L., & Marimin, N. (2024). Enhancing land use planning through integrating landscape analysis and flood inundation prediction Bekasi City's in 2030. *Geomatics Natural Hazards and Risk*, 15(1). https://doi.org/10.1080/19475705.2024.2360623

Global News. (2019). 18 Homes Evacuated as Ottawa River floods, Chaudière Bridge Closing due to High Water Levels. https://globalnews.ca/news/5211800/city-of-ottawa-to-provide-saturday-update-on-flood-operations/

Gomez, F. J., Jafarzadegan, K., Moftakhari, H., &Moradkhani, H. (2024). Probabilistic flood inundation mapping through copula Bayesian multi-modeling of precipitation products. *Natural Hazards and Earth System Sciences*, 24(8), 2647–2665. https://doi.org/10.5194/nhess-24-2647-2024

Grégoire, G., Fortin, J., Ebtehaj, I., & Bonakdari, H. (2023). Forecasting pesticide use on golf courses by integration of deep learning and decision tree techniques. *Agriculture*, 13(6), 1163. https://doi.org/10.3390/agriculture13061163

Huţanu, E., Mihu-Pintillie, A., Urzica, A., Paveluc, L. M., Stoleriu, C. C., & Grozavu, A. (2020). Using 1D HEC-RAS modeling and LiDAR data to improve flood hazard maps accuracy: A case study from Jijia Floodplain (NE Romania). *Water*, 12(6), 1624. https://doi.org/10.3390/w12061624

Idfi, G., Wahyono, I.D., Yulistyorini, A., & Khomsiati, N.L. (2019). The comparative study of flood modelling with the unsteady and the steady flow on Ngotok River. *IOP Conference Series Materials Science and Engineering*, 669(1), 012018. https://doi.org/10.1088/1757-899x/669/1/012018

Karamouz, M., Noori, N., Moridi, A., & Ahmadi, A. (2010). Evaluation of floodplain variability considering impacts of climate change. *Hydrological Processes*, 25(1), 90–103. https://doi.org/10.1002/hyp.7822

Kim, Y., & Newman, G. (2020). Advancing scenario planning through integrating urban growth prediction with future flood risk models. *Computers Environment and Urban Systems*, 82, 101498. https://doi.org/10.1016/j.compenvurbsys.2020.101498

Kirchmeier-Young, M.C., Wan, H., & Zhang, X. (2021). Anthropogenic contribution to the rainfall associated with the 2019 Ottawa River flood. *Bulletin of the American Meteorological Society*, 102, S33–S38. https://doi.org/10.1175/BAMS-D-20-0191.1

Lago, C.A.D., Giacomoni, M.H., Bentivoglio, R., Taormina, R., Gomes, M.N., & Mendiondo, E.M. (2023). Generalizing rapid flood predictions to unseen urban catchments with conditional generative adversarial networks. *Journal of Hydrology*, 618, 129276. https://doi.org/10.1016/j.jhydrol.2023.129276

Letessier, C., Cardi, J., Dussel, A., Ebtehaj, I., & Bonakdari, H. (2023). Enhancing flood prediction accuracy through integration of meteorological parameters in river flow observations: A case study Ottawa River. *Hydrology*, 10(8), 164. https://doi.org/10.3390/hydrology10080164

Mehrankia, A., MollakhaliliMeybodi, M.R., & Mirzaie, K. (2022). Prediction of heart attacks using biological signals based on recurrent GMDH neural network. *Neural Processing Letters*, 54(2), 987–1008. https://doi.org/10.1007/s11063-021-10667-8

Mishra, A., Mukherjee, S., Merz, B., Singh, V.P., Wright, D.B., Villarini, G., Paul, S., Kumar, D.N., Khedun, C.P., Niyogi, D., Schumann, G., & Stedinger, J. R. (2022). An overview of flood concepts, challenges, and future directions. *Journal of Hydrologic Engineering*, 27(6). https://doi.org/10.1061/(asce)he.1943-5584.0002164

Moriasi, D.N., Arnold, J.G., Van Liew, M.W., Bingner, R.L., Harmel, R.D., & Veith, T.L. (2007). Model evaluation guidelines for systematic quantification of accuracy in watershed simulations. *Transactions of the ASABE*, 50(3), 885–900. https://doi.org/10.13031/2013.23153

Muñoz-Carpena, R., Carmona-Cabrero, A., Yu, Z., Fox, G., &Batelaan, O. (2023). Convergence of mechanistic modeling and artificial intelligence in hydrologic science and engineering. *PLOS Water*, 2(8), e0000059. https://doi.org/10.1371/journal.pwat.0000059

Noori, A., & Bonakdari, H. (2023). A GIS-based fuzzy hierarchical modeling for flood susceptibility mapping: A case study in Ontario, Eastern Canada. *Environmental Sciences Proceedings*, 25(1), 62. https://doi.org/10.3390/ECWS-7-14242

Piri, J., Kahkha, M.R.R., & Kisi, O. (2024). Hybrid machine learning approach integrating GMDH and SVR for heavy metal concentration prediction in dust samples. *Environmental Science and Pollution Research*, 1–20. https://doi.org/10.1007/s11356-024-34795-5

Priya, R., Kumar, R.S., Naiklal, S.K., & Kumar, A. (2020). Flood inundation modelling for Tungabhadra basin using HEC-RAS: A case study for Haralahalli discharge site. *Geographical Analysis*, 9(2), 33–41. https://doi.org/10.53989/bu.ga.v9i2.8

Safari, M.J.S., Ebtehaj, I., Bonakdari, H., & Es-haghi, M.S. (2019). Sediment transport modeling in rigid boundary open channels using generalize structure of group method of data handling. *Journal of Hydrology*, 577, 123951.

Sathya, A., Thampi, S.G., & Chithra, N. (2023). Development of a framework for sand auditing of the Chaliyar River basin, Kerala, India using HEC-HMS and HEC-RAS model coupling. *International Journal of River Basin Management*, 21, 67–80. https://doi.org/10.1080/15715124.2021.1909604

Scott, D.E., & Hutchinson, C.E. (2020). An application of the GMDH algorithm to economic modeling. In: *Self-Organizing Methods in Modeling* (pp. 243–255). CRC Press. https://www.taylorfrancis.com/chapters/edit/10.1201/9781003065227-13/application-gmdh-algorithm-economic-modeling-donald-scott-charles-hutchinson

Shah, M., Rahman, A., & Chowdhury, S. (2015). Challenges for achieving sustainable flood risk management. *Journal of Flood Risk Management*, 11(S1). https://doi.org/10.1111/jfr3.12211

Sharma, V.C., & Regonda, S.K. (2021). Two-Dimensional flood inundation modeling in the Godavari River Basin, India – Insights on model output uncertainty. *Water*, 13(2), 191. https://doi.org/10.3390/w13020191

Soltani, K., Ebtehaj, I., Amiri, A., Azari, A., Gharabaghi, B., & Bonakdari, H. (2021). Mapping the spatial and temporal variability of flood susceptibility using remotely sensed normalized difference vegetation index and the forecasted changes in the future. *Science of the Total Environment*, 770, 145288. https://doi.org/10.1016/j.scitotenv.2021.145288

Teufel, B., Sushama, L., Huziy, O., Diro, G., Jeong, D., Winger, K., Garnaud, C., De Elia, R., Zwiers, F., & Matthews, H. (2019). Investigation of the mechanisms leading to the 2017 Montreal flood. *Climate Dynamics*, 52, 4193–4206. https://doi.org/10.1007/s00382-018-4375-0

Yazdan, M.M.S., Ahad, M.T., Kumar, R., & Mehedi, M.A.A. (2022). Estimating flooding at river spree floodplain using HEC-RAS simulation. *J: Multidisciplinary Scientific Journal*, 5(4), 410–426. https://doi.org/10.3390/j5040028

Zahmatkesh, Z., Han, S., & Coulibaly, P. (2021). Understanding uncertainty in probabilistic floodplain mapping in the time of climate change. *Water*, 13(9), 1248. https://doi.org/10.3390/w13091248

Zeiger, S.J., & Hubbart, J.A. (2021). Measuring and modeling event-based environmental flows: An assessment of HEC-RAS 2D rain-on-grid simulations. *Journal of Environmental Management*, 285, 112125. https://doi.org/10.1016/j.jenvman.2021.112125

Zellou B, & Rahali H. (2016). Assessment of reduced-complexity landscape evolution model suitability to adequately simulate flood events in complex flow conditions. *Natural Hazards*, 86(1), 1–29. https://doi.org/10.1007/s11069-016-2671-8

15 *k*-Means Clustering
Geometric Design of Double-Arch Dams

Enrico Zacchei and José Luis Molina

15.1 INTRODUCTION

Dams are hydraulic structures mainly used for water management. Their long-term safety is a focus of social concern; therefore, they must be well designed by using not only classical and conservative analyses but also more modern analyses (Fenves and Chopra, 1983; Zacchei and Molina, 2021).

The geometric shape of the dam is a factor that strongly influences the balance between the external loads and internal stresses, also affecting the cost of the global project (Zacchei et al, 2019; Fan et al., 2015).

In the literature, there are several studies regarding these aspects. For instance, new methods have been developed to find the optimum shape of arch dams by combining stochastic approximations with genetic algorithm methods (Seyedpoor et al., 2011), and to determine the material parameters, the deformation dam modulus, the dam foundation, and the reservoir basin (Gu et al., 2015). In Seifollahi et al. (2022), the geometry dimensions of a gravity dam have been optimized by alternative algorithms. In Khatibinia and Khosravi (2014), the optimal shape including the dam-water-foundation interaction has also been studied.

To design dams, there are some limitations imposed by topographic and geologic conditions. In Zacchei and Molina (2019, 2020), Bartoli et al. (2017) Bayesian theorem has been used to calculate the key geometrical parameters (e.g., dam height, crown crest, crest length, concrete volume) to define the optimum geometry of double-arch dams. These parameters deserve more attention, and for this, the *k*-mean clustering algorithm is adopted in this paper. It can be considered a tool for artificial intelligence (AI) (Turco et al., 2021).

After the 20th century, with the gradual development of AI and its promotion, artificial intelligence algorithms are used to simulate the input-output relationship for several high-precision analyses. In particular, *k*-means is useful for this purpose due to its simplicity, efficiency, and versatility. In fact, it is considered unsupervised machine learning that belongs to AI in the great area of computer science (Turco et al., 2021).

In a more general approach, the clustering analyses have been used to detect anomalies in the dam body (Ji et al., 2022) and to develop models to estimate its deformations (Wang et al., 2020, 2021). In particular, *k*-means algorithm has been

DOI: 10.1201/9781003547099-18

used to model the concrete volume (Kucukerdem et al., 2019), to predict the deformations during phase construction (Hu and Ma, 2021), and to detect dam deformation by focusing on the elimination of outliers (Song et al., 2021).

To the best of the authors' knowledge, there are not sufficient studies on the intercorrelations of the parameters involved in the design layout for double-arch dams by using *k*-mean algorithm. Thus, in this paper, as a main goal, each parameter has been disaggregated with respect other one, allowing to understand their weights for the layout design of double-arch dams.

15.2 MATERIALS AND METHODS

15.2.1 Materials

The materials (i.e., data, parameters, values, etc.) collected for this study have been retrieved from official national databases. In Spain, there are about 70 double-arch dams, of which for 37 dams (36 large dams with height $H > 15.0$ m) complete technical data are available (SNCZI, 2022). The collected data can be considered homogenous since they regard very similar structures designed under the same national codes and probably the same general recommendations. Thus, by considering 8 (= 6+2) main categories, 296 (= 37×8) total data have been collected (data retrieved from SEPREM, 2022; SNCZI, 2022).

These categories regard the following geometrical parameters: The dam crest, T_c, the volume of the dam concrete, V_d, the straight-line distance measured between abutments excavated to assumed foundation rock at the crest elevation, L_1, the radius of the dam axis, R_{axis}, the crest length of the dam, L_c, and the reservoir capacity, V_r. Table 15.1 lists these categories showing the maximum/minimum values, the mean value, μ_0, and the standard deviation, σ_0.

TABLE 15.1
Parameters and Their Values

Parameter (or Category)	Maximum Value	Minimum Value	μ_0	$\pm\sigma_0$
Dam height, H (m)	202.0	14.0	83.90	43.58
Crest base of the dam, T_c (m)[a]	10.0	0.55	4.44	2.48
Crest length of the dam, L_c (m)	567.23	32.50	242.94	146.94
Dam volume, V_d ($\times 10^3$ m^3)	2186.0	1.20	262.34	407.96
Straight-line distance, L_1 (m)[b]	492.42	28.21	210.90[c]	127.56[c]
Radius of the dam, R_{axis} (m)	295.45	16.93	126.54	76.53
Reservoir capacity, V_r ($\times 10^6$ m^3)	2648.64	0.06	223.01	505.18

[a] Some values were not available; thus, they have been estimated by Equation 15.2.
[b] Estimated by inverse Equation 15.1, where $R_{axis} = L_c/\alpha$ with $\alpha = 110.0°$.
[c] Used for some preliminary intercorrelations (see Figure 15.2).

Note that these values represent the non-clustered values (i.e., raw values), which provide a very large range between its minimum and maximum, resulting thus high errors σ_0. This could show in a clearer way the utility of the clustered analyses to reduce these errors and to find new intercorrelations among all parameters.

15.2.2 Methodology

The methodology of the analyses of this paper consists basically of two steps. The first step regards the individuation of analytical relations to design the geometry of a double-arch dam, whereas the second step regards the implementation of the *k*-means algorithm.

In the international USACE (1994) manual, it is possible to find some relations for the preliminary double-arch dam design in geometric terms. Figure 15.1 shows some fundamental dimensions in plan and transversal view (some parameters have already been defined. The blue arrow indicates the water flow).

The radius of the dam axis, R_{axis}, can be estimated by:

$$R_{\text{axis}} = 0.60\ L_1 \tag{15.1}$$

where L_1 has already been defined. The radius R_{axis} is correlated to the central angle for the top arch, α, which usually varies between 100.0° and 120.0°. The crown cantilever can be defined by the crest, T_c, the base T_b, and the thickness at 0.45 H above the base, $T_{0.45}$, of the dam, respectively:

$$\begin{cases} T_c = 0.01(H + 1.2\ L_1) \\ T_b = \sqrt[3]{0.0012\ H\ L_1 L_2 \left(\dfrac{H}{400}\right)^{H/400}} \\ T_{0.45} = 0.95\ T_b \end{cases} \tag{15.2}$$

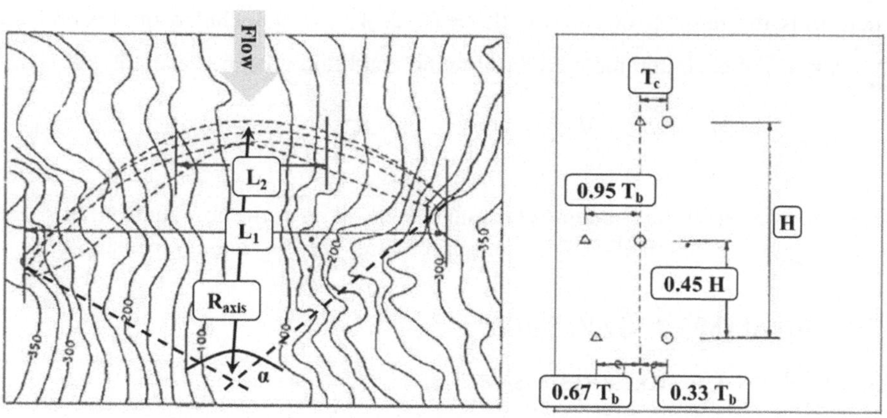

FIGURE 15.1 Example of geometry of a double-arch dam: (a) plan view and (b) transversal view. (Modified from USACE, 1994.)

where H is the dam height and L_2 is the straight-line distance at 0.15 H above the base. In addition, upstream (US) and downstream (DS) projections of the extrados and intrados faces, respectively, can be estimated empirically as $US_{base}=0.67\ T_b$, $US_{0.45H}=0.95\ T_b$, and $DS_{base}=0.33\ T_b$.

The volume of the dam concrete, V_d, can be estimated by:

$$V_d = \left[\left(2.0\times10^{-6}\right) H^2 L_2 \left(\frac{H + 0.80\ L_2}{L_1 - L_2}\right)\right] + \left[\left(4.0\times10^{-4}\right) \mathrm{H}\, L_1 \left(\mathrm{H} + L_1\right)\right]. \quad (15.3)$$

Equations 15.1–15.3 were based on historical data compiled from existing dams; thus, they could be used for a preliminary layout. For this, here the focus is to improve these parameters by the k-means algorithm under more historical data from 37 Spanish double-arch dams. In this sense, the parameters involved in Equations 15.1–15.3 represent the inputs of this study.

In the k-means algorithm, the most used Euclidean distance, dE, and thus the squared Euclidean distance (square error), dE^2, is adopted (Zacchei and Brasil, 2024):

$$\mathrm{dE} = \sqrt{\sum_i^k \left(c_i - p_i\right)^2} \rightarrow \mathrm{dE}^2 = \sum_i^k \left(c_i - p_i\right)^2 \quad (15.4)$$

where $c_i(x, y)$ is the center of the cluster k, $p_i(x, y)$ is the point to be compared, and $i=\{1, 2, 3, \ldots, k\}$ is the dimension of p_i and c_i. For a set, t, of cases $\Pi=\{p_1, p_2, p_3, \ldots, p_n\} \in \mathrm{R}^d$, where R^d is the data space of d dimensions, the k-means algorithm tries to find a set of k cluster centers $C=\{c_1, c_2, c_3, \ldots, c_k\} \in \mathrm{R}^d$ that is a solution of the minimization problem:

$$\min_n \{E\} = \min_n \left\{\sum_i^k \sum_j^{n_i} \left\|p_{ij} - c_i\right\|^2\right\} \quad (15.5)$$

where n_i is the number of cases with $j=\{1, 2, 3, \ldots, n_i\}$ included in cluster k and $\sum_i^k n_i = n$. For each iterated c_i, the following condition must be verified:

$$\frac{\partial E}{\partial c_i} = 0 \rightarrow c_i^{(t+1)} = \frac{1}{\left|n_i^t\right|} \sum_j^{n_i^t} p_{ij} \quad (15.6)$$

Thus, the average of the elements of each group is taken as the new centroid (for more details regarding this methodology and procedure, see Zacchei and Brasil, 2024).

15.3 ANALYSES AND RESULTS

15.3.1 Non-Clustered Results

From all collected data, some preliminary results can be obtained. Figure 15.2 shows, for brevity, only one combination, i.e., V_d vs H, which are directly correlated to each other as shown in Equation 15.3. Note that this combination regards non-clustered data.

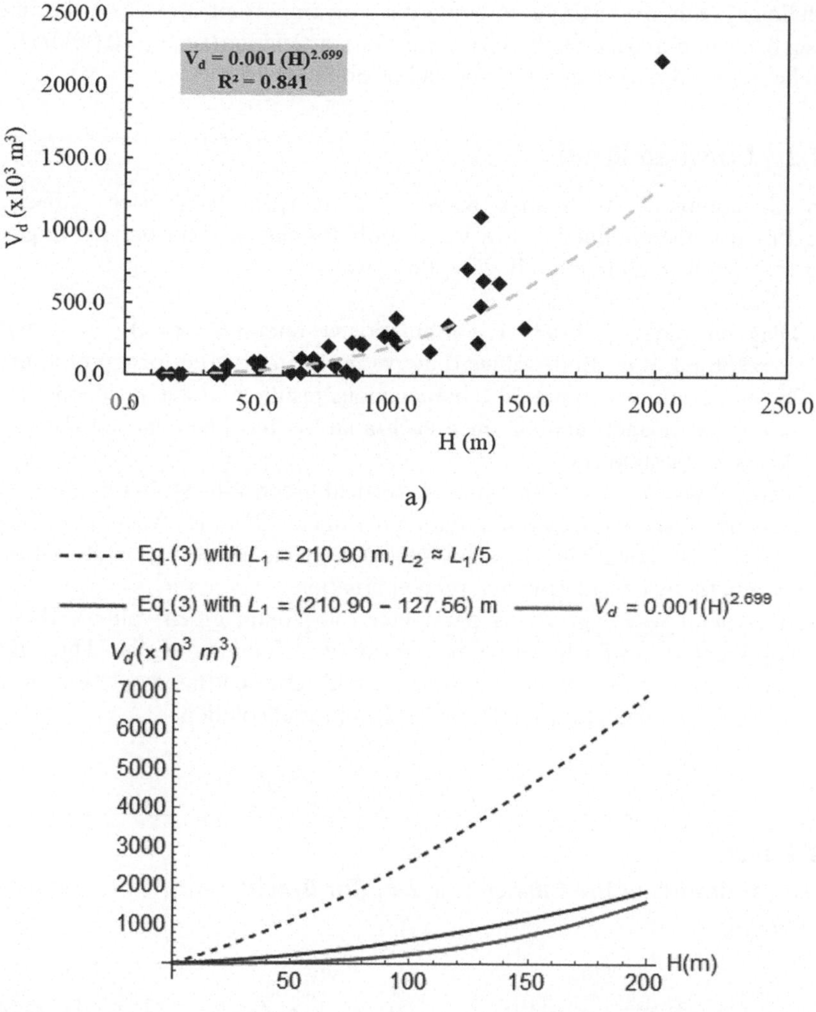

FIGURE 15.2 Correlations between non-clustered V_d and H: (a) collected data and (b) analytical trends.

Figure 15.2a shows a good relation between both parameters since $R^2 = 0.84$; thus, it is possible to estimate a new relation (in yellow) to be compared with classical Equation 15.3. In Figure 15.2b, it is possible to see that Equation 15.3 (dashed black line) overestimates V_d values with respect to those obtained by this new relation (blue line). This overestimation can be quantified as about $\mu_{0-}\sigma_0$ (plotted by solid black line). The used values for L_1 have been shown in Table 15.1, whereas the approximation $L_2 \approx L_1/5$ has been retrieved from a practical example illustrated in literature (Zacchei and Molina, 2019).

Obviously, this overestimation could have an important impact on the internal stresses of the dam and its weight. Therefore, this new relation (i.e., $V_d = 0.001 \times (H)^{2.699}$) could be more valid to estimate V_d instead of Equation 15.3.

15.3.2 Clustered Results

From all combinations (matrix 8×8), 192 analyses have been carried out. Table 15.2 lists the obtained $\mu_1 \pm \sigma_1$ values only for cluster 1, for brevity. In general, Table 15.2 can be read in the following three ways:

i. Diagonal way, e.g., for $H - T_c$ combination we obtain $H = 84.51 \pm 24.70$ m for $T_c = 3.76 \pm 1.21$ m. Both values (blue values) can be used for a preliminary layout design, which provides homogenous results in a statistical way. This is a more correct reading since each analysis has been carried out for a couple of parameters.
ii. Vertical way, e.g., for H parameter (vertical green values). In this case, it is possible to see the trend of a unique parameter (here H) weighed in function of other ones. This is useful to obtain a general estimation of a certain parameter by considering several combinations.
iii. Horizontal way, e.g., for H parameter (horizontal green values). Here it is possible to evaluate all parameters only in function of one. Thus, if H parameter is fixed, other parameters could represent good solutions to be adopted; however, they could not be the optimal solution.

TABLE 15.2
Statistical Results of the Cluster 1, $\mu_1 \pm \sigma_1$ (for Brevity Only Six Categories Are Shown)

	H (m)		**T_c (m)**		**L_c (m)**	
H	83.59	±13.18	3.76	±1.21	272.89	±56.64
T_c	84.51	±24.70	2.13	±0.85	242.27	±52.18
L_c	97.37	±23.52	4.04	±1.18	267.41	±56.06
V_d	75.03	±18.47	4.01	±1.23	261.77	±56.06
V_r	69.14	±15.85	3.57	±1.13	231.81	±71.81
R_{axis}	97.37	±23.52	4.04	±1.18	267.41	±56.06
	V_d ($\times 10^3$ m^3)		V_r ($\times 10^6$ m^3)		R_{axis} (m)	
H	125.41	±91.86	34.29	±30.67	142.14	±29.50
T_c	211.52	±197.55	167.81	±353.98	126.19	±27.17
L_c	189.73	±103.40	77.56	±149.67	139.28	±29.20
V_d	145.39	±81.54	47.52	±36.83	136.34	±29.20
V_r	172.75	±126.90	45.51	±37.06	131.39	±42.39
R_{axis}	189.73	±103.40	162.54	±334.03	139.28	±29.20

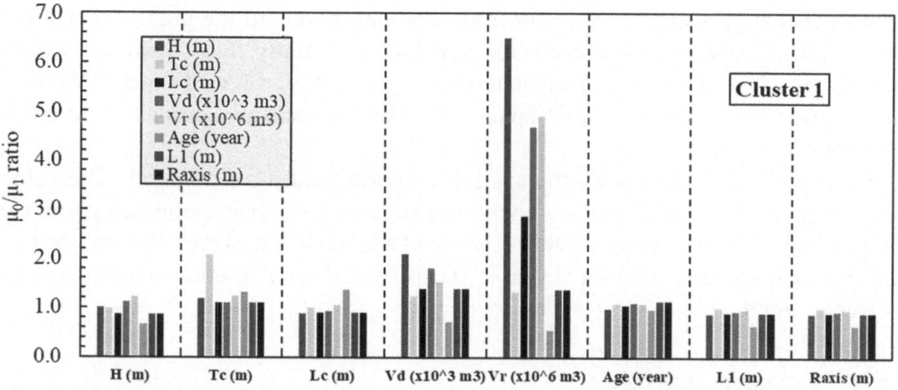

FIGURE 15.3 μ_0/μ_1 ratio of all parameters and combinations for cluster 1. The individual bars, from left to right, represent the key parameters: height of the dam (H), dam crest (T_c), crest length (L_c), volume of the dam (V_d), reservoir capacity (V_c), age, straight-line distance (L_1), and radius (R_{axis}). The bar graph shows a prominent peak for V_r and moderate variation for T_c and V_d, with other parameters remaining near 1 from left to right.

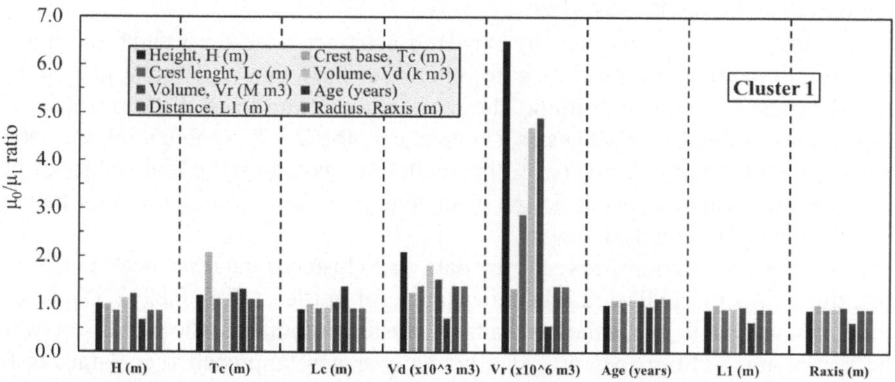

FIGURE 15.4 Spatial-temporal clustering for some parameters from cluster 1 results. The bar chart presents 3D bar plots of key parameters (H, R_{axis}, L_c, V_d, and V_r) over different years within Cluster 1, highlighting variations and trends in their values across the spatial-temporal domain.

At this stage, it is interesting to do a comparison between non-cluster data (Table 15.1) with clustered data (Table 15.2). Figure 15.3 shows, by histograms, the μ_0/μ_1 ratio of cluster 1. Here it is possible to see the utility of the cluster algorithm. In general, when $\mu_0/\mu_1 \approx 1.0$, the preliminary non-clustered value does not suffer alterations, therefore it could be adopted during the layout design of the dam. On the other hand, when $\mu_0/\mu_1 > 1.0$ and/or $\mu_0/\mu_1 < 1.0$, the preliminary value could be over- or under-estimated, respectively.

The very high values of the μ_0/μ_1 ratio are mainly due to the great dispersion of the data (i.e., great error) in a stochastic distribution. Usually, this is verified when the coefficient of variation, CV_0 (calculated by σ_0/μ_0), is $CV_0 > 1.50$. In fact, $CV_0 = 1.56$ for V_d, and $CV_0 = 2.27$ for V_r (see Table 15.1). These effects also occur for a very low μ_0/μ_1 ratio.

Finally, Figure 15.4 shows the spatial-temporal values for cluster 1. The values of each parameter are shown in function of the age [1]and their combined parameter (in brackets). This could be another reading of the results; however, the studied time interval is short (i.e., 44.0–51.0 years); thus, it is difficult to understand a possible universal trend.

15.4 CONCLUSIONS

In this paper *k*-means algorithm has been implemented for the parameters involved in the geometric design of double-arch dams. Analytic and numerical analyses have been carried out.

From the database, 296 parameters have been collected and divided into eight categories. These parameters, regarding 37 Spanish dams, mainly treat geometrical values. Preliminary analyses show a good relation between V_d and H, indicating that the classic relation could overestimate the V_d values. For this, a new relation has been proposed, i.e., $V_d = 0.001 \times (H)^{2.699}$.

The disaggregation analyses, by clustered data, provided the weight and impact of a certain parameter with respect to other one. This shows the great potential of the AI by the *k*-means algorithm. The more correct way to read the results could consist in considering the values as a couple, e.g., for $H - T_c$ combination, we obtain $H = 84.51 \pm 24.70$ m for $T_c = 3.76 \pm 1.21$ m (called "diagonal way"). Both values can be used for a preliminary layout design of double-arch dams, which provides homogenous results in a statistical way.

Finally, a comparison between raw data and clustered data has been made. For this, the μ_0/μ_1 ratio indicated a better balance for cluster 1 (i.e., 0.89–2.95). In this sense, it is advisable to use the results from cluster 1 to analyze the weights between the parameters. This paper provides not only a new approach (i.e., *k*-means for double-arch dams) but also practical values for more accurate analyses.

ACKNOWLEDGMENTS

The first author acknowledges the Itecons Institute, Portugal, for the Wolfram Mathematica license and the University of Salamanca (USAL), Spain, for paying the rights (when applicable) to completely download all papers in the references.

---

[1] The age is another parameter (or category) considered in this study. Given that dams, like all large-scale constructions, are built rarely and over long periods of time, this parameter has not been introduced up to now. However, for curiosity, its estimated non-clustered range was: $\mu_0 \pm \sigma_0 \approx 50.0 \pm 18.0$ years.

DECLARATION OF GENERATIVE AI IN SCIENTIFIC WRITING

The authors declare that they have not used generative AI in scientific writing.

REFERENCES

Bartoli, G., Bett, M., Facchini, L., Marra, A.M., Monchetti, S., Bayesian model updating of historic masonry towers through dynamic experimental data. *Proc Eng,* 199, 1258–1263, 2017.

Fan, Q., Zhou, S., Yang, N., Optimization design of foundation excavation for Xiluodu super-high arch dam in China. *J Rock Mech Geotech Eng,* 7, 120–135, 2015

Fenves, G., Chopra, A.K., Effects of reservoir bottom absorption on earthquake response of concrete gravity dams. *Earthq Eng Struct Dyn*, 11, 809–829, 1983.

Gu, H., Wu, Z., Huang, X., Song, J., Zoning modulus inversion method for concrete dams based on chaos genetic optimization algorithm. *Math Probl Eng,* 2015, 1–9, 2015.

Hu, J., Ma, F., Comparison of hierarchical clustering based deformation prediction models for high arch dams during the initial operation period, *J Civil Struct Health Monit*, 11, 897–914, 2021.

Ji, L., Zhang, X., Zhao, Y., Li, Z., Anomaly detection of dam monitoring data based on improved spectral clustering. *J Internet Technol* 23, 1–11, 2022.

Khatibinia, M., Khosravi, S., A hybrid approach based on an improved gravitational search algorithm and orthogonal crossover for optimal shape design of concrete gravity dams. *Appl Soft Comput* 16, 223–233, 2014.

Kucukerdem, T.S., Kilit, M., Saplioglu, K., Determination of the number of clusters used in fuzzy inference systems by means of *k*-means and modeling of dam volume: Kestel dam example. *Pamukkale Univ J Eng Sci*, 25, 962–967, 2019

Seifollahi, M., Abbasi, S., Abraham, J., Norouzi, R., Daneshfaraz, R., Lotfollahi-Yaghin, M.A., Alkan, A., Optimization of gravity concrete dams using the grasshopper algorithm (Case study: Koyna Dam), *Geotech Geol Eng*, 40, 5481–5496, 2022.

SEPREM, 2022. Spanish Association of Dams and Reservoirs (SEPREM), Database. https://www.seprem.es/index.php

Seyedpoor, S.M., Salajegheh, J., Salajegheh, E., Gholizadeh, S., Optimal design of arch dams subjected to earthquake loading by a combination of simultaneous perturbation stochastic approximation and particle swarm algorithms. *Appl Soft Comput*, 11, 39–48, 2011.

SNCZI, 2022. Inventory of Dams and Reservoirs (SNCZI), Database. https://sig.mapama.gob.es/snczi/

Song, J., Zhang, S., Tong, F., Yang, J., Zeng, Z., Yuan, S., Outlier detection based on multivariable panel data and *k*-means clustering for dam deformation monitoring data. *Adv Civil Eng*, 2021, 3739551, 2021.

Turco, C., Funari, M.F., Teixeira, E., Mateus, R., Artificial neural networks to predict the mechanical properties of natural fibre-reinforced compressed earth blocks (CEBs). *Fibers*, 9, 1–21, 2021.

USACE, 1994. *Arch Dan Design*, Manual No. 1110-2-2201, US Army Corps of Engineers (USACE), Washington, DC, USA, p. 240.

Wang, S., Xu, C., Gu, C., Su, H., Hu, K., Xia, Q., Displacement monitoring model of concrete dams using the shape feature clustering-based temperature principal component factor. *Struct Control Health Monit*, 27, 1–21, 2020.

Wang, S., Xu, C., Liu, Y., Wu, B., Mixed-coefficient panel model for evaluating the overall deformation behavior of high arch dams using the spatial clustering. *Struct Control Health Monit*, 28, 1–19, 2021.

Zacchei, E., Brasil, R., *k*-means for earthquakes: disaggregation analyses of small events by considering wave components and soil types. *Arab J Geosci* 17, 302, 2024.

Zacchei, E., Molina, J.L., Brasil, R., Nonlinear degradation analysis of arch-dam blocks by using deterministic and probabilistic seismic input. *J Vib Eng Technol,* 7, 301–309, 2019.

Zacchei, E., Molina, J.L., Introducing importance factors (IFs) to estimate a dam's risk of collapse produced by seismic processes. *Int J Disaster Risk Red,* 60, 1–13, 2021.

Zacchei, E., Molina, J.L., Optimization of geometric parameters for double-arch dams through Bayesian implementation. *J Struct Eng,* 146, 1–11, 2020.

Zacchei, E., Molina, J.L., Shape optimization of double-arch dams by using parameters obtained through Bayesian estimators. *Iranian J Sci Technol Trans Civil Eng,* 43, 649–662, 2019.

16 Water Security in a Changing World

Harnessing AI to Mitigate Climate Change Impacts

Joan Cecilia C. Casila, Luisito C. Abueg, Liliane M. Suarez, Sherwin P. De Jesus, Claudette Joyce B. Mendina, and Jerome P. Romero

16.1 INTRODUCTION

Water security and climate change are fundamentally connected since unpredictable and highly variable weather events affect the water resources (United Nations, 2024b). Water security is described as the capability a population to ensure sustainable access to adequate quality and quantities of water for livelihoods, human well-being, and ecosystems (United Nations, 2013). Climate change is one of the most pressing global issues that poses significant threat to water security as it caused several water-related hazards and weather events such as droughts, extreme typhoons due to rising sea-level, unpredictable rainfall patterns, increased global temperature, and melting of glaciers (Holden & Marshall, 2018). According to the World Meteorological Organization (WMO) in 2021, because of global warming and climate change, weather-related disasters have increased five times in the last 50 years. Climate change is alarmingly affecting the usable and available water resources. The status of regional and global water supply and how water security and global net-zero emissions goal can be attained given the emergence of artificial intelligence (AI) will be discussed in the succeeding sections.

16.2 REGIONAL AND GLOBAL WATER SUPPLY AND SECURITY

Every region in the world faces water security challenges. According to the United Nations (2023) report on the Global Water Security Assessment, about 6.3 billion people live in countries with high and critical water stress levels. Of these, 4.3 billion people are in the Asia-Pacific region and 1.4 billion live in Africa. Europe and the Americas are more water secure than other global regions.

According to the2030 Water Resources Group (2012), one-third of the world's population will live near water basins where the water deficit will be larger than

DOI: 10.1201/9781003547099-19

50% by 2030. Several countries in Western United States, northern Africa, southern Africa, the Middle East, Australia and parts of south Asia and China are already experiencing high "water stress"—when the annual renewable freshwater supplies are below 1,700 cubic meters per person per year. As global demands for freshwater increase, the supply of fresh water will not keep pace with the demand due to the absence of more effective water resources management. The annual global water requirements will reach 6,900 billion cubic meters in 2030 which is 40% above the current sustainable water supply.

Every country, regardless of its level of economic development and infrastructure, will be vulnerable to water disasters exacerbated by climate change. These disasters can have a profound impact on human life, leading to significant losses and damages. Additionally, inefficient water consumption and changing land-use practices, such as deforestation and soil degradation, will further diminish water supply and security (Nguyen et al., 2023). Developing countries are particularly affected and struggle to cope due to limited resources and infrastructure insufficiency (United Nations, 2024a; Zhang et al., 2019; Mbow et al., 2019). Industries that rely heavily on water for its operations could have significant economic losses with water supply disruptions.

16.3 IMPACTS OF CLIMATE CHANGE ON WATER SECURITY

Water security is crucial to sustain life, ecosystems, industries, and agriculture. It includes not only the availability of freshwater resources but also their quality and accessibility (Mishra et al., 2021). Climate change poses a significant threat to water since higher global temperatures are expected to alter precipitation patterns, and increase extreme weather events (Seneviratne et al., 2021). Climate change appears not only as an environmental issue but also as a social and economic challenge that requires urgent action.

16.3.1 Variability in Rainfall Patterns and Increased Flood and Drought Occurrences

The continuous cycle of water in the Earth's surface, ocean, and atmosphere with the exchange in mass, matter and energy is referred to as the hydrological cycle. It plays a vital role in the ecosystem, energy budget, and climate. The changes occurring in the climatic system affects the water system (availability and distribution) and vice versa (Olaoluwa et al., 2022).

The rise in temperatures due to climate change intensifies the Earth's water cycle, as projected by current climate models. An increased rate of evaporation leads to more persistent and powerful storms, which heightens the risk of flooding. Conversely, areas located away from storm tracks face a greater risk of drought (NASA, 2024). These variabilities can lead to reduced water quality, increased evaporation rates, and altered hydrological cycles, adding stress on already vulnerable water systems (Kundzewicz, 2008).

Since the 1970s, many regions around the world, especially in the tropics and subtropics, have intensified precipitation variability and experienced longer droughts. In land areas, the frequency of heavy precipitation has also increased (Casila, 2022;

IPCC, 2007). Recent studies forecast that rainfall variability and drought will continue to increase in the future (Kundzewicz et al., 2006).

16.3.2 Glacier Melting, Sea-Level Rise and Salinization

Glaciers worldwide are considered scientific indicators and records of how the global climate changed over time (World Wildlife Fund, 2024). According to Glick (2024), arctic sea ice melts down over the past century and has declined by about 10% in the past 30 years. For instance, the famous snow on Mt. Kilimanjaro in Tanzania has melted by more than 80% since 1912. Additionally, the eastern and central glaciers of the Garwhal Himalayas are predicted to virtually disappear by 2035. It is known that an increase in temperature results in the melting of ice and water flows to the oceans, resulting in warming of ocean water and expansion in volume (Fox-Kemper et al.,2021). If the West Antarctic ice sheet were to melt, it could raise the sea levels by nearly 20 feet (Glick, 2024).

16.3.3 Overexploitation of Groundwater

Groundwater reliance has grown due to prolonged periods of drought and the unavailability of safe surface water. This entails the need for groundwater to play a more central role in climate discussions (International Groundwater Resources Assessment Centre (IGRAC), 2024), as various issues are now threatening the availability of groundwater supplies. One of the issues of groundwater exploitation is land subsidence where excessive water extraction from aquifers makes the ground surface sink (Han, 2024). The depletion of groundwater around the globe is happening at an accelerating speed, and it was found that water levels are dropping in 70% of the aquifers (Tasoff, 2024). This will not only affect the availability and accessibility of water but will also contribute to ecosystem disruption (Dao et al., 2024). It is expected to intensify in the future due to the increasing demands of the population, rapid industrialization, amplified by the impact of climate change (Bagheri-Gavkosh et al., 2021).

In addition to this, solid waste landfills, agricultural activities, and industrial processes are causing groundwater pollution (Igwegbe et al., 2024). The landfill leachate carries harmful substances such as heavy metals, organic matter, and pathogens that percolate into the subsurface. While it affects the quantity of usable groundwater, it also poses public health risks (Igwegbe et al., 2024). Given these few factors and issues faced regarding groundwater that cause severe environmental and social issues. Policymakers, communities, and other significant stakeholders must collaborate to manage the finite groundwater resource (Megdal, 2018). Strategic management of groundwater is vital for ensuring global water and food security in a changing climate (International Groundwater Resources Assessment Centre (IGRAC), 2024).

16.3.4 Physical and Socioeconomic Impact of Water Scarcity

Water scarcity significantly impedes the achievement of the global sustainable development goals (SDGs) (United Nations, 2024a). It has profound physical and socioeconomic impacts that affect communities and ecosystems (Ahn & Juraev, 2024).

Areas with water scarcity and irregular water supply experience reduced agricultural productivity, which directly affects food availability and accessibility, making them particularly vulnerable to food poverty (FAO, 2015). Poor consumers are disproportionately harmed by rising food prices driven by the impacts of climate change. This necessitates enhanced social protection measures to support those who are most at risk (Hoddinott, 2006).

16.4 NET-ZERO EMISSIONS INITIATIVES ON WATER SECURITY

To address the pressing impacts of climate change and achieve a sustainable and water-secure future, urgent actions and innovations are necessary. Emissions of greenhouse gases (GHGs) such as carbon dioxide, methane, and nitrous oxide must also be monitored and mitigated to achieve net-zero emissions.

Net-zero emissions is a global target which aims to balance the GHG emissions produced and taken out of the atmosphere (Climate Council, 2023). A total of 197 countries declared to meet the 1.5°C limit, and with 50% probability, it will translate to a remaining carbon budget of 400–800 gigatons of CO_2 ($GtCO_2$). Maintaining this limit will require carbon dioxide emissions to peak before 2030 and achieve the net-zero target in 2050. Achieving this will aid in securing water supply and improved water quality as it will help mitigate global warming effects and impacts of climate change (Fankhauser et al., 2022). However, working toward net-zero remains challenging due to continuous industrialization. Accurately tracking emissions and pollutants in water and wastewater from different sectors and industries is very complex but could be addressed through the utilization and integration of AI (Olawade et al., 2024).

16.5 THE ROLE OF AI IN MITIGATING WATER RESOURCE IMPACTS

AI presents significant potential in mitigating water challenges amplified by climate change. AI technologies can improve water management practices through enhanced forecasting, monitoring, and decision-making (Lundberg & Lee, 2017). The powerful capacity of AI enables the processing and analysis of an immense range of water, socioeconomic, and climatic data (Alpan & Sekeroglu, 2020), which can be very helpful for securing efficient water use, particularly in areas that are more vulnerable to increasing water stress.

16.5.1 AI on Climate Modeling

AI can be utilized in modeling and simulating hydrological systems under varying climate conditions, allowing policymakers to make informed decisions about resource allocation and management (Chang et al., 2023). By integrating AI with current water management frameworks, stakeholders can develop adaptive strategies that are efficient and effective in mitigating the effects of climate change.

Rampal et al. (2024) showed that climate downscaling is a crucial process in climate science that aims to enhance the resolution of climate projections, allowing better understanding of local climate variability and change. Recent advancements in ML show an improvement in the performance of empirical downscaling methods. Techniques such as Convolutional Neural Networks (CNNs) and Generative Adversarial Networks (GANs) are increasingly used to enhance the efficiency and accuracy of downscaling processes. ML also significantly enhances climate projections by improving the efficiency and accuracy of downscaling methods. ML techniques like GANs and diffusion models have outperformed traditional algorithms in reproducing present-day climate conditions and extremes.

Moreover, the new atmospheric model named neural GCM outperforms some of the existing weather and climate prediction models (Fuckar, 2024). It shows a high potential for large savings in computational cost over conventional dynamic models. Kochkov et al. (2024) indicated that GCMs have undergone continuous improvement for the past 70 years, benefiting from enhanced numerical techniques and more detailed physical models. These advancements have been supported by increasing computational power of computers, allowing for higher resolution simulations. Within GCMs, physical processes that cannot be directly resolved, such as cloud formation, radiation, and precipitation, are represented through semi-empirical parameterizations. NeuralGCM performs competitively with ML models for forecasts ranging from 1 to 10 days and matches the performance of the European Centre for Medium-Range Weather Forecasts ensemble predictions for forecasts up to 15 days. When provided with sea surface temperature data, NeuralGCM is capable of accurately tracking climate metrics over several decades. Additionally, climate forecasts at a resolution of 140 kilometers reveal emergent phenomena, including realistic tropical cyclone frequencies and trajectories. In both weather and climate predictions, this method achieves significant computational savings compared to traditional GCMs.

16.5.2 AI in Drought and Flood Prediction

Extreme events such as widespread floods and severe droughts have been thoroughly observed, analyzed, and forecasted. Numerous research studies have been done to minimize the consequences and damages caused to human lives and socioeconomic development (Arduino et al., 2005). In arid and semi-arid regions, decision-making in the water sector is effectively guided by a clear forecasting model, providing crucial information to relevant stakeholders and in the layout of contingencies (Osman et al., 2020)

Researchers utilize both probabilistic and deterministic approaches to develop mathematical models for flood forecasting (Xiang et al., 2024). AI has reinforced weather forecasting using powerful statistics in analyzing a wide range of climate data and patterns. It also utilizes low cost of resources compared to the traditional methods used in weather predictions. AI quantifies extreme drought, flood, and other weather characteristics such as magnitude and frequency. These aid in data-driven decision-making of authorities, as well as in creating better plans and improving resilience in communities at risk (Cancelliere et al., 2007). To highlight the application of AI in flood and drought forecasting, a research study is presented in the following subsection.

16.5.2.1 Case Study of AI-Driven Flood and Drought Forecasting

Adikari et al. (2021) evaluated AI models for flood and drought forecasting in arid and tropical regions. Comparisons of three major accepted AI-based approaches in flood and drought forecasting were conducted, namely CNN, long short-term memory (LSTM) network, and wavelet decomposition functions combined with the Adaptive Neuro-Fuzzy Inference System (WANFIS). CNN and LSTM addressed substantial linear regression problems. These techniques are recognized as major deep learning techniques with high accuracy.

CNN is designed to recognize patterns in one-dimensional or two-dimensional data. They consist of convolutional layers that capture spatial or temporal features, followed by pooling layers to reduce dimensionality, and fully connected layers for prediction. LSTM is a type of Recurrent Neural Network (RNN) designed to handle sequential data by maintaining a memory of previous outputs across long time periods. This uses "gates" to control the flow of information, allowing the model to retain or discard information as needed. This is particularly effective for time-series forecasting, where long-term dependencies in data are important such as weather prediction and hydrological patterns. Lastly, the WANFIS is effective in handling single-input features like the Standard Precipitation Index (SPI) for drought forecasting. The wavelet decomposition enhances the model's ability to manage nonlinear and noisy data, improving prediction accuracy.

In the study, fluvial floods were measured by runoff changes in rivers while droughts were assessed using the SPI. Results showed that, deep learning models like CNN were more effective for complex, multivariate forecasting tasks, ML models like WANFIS excel in univariate and nonlinear tasks such as drought prediction. Adikari et al. (2021) highlighted the importance of selecting the appropriate AI model based on the task, which in this case is flood and drought forecasting, and the nature of the input data, whether multiple features or single feature.

16.5.3 AI on Soil Erosion and Water Runoff Predictions

ML techniques have also emerged as effective tools for modeling and mapping water-induced soil erosion, including surface, gully, and interrill erosion (Zhao et al., 2020). Sahour et al. (2021) employed three ML algorithms, namely boosted regression trees (BRT), deep learning (DL), and multiple linear regression to establish relationships between various predictors such as vegetation canopy percentage, slope and soil erosion rates, BRT utilizes a tree-based method that combines multiple decision trees to improve prediction accuracy, effectively handling complex interactions and nonlinear relationships among predictors. In contrast, DL employs a neural network architecture with multiple layers, allowing it to capture intricate patterns in data through backpropagation. Multiple linear regression, on the other hand, establishes a linear relationship between the target variable, which in this case is soil erosion and independent variables.

Utilizing these techniques significantly enhanced soil conservation planning by accurately predicting erosion rates and spatial distribution. The ML techniques mentioned can enhance the accuracy of predicting soil erosion rates by establishing complex relationships between erosion and its controlling factors, such as vegetation cover and slope characteristics. These models can process large datasets and support better spatial mapping of erosion across extensive areas, which is often challenging with traditional methods. AI-driven predictions can facilitate timely interventions in soil conservation projects, and ultimately contribute to sustainable land management practices (Kumar et al., 2024).

16.5.4 AI on Water Monitoring and Tracking

Monitoring and tracking water quality is vital in achieving water security. Durgun (2024) integrated AI into a sensor system that utilized spectroscopic analysis for real-time water quality assessment. The system effectively distinguished between clean, contaminated, and UV-disinfected water, which enhanced the reliability of monitoring efforts. Specifically, Random Forest, Support Vector Machines (SVM), and Neural Network (NN) are used for the processing and classifying spectral data collected from water samples, demonstrating high accuracy in real-time classification of water quality variations.

Random Forest is a widely used ML algorithm, developed by Leo Breiman and Adele Cutler, that merges the predictions of multiple decision trees to generate final result. The popularity of this ML algorithm stems from its simplicity and versatility, making it suitable for both classification and regression tasks (IBM, 2023). On the other hand, SVMs are frequently used for classification tasks. They work by identifying the optimal hyperplane that separates two classes, maximizing the margin between the nearest data points from each class. SVM is widely utilized in water quality monitoring given its ability to predict effectively various water quality parameters (WQPs). SVM models analyze time-series data, allowing for the possible identification of complex relationships between input variables and water quality outcomes over time (Jamshidzadeh et al., 2024). In the context of water monitoring and security, NN can analyze complex datasets to predict effectively water quality and consumption behavior (Rustam et al., 2022). By leveraging datasets from various platforms, neural networks can identify patterns and trends in water quality metrics, enabling timely interventions to ensure water safety and availability. This approach not only enhances predictive capabilities but also supports decision-making in water resource management, contributing to overall water security.

16.5.5 AI for Water Resource Optimization in Agriculture

AI is transforming agriculture in terms of water systems, soil moisture monitoring, and precision agriculture, given that water is the primary driver of most agricultural activities and is essential for sustainable development (He et al., 2024). However, this resource is not being optimized, since regions across the globe are facing water shortage problems. If not managed properly, it can result in a more severe scenario (Lakhiar et al., 2024), leading to more losses and less agricultural production. AI can more than resolve the management issues in water resources; it can also enhance management practices that can result in sustainable farming (Ashoka et al., 2024). By integrating AI into smart irrigation systems in terms of monitoring variables such as air temperature, rainfall, evapotranspiration, and solar radiation, systems can predict the appropriate amount of water and time to water the crops, improving water use efficiency, increasing yield, reducing fertilizer application, reducing labor costs, and saving energy. Optimizing water usage ensures that crops receive precise water needed while minimizing waste. By leveraging real-time data and predictive analytics, AI-driven smart irrigation systems enhance agricultural productivity and contribute to more sustainable water management practices.

AI can be utilized in soil moisture monitoring to enhance irrigation scheduling, as well as to predict the exact amount of water needed by the crops, depending on weather patterns and growth stages. Ahmed et al. (2023) argued that soil moisture monitoring is an essential approach to irrigation scheduling. Including sensors and machine learning algorithms in soil moisture monitoring can reduce water consumption by up to 30% compared to conventional irrigation practices (Elshaikh et al., 2024a). Many innovative smart irrigation approaches were introduced, but Ahmed et al. (2023) categorized them into three. First, we have the model predictive irrigation systems, which are employed to simulate water movement within canals, maintaining specific water levels at various locations, and regulating the flow of water that influences these levels (Menon et al., 2020; Wang & Riaz, 2022; Zheng et al., 2019), second, variable-rate irrigation (VRI) systems are used to apply irrigation at different rates across areas within a field in an optimized manner (Bhakta et al., 2019). Third, unmanned aerial vehicles (AUVs) or drones, which are used for irrigation management.

AI and precision agriculture focus on making systems more efficient, reducing water consumption while increasing crop yields. Technological innovations are fundamental elements of AI and precision agriculture. Specific examples are ground-based sensor networks, AUVs, and VRI technologies (Chithra et al., 2024). Researchers have developed real-time irrigation management (Pierre et al., 2023) and precision agriculture (Pattnaik et al., 2020), an AI-based decision support system that utilizes data from sensors, crop models, and weather stations to improve irrigation schedules. The system could increase yield and reduce water use, thus reducing wastage (Elshaikh et al., 2024a). Using sensors, predictive algorithms, and data analytics, AI enhances irrigation systems, soil monitoring, and resource management to optimize water use, improving agricultural productivity and sustainability while considering some SDGs.

16.5.6 AI for Optimization of Water Distribution Networks

AI has been used in many water distribution network systems. Studies have described systems that combine self-learning AI techniques, statistical data analysis, and tools to detect events such as pipe bursts and leaks, as well as equipment and other failures in the network. The AI-based systems learn from historical events to improve future events and malfunction detection. This system, which was developed for United Utilities in northwest England that significantly improved the ability to deal with such events (IAHR, 2020).

One significant ML approach is the logic and machine learning integration. Combining logic-based algorithms with ML techniques allow for effective early leak detection. The local outlier factor (LOF) model has shown superior performance in identifying anomalies in water distribution systems, thereby minimizing water loss and associated economic costs (Joseph et al., 2024).

Another approach is using DL techniques. Komba et al. (2024) used a convolutional neural network (CNN) framework to utilize pressure monitoring data that are being transformed into images, achieving a true positive rate of 98% for leak detection. Support Vector Machines and Random Forest (SVM-RF) algorithms merge these two ensembles to achieve a leak detection accuracy of 97% by analyzing flow rate, pressure, and temperature data (Komba et al., 2024).

Despite the significant improvements in leak detection and efficiency enabled by AI-driven methods, there are still challenges in integrating these technologies into existing infrastructures and maintaining consistent performance in diverse operational conditions.

In terms of customer data interpretation, AI has the capacity to apply artificial neural networks (ANNs) on data collected from smart water meters to predict water consumption and enhance the operational effectiveness of water supply systems (IAHR, 2020). Providing near-real-time, personalized information and feedback about water usage could significantly enhance customer satisfaction. Affordable smart water meters can now be installed to gather the essential data needed to analyze and share individual customer information in a proactive and impactful manner.

16.5.7 AI on Smart Water Systems for Resource Conservation

Smart water systems leveraging AI and IoT technologies are pivotal in addressing water scarcity by enhancing access and minimizing waste. These systems facilitate real-time monitoring, predictive analytics, and efficient resource management, ultimately leading to sustainable water conservation practices. Smart water management systems utilize IoT sensors for continuous monitoring of water usage and quality, achieving up to 95% accuracy in consumption data (Suryavanshi et al., 2024). There are also applications in predictive analytics and resource optimization, where AI algorithms can analyze historical data to forecast water demand and optimize distribution, ensuring efficient allocation of resources (Tyagi et al., 2024). Dynamic adjustment systems can also modify water levels based on real-time data, potentially conserving over 15% of water resources (Chen et al., 2023). While these technologies offer great advantages, challenges such as high costs and the need for strong regulatory frameworks may hinder their widespread use, requiring more research and collaboration among stakeholders (Elshaikhet al., 2024b). Water businesses were found to be slow in changing paradigms and adopting digital technologies. Incremental steps are taken, often acting reactively to water scarcity conditions (IAHR, 2020).

A case study focuses on using AI technology to optimize water network performance in terms of pressure and water quality (Kim et al., 2024). Self-learning AI technologies analyze network data and patterns in real-time to control networks, increasing process efficiency in terms of operations and energy. An AI installation in the Netherlands improved energy efficiency in operations between 5% and 10%. These technologies optimize pressure, resulting in a reduced number of bursts and leaks. In Kim et al. (2024), there is ML methodology that improves the detection of pressure variations in WDNs, enabling rapid responses to incidents like pipe bursts, was developed. This system utilizes advanced algorithms, including persistent short-term memory and generative adversarial networks, to dynamically adjust data acquisition frequencies, significantly enhancing anomaly detection capabilities. Lastly, an IoT-based water management system integrates DL techniques, such as long short-term memory (LSTM), to predict consumer water demand and optimize distribution. This system aims to minimize water loss during transport while ensuring quality, showcasing the potential of smart technologies in urban water management (Meena & Dhanalakshmi, 2023).

16.5.8 AI for Water Treatment and Desalination

Adaptation to climate variabilities that resulted in freshwater shortages has prompted authorities to promote wastewater treatment and seawater desalination, especially in countries and regions already facing climate-induced water scarcity issues (Estrela, 2024).

Desalination technologies are utilized to increase freshwater by extraction from seawater (Zheng et al., 2020). This technology is an effective strategy that aids social and economic development (Liu et al., 2021). Desalination technologies have two categories: membrane desalination (energy is employed to generate pressure and electric gradients) and thermal (energy is used as heat to evaporate water—phase change) (Do Thi et al., 2021; Herrero-gonzalez & Ibañez, 2021). AI is vital in transforming desalination processes by optimizing energy use and enhancing overall process efficiency. According to Ai et al. (2022), the integration of AI techniques paves the way for a vast volume of data that enhances the understanding of the behavior of desalination systems. AI can recognize the intricate connections and patterns of desalination processes by processing and analyzing data (Chekifi et al., 2024). By integrating AI-driven algorithms with desalination technologies, industries using this technology can reduce energy consumption, monitor operational conditions in real-time, and improve predictive maintenance, making desalination technology a more sustainable solution for addressing global water scarcity.

Wastewater treatment plants remove contaminants and minimize pollutant discharges, ensuring a safe environment (Cairone et al., 2024) and protecting public health. AI has three critical aspects concerning wastewater treatment: the prediction of removal efficiency for both organic and inorganic pollutants, real-time monitoring of important water quality parameters (including pH, COD, BOD, turbidity, TDS, and conductivity) to ensure accurate data and assess the quality and safety of water, and detection of malfunction and inefficiencies in wastewater treatment processes and appurtenant equipment to ensure effective operation (Nagpal et al., 2024). AI transforms water treatment systems by enhancing process efficiency, thus reducing energy consumption and minimizing emissions.

AI is revolutionizing desalination technologies and wastewater treatment systems by optimizing energy use, improving process efficiency, and reducing emissions. Through real-time monitoring and predictive analytics, AI enhances the operation of desalination plants and wastewater treatment systems, making them more sustainable solutions for addressing global water scarcity and reducing carbon footprint while complying with SDG nos. 6 and 13.

16.5.9 AI for Watershed Management

The critical role of watershed management lies in its ability to mitigate climate change, enhance water security, and promote the sustainable use of water resources while reinforcing ecosystem resilience. Effective management practices, particularly community-based approaches, can significantly improve watershed health, which is intrinsically linked to water security.

Water security involves ensuring the availability of an acceptable quantity and quality of water for health, livelihoods, ecosystems, and production (Water Security &NbS, 2024). Watershed management contributes to water security by protecting water sources with sustainable land management practices, enhancing resilience, and supporting ecosystems, which is essential for human well-being and socioeconomic development.

Watershed management plays a vital role in climate change mitigation, such as how forests and wetlands help in carbon sequestration and the use of nature-based solutions, such as reforestation, to enhance watershed resilience and mitigate climate change impacts. AI technologies are increasingly being integrated into sustainable watershed restoration efforts, enhancing decision-making and ecological management. AI plays a transformative role in sustainable watershed restoration by providing advanced tools and technologies to monitor, analyze, and manage water resources effectively.

AI contributes to these efforts through intelligent monitoring systems, predictive modeling, and ecological integration. AI systems utilize sensor networks and mobile devices for comprehensive environmental monitoring, addressing data collection deficiencies in traditional methods (Zhang et al., 2024). These systems can represent and manage heterogeneous data, enabling better-informed decision-making for watershed management. AI models, such as random forest regression, have also shown high accuracy in predicting streamflow, which is crucial for effective water resource management (Raja Shekar & Mathew, 2024). The ability to simulate various scenarios allows for improved planning and management strategies, ultimately leading to sustainable watershed practices. Finally, AI tools can enhance ecological restoration projects by optimizing land use and identifying ecological corridors, although they must be applied with an understanding of underlying values and potential biases (Wang et al., 2023b; Woodrich & Pape, 2024). The integration of AI in restoration efforts can lead to more effective outcomes, provided that practitioners are aware of the complexities involved.

16.6 AI FOR CLIMATE-RESILIENT WATER MANAGEMENT AND GOVERNANCE

In a continuously developing world, post-pandemic and during the era of the fourth industrial revolution, big data and technology have become a staple of social, political, and economic spheres of human life. These aspects made technology ever important and a staple of the conduct of human activities and life. More particularly, the COVID-19 pandemic has accelerated the migration of activities to cyberspace and required drastic adjustments to suit the needs of the health crisis and leading to remodeling of existing modes and modalities of economy and society (Abueg, 2020, 2021).

While these changes are drastic and abrupt and require immediate change and adoption, the advances in technology, even before the pandemic, paved for the birth of big data, which led to the discipline of data science and the birth of AI. These technological advances became equally important not only in economic, social, and political aspects of human life, but also applied in natural resource management and conservation. In particular, AI and related technological advances proved essential and necessary in water resource utilization (both for the household and other socioeconomic sectors), as well as in management (e.g., water security and water quality monitoring).

Given that there is world recognition of water being necessary for human life and essential in various socioeconomic aspects, the UN, through its SDGs, highlighted and made explicit through SDG nos. 6, 13, and 14 and related as well to SDG nos. 3 and 17 (expanding the former MDGs, particularly MDG no. 7). These initial goals outlined by UN-member countries through the former MDGs and now SDGs have highlighted the need for cooperation and coordination on efforts to promote these goals. In the context of water resource management, the role of AI has also been given utmost importance in the need to ensure the realization of these goals, as well as to make water use more sustainable and backed with data-driven decision-making of stakeholders (Jenny et al., 2020). In addition to sustainability, the uncertainty brought forth by global warming and now climate change made data more important and necessary than before in ensuring a more sustainable use of water resources via time-series and spatial data (Xiang et al., 2021). These are some of the important facets of water resource management recognized by literature and stakeholders around the world, given the current conditions of global climate and human population. For these reasons, water resource management initiatives that cut across political borders and regions see the need for adequate data and AI and technology-related advancements to improve water security, management, and utilization, among others (Atkisson, 2024; Kamyab et al., 2023).

However, given the promise of blending AI in integrated water resource management, there have been challenges in implementing these developments, as well as the degree of reception, especially by vulnerable groups and sectors across the world, and the developing nations at large. For one, the mere migration of social and economic functions to the cyberspace requires investments in technology, some degree of technical literacy, and the reception of the stakeholders (e.g., Abueg, 2020, 2021). Adoption to technology is not only a blessing, but also a "curse", depending on the ability to adapt to these new advancements—a problem of developing countries having limited resources and even struggling with water resources. The fundamental problems of financial resource constraints and skills improvement are seen in a lot of countries in the world, especially those that are greatly affected by the pandemic (Abueg, 2023, 2024). Having these constraints and problems recognized in the literature of water resource management, it is therefore imperative for support not only from international organizations, but also to stakeholders who can provide any support, no matter how small these initiatives are.

Recognizing the need for investments in technology and AI in water resource management, developing economies and vulnerable sectors of society may be supported via shared access to competent resource personnel and experts, subsidies, and support in acquiring and learning AI and related technology for integrated water resource management. We also recognize the need for preparing the target communities, groups, and sectors to be receptive to learning, adopting, and applying AI technology to improve water in terms of quality and resource management (e.g., Doorn, 2021).

For example, one problem of water resource management, even before the COVID-19 pandemic, is the increasing amount of pollution in oceans across the globe, which was further exacerbated by the pandemic itself (Abueg, 2019; Casila et al, 2025). Water pollutants also provide indications of human activity that may help provide insights into appropriate policy action for policymakers, stakeholders, and society at large (Casila et al., 2023). These may be aided by ample and appropriate technology, now enhanced with big data and AI (UNESCO, 2024).

Not only is pollution an issue of water resource management, but also the concern about safe and potable drinking water for human survival and health (Maroju et al., 2023). These also have implications on water resource management across borders (e.g., Coesnon et al., 2021), which have similar implications as in pollution. These concerns and developments highlight that water "knows no boundaries": water is a shared resource and beneficial to everyone, so its sustainable use and management are also a shared responsibility of everyone. Moreover, these needed enhancements should also keep pace with the increasing demand of the world, to keep water use and management sustainable, prevent risks to human health and environment, and to improve management by stakeholders and policymakers (Jha, 2023).

16.7 FUTURE OF AI IN WATER SECURITY AND CLIMATE ADAPTATION

16.7.1 Emerging AI Trends in Water Resource Management

Emerging trends in AI for water resource management highlight the integration of advanced technologies to enhance efficiency and sustainability. For example, LSTM networks are used for simulating daily reservoir inflow, effectively capturing seasonal variations (Lin et al., 2024). Additionally, multilayer perceptron (MLP) and gated recurrent unit (GRU) models are employed for estimating agricultural and public water demand, integrating hydrological and socioeconomic factors (Lin et al., 2024). ML and data analytics improve projections of water availability, aiding real-time decision-making (Saxena et al., 2023). Furthermore, smart water management systems combine AI and IoT to enhance operational efficiency, predictive maintenance, and resource conservation (Dada et al., 2024). AI-driven Decision Support Systems (DSSs) also facilitate better monitoring and optimization of water resources, providing essential support for decision-making processes (Elshaikh, et al., 2024c).

16.7.2 The Role of AI in Achieving Global Water Security Goals

AI plays a pivotal role in achieving global water security goals by enhancing water resource management, optimizing water distribution, and improving water quality monitoring. AI technologies, including ML and data analytics, provide innovative solutions to address water scarcity, improve water infrastructure, and ensure sustainable water management practices. These advancements are crucial for meeting the United Nations' SDGs. The integration of AI in water management is transforming traditional practices and paving the way for a more resilient and water-secure future.

AI in water resource management enhances the precision of water availability projections, facilitating better resource distribution and emergency prevention (Saxena et al., 2024). It also enables predictive maintenance of water infrastructure and forecasting of water demand and consumption, leading to more efficient water use (Mehmood et al., 2020). In agriculture, AI-driven systems optimize irrigation scheduling and water application, promoting water efficiency and sustainable farming practices (Ashoka et al., 2024). Additionally, AI technologies are used for real-time water quality monitoring, impacting both small samples and large water bodies, which significantly improves water safety, while its ability to forecast pre-water-related disasters allows for better post-disaster management (Mehmood et al., 2020). Furthermore, AI fosters community engagement and innovation through initiatives like Intelligent Water Hackathons, which develop solutions for water stress in high-risk regions such as Middle East and North Africa (MENA), encouraging collective knowledge sharing and improved water resource outcomes (Batisha, 2023).

16.7.3 Opportunities for AI in Climate Adaptation

The application of AI in water management is expected to increase further due to the ongoing catastrophic impacts of climate change worldwide (Ahmed et al., 2024). It was proven that AI can be a transformative tool for climate adaptation. AI can predict, prepare, and respond to the challenges of the changing climate (Anderson, 2024).

Effective climate adaptation requires resilient solutions that can withstand intense climatic phenomena and must also recover abruptly from the disruptions (Wang et al., 2023a). The integration of AI in different predictive models and simulations guides the urban planners, policy makers, scientists, and experts to design and create climate change resilient solutions (Tank, 2024).

The limitless power of AI with ethical considerations can be diverted more toward enhancing the existing predictive models, warning systems, and creating proactive measures to reduce the devastating impacts of climate-induced disasters (Murari, 2024). Effective natural resources management with AI and building climate-resilient infrastructure that can withstand intense climate stress can optimize water and energy usage ensuring responsible consumption and waste reduction (Genghini, 2023).

16.7.4 Future of AI in Mitigating Water Security Impacts

Water security and enhanced resilience to hydrological extremes can only be achieved through a comprehensive understanding of water resource dynamics, and the presence of AI applications is helping the world to deal with several concerns related to water. The knowledge to address these concerns must be according to the large span of data and observations (Sheffield et al., 2018). AI enables intelligent decision-making through data integration and smart visualizations (Vekaria & Sinha, 2024).

AI can be a vital asset for managing and optimizing irrigation and reduce water wastage by determining the appropriate to exact amount of water for agricultural production (Alam, 2023). This will allow the farmers to adapt irrigation strategies and distribution patterns despite the high variability of precipitation caused by climate change.

Moreover, AI algorithms and ML models aid in identifying highly vulnerable areas and communities to mitigate risks of water insecurity. These models and simulations are important in crafting adaptive strategies. Moreover, using satellite imagery, IoT sensors, and historical records, water resource management will be more comprehensive and effective (Sheffield et al., 2018).

Aside from this, the integration of AI in remote sensing technologies strengthens the groundwater monitoring, which still remains as the primary water source globally, particularly in arid and semi-arid areas where surface water is scarce (Ibrahim et al., 2024). AI-driven groundwater monitoring has an advantage in integrating data from multiple sources, which provides a comprehensive record of current groundwater conditions and trends. This will help experts to make accurate and data-driven decisions in optimizing the groundwater resources (Schwefel, 2024).

In addition to monitoring and predicting the water supply and storage, AI can be utilized in managing water demand by analyzing the usage and behavior patterns of the population. Depending on the nature and conditions, AI can formulate tailored water conservation strategies for different sectors. These strategies may focus on the recommendation of water-efficient equipment and machinery and identifying leaks and unnecessary water wastage (Al Alahmed & Bahnasawi, 2024).

Through the innovative applications of AI, achieving global water security is increasingly within our reach. The capability of AI in data integration, predictive analytics, optimization, and smart decision-making can address the serious impacts of climate change on water security. The proactive management of water through AI secures the fundamental right of access and availability of the population to water. These developments in AI foster the global sustainability and resiliency.

16.8 CONCLUSION AND RECOMMENDATIONS

Climate change has developed into a major constraint in ensuring water security across the world. By water security, we do not only mean adequate supply of water for various sectors of economies or stakeholders of societies worldwide but also the need for potable quality for human use and health complementation. Given the increasing demand for water resources due primarily to the increasing global population, the uncertainty brought forth by climate change has made the problem worse and troublesome for developing countries and vulnerable social groups. Poverty has also exacerbated the problems of water security in these areas, contributing as well to degrading environment and increase in pollution, affecting water access and quality.

One important facet of the developments in technology is improvements in computing, data generation and storage, and connectivity, which led to the development of data science, machine learning, and AI. These developments as been shown and highlighted to be necessary and important in mitigating problems concerning water security, as well as ensuring the sustainability of water use and management. While these technological solutions show promise in addressing the increasing magnitude and complexity of problems in integrated water resource management, acquisition and adoption have also been equally problematic in identified vulnerable strata of the global population. We have identified the importance of support to acquire this technology (as these would have significant financial costs), complementation of competent manpower and expertise, and also investing in the targeted groups to be receptive and ensuring learning, adoption, and continued use. These constraints have been the directions for research and policy formulation in water resource management, and have also become the themes for sustainability and resiliency. Given the lessons of the COVID-19 pandemic, global catastrophic problems of frequent typhoons and flooding, revisiting the definitions of sustainability and resiliency has been recalibrated to include data science, machine learning, and AI in the context of water resource management.

Finally, in moving forward to achieve the goals of local, regional, and international organizations and groups, we highlight the need for continued coordination and cooperation of efforts, no matter how small or big they are. We see the need for participation of both the public and private sectors in this AI-improved water resource management frontier.

REFERENCES

2030 Water Resources Group. (2012). *The Water Resources Group: Background, Impact and the Way Forward.* https://www.weforum.org/publications/water-resources-group-background-impact-and-way-forward/

Abueg, L. (2019). *A Survey of the Ocean's Plastic Waste Problem, and Some Policy Developments of the Philippines.* https://mpra.ub.uni-muenchen.de/96263/

Abueg, L. C. (2020). *Silver Linings in Philippine History and Macroeconomics of the COVID-19 Pandemic Response: Beyond the Longest Lockdown.* https://www.researchgate.net/publication/347399139_Silver_linings_in_Philippine_history_and_macroeconomics_of_the_Covid-19_pandemic_response_beyond_the_longest_lockdown

Abueg, L. C. (2021). What can gender economics learn from the Pinoy BL genre of the COVID-19 pandemic? *Review of Women's Studies*, *31*(1). https://www.researchgate.net/publication/352092043_What_can_gender_economics_learn_from_the_Pinoy_BL_genre_of_the_COVID-19_pandemic

Abueg, L. C. (2023). Stretching Filipino "resilience" amidst a pandemic-induced recession with market and government failures: Insights from community pantries and online barter. *Graduate Inequality Review*, 2, 68–76.

Abueg, L. C. (2024). When poverty lines intersect ethical boundaries: some reflections on the spectrum on the dynamics of poverty from a multi-character participatory perspective. *Graduate Inequality Review*, 3, 16–28.

Adikari, K. E., Shrestha, S., Ratnayake, D. T., Budhathoki, A., Shanmugam, M. S., & Dailey, M. N. (2021). Evaluation of artificial intelligence models for flood and drought forecasting in arid and tropical regions. *Environmental Modelling & Software*, *144*, 105136.

Ahmed, Z., Gui, D., Murtaza, G., Yunfei, L., & Ali, S. (2023). An overview of smart irrigation management for improving water productivity under climate change in drylands. Agronomy, *13*(8), 2113.

Ahmed, A. A., Sayed, S., Abdoulhalik, A., Moutari, S., & Oyedele, L. (2024). Applications of machine learning to water resources management: A review of present status and future opportunities. *Journal of Cleaner Production*, *441*, 140715. https://doi.org/10.1016/J.JCLEPRO.2024.140715

Ahn, Y., & Juraev, Z. (2024). *Water Scarcity and Its Economic-Social Consequences: The Case of Central Asia.* https://Communities.Springernature.Com/Posts/Water-Scarcity-and-Its-Economic-Social-Consequences-the-Case-of-Central-Asia. https://communities.springernature.com/posts/water-scarcity-and-its-economic-social-consequences-the-case-of-central-asia#:~:text=Water%20scarcity%20in%20Central%20Asia%20is%20primarily%20due%20to%20geographical,actually%20suffering%20from%20water%20scarcity

Ai, Z., Ishihama, F., &Hanasaki, N. (2022). Mapping current and future seawater desalination plants globally using species distribution models. Water Resources Research. https://doi.org/10.1029/2021WR031156

Al Alahmed, Y., & Bahnasawi, H. (2024). Harnessing the power of AI for sustainable water consumption. *Tuijin Jishu/Journal of Propulsion Technology*, 45, 1001–4055.

Alam, R. (2023). *Five Ways Artificial Intelligence is Going to Shape the Future of Water and Resilient Infrastructure.* https://www.wsp.com/En-Ph/Insights/2023-Artificial-Intelligence-Shaping-Future-of-Water

Alpan, K., & Sekeroglu, B. (2020). Prediction of pollutant concentrations by meteorological data using machine learning algorithms. *International Archives of the Photogrammetry, Remote Sensing and Spatial Information Sciences – ISPRS Archives.* https://doi.org/10.5194/isprs-archives-XLIV-4-W3-2020-21-2020

Anderson, K. (2024). *How can Artificial Intelligence Help Tackle Climate Change?* https://Greenly.Earth/En-Us/Blog/Ecology-News/How-Can-Artificial-Intelligence-Help-Tackle-Climate-Change

Arduino, G., Reggiani, P., & Todini, E. (2005). Recent advances in flood forecasting and flood risk assessment. *Hydrology and Earth System Sciences*, 9, 280–284. https://doi.org/10.5194/hess-9-280-2005

Ashoka, P., Devi, B. R., Nilesh, S., Behera, M., Gautam, A., Jha, A., & Sinha, G. (2024). Artificial intelligence in water management for sustainable farming: A review. *Journal of Scientific Research and Reports.* https://doi.org/10.9734/jsrr/2024/v30i62068

Atkisson, A. (2024). *The AI Imperative: Accelerating Water Management Processes in the Era of Climate Change.* Global Water Partnership.

Bagheri-Gavkosh, M., Hosseini, S. M., Ataie-Ashtiani, B., Sohani, Y., Ebrahimian, H., Morovat, F., & Ashrafi, S. (2021). Land subsidence: A global challenge. *Science of the Total Environment*, 778. https://doi.org/10.1016/j.scitotenv.2021.146193

Batisha, A. (2023). A lighthouse to future opportunities for sustainable water provided by intelligent water hackathons in the Arabsphere. Humanities and Social Sciences Communications. https://doi.org/10.1057/s41599-023-02146-3

Bhakta, I., Phadikar, S., & Majumder, K. (2019). State-of-the-art technologies in precision agriculture: a systematic review. *Journal of the Science of Food and Agriculture*, 99, 4878–4888. https://doi.org/10.1002/jsfa.9693

Cairone, S., Hasan, S. W., Choo, K.-H., Lekkas, D. F., Fortunato, L., Zorpas, A. A., Korshin, G., Zarra, T., Belgiorno, V., & Naddeo, V. (2024). Revolutionizing wastewater treatment toward circular economy and carbon neutrality goals: Pioneering sustainable and efficient solutions for automation and advanced process control with smart and cutting-edge technologies. *Journal of Water Process Engineering*, 63, 105486.

Cancelliere, A., Mauro, G. Di, Bonaccorso, B., & Rossi, G. (2007). Drought forecasting using the standardized precipitation index. Water Resources Management. https://doi.org/10.1007/s11269-006-9062-y

Casila, J. C. C., Abueg, L. C., Khajuria, A., Memon, S., Haddout, S., Priya, K. L. (2025). Achieving net-zero, protecting the oceans, and embracing circular economy: the urgency of addressing climate change. In: Casila, J. C., Khajuria, A., Memon, S. (eds) *Proceedings of The 3rd International Conference on Climate Change and Ocean Renewable Energy*. CCORE 2024. Springer Proceedings in Earth and Environmental Sciences. Springer, Cham. https://doi.org/10.1007/978-3-031-93887-0_1

Casila, J. C., Dimapilis, D. A., Limbago, J. S., Delos Reyes, A. A., Casila, E. B., & Haddout, S. (2023). Characterization and quantification of surficial sediment microplastics and its correlation with heavy metals, soil texture, and flow velocity. *Analytical Letters*, *57*(7), 1092–1105. https://doi.org/10.1080/00032719.2023.2241090

Casila, J. C., Ureta, J. C., Haddout, S., & Priya, K. L. (2022). GHG emissions, environmental technologies, and renewable energy status in the Philippines. In *Towards Net Zero Emissions: Policy and Practice. Hanoi Geoengineering 2022: Innovative Geosciences, Circular Economy and Sustainability* (pp. 1–8). Hanoi University of Science.

Chang, F. J., Chang, L. C., & Chen, J. F. (2023). Artificial intelligence techniques in hydrology and water resources management. *Water*, 15. https://doi.org/10.3390/w15101846

Chekifi, T., Benmoussa, A., & Boukraa, M. (2024). Desalination powered by renewables: A challenge and an AI opportunity. Water Resources Management. https://doi.org/10.1007/s11269-024-03935-4

Chen, S. L., Chou, H. S., Huang, C. H., Chen, C. Y., Li, L. Y., Huang, C. H., Chen, Y. Y., Tang, J. H., Chang, W. H., & Huang, J. S. (2023). An intelligent water monitoring IoT system for ecological environment and smart cities. Sensors, *23*. https://doi.org/10.3390/s23208540

Chithra, R., Vasantheeswaran, R., A.p, T., & Thangavel, G. (2024). Enhancing agricultural efficiency and sustainability through advanced IoT and AI-driven precision farming technologies. In *1st International Conference on Electronics, Computing, Communication and Control Technology, ICECCC 2024*. https://doi.org/10.1109/ICECCC61767.2024.10593925

Climate Council. (2023). *What does Net Zero Emissions Mean?* https://www.climatecouncil.org.au/resources/what-does-net-zero-emissions-mean/

Coesnon, M., Dasgupta, D., Schillinger, J., & Zmak, E. (2021). Water across borders: Youth perspectives on transboundary water management. In T. Chen, S. Khatri, A. Quintanilla, & J. Schillinger, (Eds.). *World Youth Parliament for Water*. https://youthforwater.org/app/uploads/2021/05/WYPW_Water-across-Borders_GA-2021.pdf

Dada, M. A., Majemite, M. T., Obaigbena, A., Daraojimba, O. H., Oliha, J. S., & Nwokediegwu, Z. Q. S. (2024). Review of smart water management: IoT and AI in water and wastewater treatment. *World Journal of Advanced Research and Reviews*. https://doi.org/10.30574/wjarr.2024.21.1.0171

Dao, P. U., Heuzard, A. G., Le, T. X. H., Zhao, J., Yin, R., Shang, C., & Fan, C. (2024). The impacts of climate change on groundwater quality: A review. *Science of the Total Environment*, 912. https://doi.org/10.1016/j.scitotenv.2023.169241

Do Thi, H. T., Pasztor, T., Fozer, D., Manenti, F., & Toth, A. J. (2021). Comparison of desalination technologies using renewable energy sources with life cycle, pestle, and multi-criteria decision analyses. Water. https://doi.org/10.3390/w13213023

Doorn, N. (2021). Artificial intelligence in the water domain: Opportunities for responsible use. *Science of the Total Environment*, 755. https://doi.org/10.1016/j.scitotenv.2020.142561

Durgun, Y. (2024). Real-time water quality monitoring using AI-enabled sensors: detection of contaminants and UV disinfection analysis in smart urban water systems. Journal of King Saud University – Science, *36*. https://doi.org/10.1016/j.jksus.2024.103409

Elshaikh, A., Elsheikh, E., & Mabrouki, J. (2024a). Applications of artificial intelligence in precision irrigation. *Journal of Environmental and Earth Sciences*, 6(2), 176–186.

Elshaikh, A., Mabrouki, J., & Mohamed, A. A. O. (2024b). Harnessing smart technologies for sustainable water resources management in water-scarce environments. In *Advancements in Climate and Smart Environment Technology* (pp. 39–55). IGI Global. https://doi.org/10.4018/979-8-3693-3807-0.ch004

Elshaikh, A., Mabrouki, J., & Mohamed, A. A. O. (2024c). The future of water management: leveraging ai for effective decision support. In *Advancements in Climate and Smart Environment Technology*. https://doi.org/10.4018/979-8-3693-3807-0.ch015

Estrela, T. (2024). Introduction to section 3. *Water International*, 49(3–4), 466–475.

Fankhauser, S., Smith, S. M., Allen, M., Axelsson, K., Hale, T., Hepburn, C., Kendall, J. M., Khosla, R., Lezaun, J., Mitchell-Larson, E., Obersteiner, M., Rajamani, L., Rickaby, R., Seddon, N., & Wetzer, T. (2022). The meaning of net zero and how to get it right. *Nature Climate Change,* 12(1), 15–21. https://doi.org/10.1038/s41558-021-01245-w

Food and Agriculture Organization of the United Nations (FAO). (2015). *Climate Change and Food Security: Risks and Responses.* https://www.fao.org/policy-support/tools-and-publications/resources-details/en/c/427091/

Fox-Kemper, B., H. T. Hewitt, C. Xiao, G. Aðalgeirsdóttir, S. S. Drijfhout, T. L. Edwards, N. R. Golledge, M. Hemer, R. E. Kopp, G. Krinner, A. Mix, D. Notz, S. Nowicki, I. S. Nurhati, L. Ruiz, J. -B. Sallée, A. B. A. Slangen, & Y. Yu, (2021). Ocean, cryosphere and sea level change. In Masson-Delmotte, V., P. Zhai, A. Pirani, S. L. Connors, C. Péan, S. Berger, N. Caud, Y. Chen, L. Goldfarb, M. I. Gomis, M. Huang, K. Leitzell, E. Lonnoy, J.B.R. Matthews, T. K. Maycock, T. Waterfield, O. Yelekçi, R. Yu, & B. Zhou (eds.), *Climate Change 2021: The Physical Science Basis. Contribution of Working Group I to the Sixth Assessment Report of the Intergovernmental Panel on Climate Change.* Cambridge University Press, Cambridge, United Kingdom and New York, NY, USA, (pp. 1211–1362), doi:10.1017/9781009157896.011..

Fuckar, N. (2024). *Improving Weather and Climate Predictions Using Machine Learning.* University of Oxford's School of Geography and the Environment. United Kingdom.

Genghini, L. (2023). *8 Ways Artificial Intelligence can Contribute to Environmental Conservation. https://2030.Builders/8-Ways-Ai-Can-Contribute-to-Environmental-Conservation/.*

Glick, D. (2024). *The Big Thaw.* National Geographic. Washington, D.C., U.S.A. https://www.nationalgeographic.com/environment/global-warming/big-thaw/

Han, D. (2024). Effects of human activities on china's coastal groundwater. In *Treatise on Estuarine and Coastal* Science (Second edition, pp. 412–433). Elsevier. https://doi.org/10.1016/b978-0-323-90798-9.00088-3

He, L., Du, Y., Yu, M., Wen, H., Ma, H., & Xu, Y. (2024). A stochastic simulation-based method for predicting the carrying capacity of agricultural water resources. Agricultural Water Management, *291,* 108630.

Herrero-gonzalez, M., & Ibañez, R. (2021). Chemical and energy recovery alternatives in SWRO desalination through electro-membrane technologies. Applied Sciences. https://doi.org/10.3390/app11178100

Hoddinott, J. (2006). Shocks and their consequences across and within households in Rural Zimbabwe. *The Journal of Development Studies*, *42*(2), 301–321. https://doi.org/10.1080/00220380500405501

Holden, W. N., & Marshall, S. J. (2018). Climate change and typhoons in the Philippines: extreme weather events in the anthropocene. In *Integrating Disaster Science and Management: Global Case Studies in Mitigation and Recovery* (pp. 407–421). Elsevier. https://doi.org/10.1016/B978-0-12-812056-9.00024-5

IBM. (2023). *What Is Random Forest?* IBM. https://www.ibm.com/topics/random-forest

Ibrahim, A., Wayayok, A., Shafri, H. Z. M., & Toridi, N. M. (2024). Remote sensing technologies for unlocking new groundwater insights: a comprehensive review. *Journal of Hydrology* X, 23. https://doi.org/10.1016/j.hydroa.2024.100175

Igwegbe, C. A., López-Maldonado, E. A., Landázuri, A. C., Ovuoraye, P. E., Ogbu, A. I., Vela-García, N., & Białowiec, A. (2024). Sustainable municipal landfill leachate management: Current practices, challenges, and future directions. *Desalination and Water Treatment,* 320. https://doi.org/10.1016/j.dwt.2024.100709

Intergovernmental Panel on Climate Change (IPCC). (2007). *Climate Change 2007: Synthesis Report*. https://www.ipcc.ch/report/ar4/syr/

International Association for Hydro-Environment Engineering and Research (IAHR) (Hg.) (2020): Hydrolink 2020/4. Artificial intelligent. Madrid: IAHR (Hydrolink, 2020/4). https://www.iahr.org/library/infor?pid=20138

International Groundwater Resources Assessment Centre (IGRAC). (2024). *Groundwater & Climate Change*. UNESCO. https://www.un-igrac.org/

Jamshidzadeh, Z., Latif, S. D., Ehteram, M., Sheikh Khozani, Z., Ahmed, A. N., Sherif, M., & El-Shafie, A. (2024). An advanced hybrid deep learning model for predicting total dissolved solids and electrical conductivity (EC) in coastal aquifers. Environmental Sciences Europe, *36*. https://doi.org/10.1186/s12302-024-00850-8

Jenny, H., Wang, Y., Alonso, E. G., & Minguez, R. (2020). *Using Artificial Intelligence for Smart Water Management Systems*. https://doi.org/10.22617/BRF200191-2

Jha, S. K. (2023). *Water Can't Wait: Accelerating the Adoption of Innovations in Water Security*. https://Blogs.Worldbank.Org/En/Water/Water-Cant-Wait-Accelerating-Adoption-Innovations-Water-Security

Joseph, K., Shetty, J., Sharma, A. K., van Staden, R., Wasantha, P. L. P., Small, S., & Bennett, N. (2024). Leak and burst detection in water distribution network using logic- and machine learning-based approaches. Water, *16*. https://doi.org/10.3390/w16141935

Kamyab, H., Khademi, T., Chelliapan, S., SaberiKamarposhti, M., Rezania, S., Yusuf, M., Farajnezhad, M., Abbas, M., Hun Jeon, B., & Ahn, Y. (2023). The latest innovative avenues for the utilization of artificial Intelligence and big data analytics in water resource management. Results in Engineering, *20*. https://doi.org/10.1016/j.rineng.2023.101566

Kim, H., Jung, K. J., Lee, S., & Jeong, E. H. (2024). Rapid response to pressure variations in water distribution networks through machine learning-enhanced data acquisition. *Aqua Water Infrastructure, Ecosystems and Society*, 73, 1358–1371. https://doi.org/10.2166/aqua.2024.030

Kochkov, D., Yuval, J., Langmore, I., Norgaard, P., Smith, J., Mooers, G., Klöwer, M., Lottes, J., Rasp, S., Düben, P., Hatfield, S., Battaglia, P., Sanchez-Gonzalez, A., Willson, M., Brenner, M. P., & Hoyer, S. (2024). Neural general circulation models for weather and climate. Nature. https://doi.org/10.1038/s41586-024-07744-y

Komba, G. M., Mathonsi, T. E., & Owolawi, P. A. (2024). Optimizing leak detection and location in water distribution networks using SVM-RF algorithm. In *2024 15th International Conference on Mechanical and Intelligent Manufacturing Technologies, ICMIMT 2024* (pp. 27–33). https://doi.org/10.1109/ICMIMT61937.2024.10585935

Kumar, S. S. M., Pandey, M., & Umamahesh, N. V. (2024). Predicting total upland sediment yield using regression and machine learning models for improved land management and water conservation. *Journal of Hydroinformatics*, 26, 2045–2058. https://doi.org/10.2166/hydro.2024.159

Kundzewicz, Z. W. (2008). Climate change impacts on the hydrological cycle. Ecohydrology and Hydrobiology. https://doi.org/10.2478/v10104-009-0015-y

Lakhiar, I. A., Yan, H., Zhang, C., Wang, G., He, B., Hao, B., Han, Y., Wang, B., Bao, R., Syed, T. N., Chauhdary, J. N., & Rakibuzzaman, M. (2024). A review of precision irrigation water-saving technology under changing climate for enhancing water use efficiency, crop yield, and environmental footprints. Agriculture, *14*(7), 1141.

Lin, H. Y., Lee, S. H., Wang, J. H., & Chang, M. J. (2024). Utilizing artificial intelligence techniques for a long–term water resource assessment in the ShihMen reservoir for water resource allocation. Water. https://doi.org/10.3390/w16162346

Liu, S., Wang, Z., Han, M., & Zhang, J. (2021). Embodied water consumption between typical desalination projects: Reverse osmosis versus low-temperature multi-effect distillation. Journal of Cleaner Production. https://doi.org/10.1016/j.jclepro.2021.126340

Lundberg, S. M., & Lee, S. I. (2017). A unified approach to interpreting model predictions. In *Advances in Neural Information Processing Systems*, 2017-*December*, 4766–4775.

Maroju, R. G., Choudhari, S. G., Shaikh, M. K., Borkar, S. K., & Mendhe, H. (2023). Application of artificial intelligence in the management of drinking water: a narrative review. Cureus. https://doi.org/10.7759/cureus.49344

Mbow, C., Rosenzweig, C., Barioni, L. G., Benton, T. G., Herrero, M., Krishnapilla, M., Liwenga, E., Pradhan, P., Rivera-Ferre, M. G., Sapkota, T., Tubiello, F. N., & Xu, Y. (2019). Chapter 5 "Food Security". In N. Benkeblia, A. Challino, A. Khan, & J. R. Porter (Eds.), *Climate Change and Land: An IPCC special report on climate change, desertification, land degradation, sustainable land management, food security, and greenhouse gas fluxes in terrestrial ecosystems* (pp. 439–550). Cambridge University Press. https://doi.org/10.1017/9781009157988.007.

Meena, K. B., & Dhanalakshmi, S. (2023). IoT water management and distribution system for smart city using artificial intelligence. *Journal of Ubiquitous Computing and Communication Technologies*, 5, 339–351. https://doi.org/10.36548/jucct.2023.4.001

Megdal, S. B. (2018). Invisible water: The importance of good groundwater governance and management. Npj Clean Water, *1*. https://doi.org/10.1038/s41545-018-0015-9

Mehmood, H., Liao, D., & Mahadeo, K. (2020). A review of artificial intelligence applications to achieve water-related sustainable development goals. In *2020 IEEE/ITU International Conference on Artificial Intelligence for Good, AI4G 2020*. https://doi.org/10.1109/AI4G50087.2020.9311018

Menon, J., Mudgal, B. V., Guruprasath, M., & Sivalingam, S. (2020). Control of an irrigation branch canal using model predictive controller. Current Science, *118*(8), 1255–1264.

Mishra, B. K., Kumar, P., Saraswat, C., Chakraborty, S., & Gautam, A. (2021). Water security in a changing environment: concept, challenges and solutions. Water. https://doi.org/10.3390/w13040490

Murari, H. (2024). *AI for Climate Change: Innovative Models for Predicting Environmental Impact - Dataversity*. https://Www.Dataversity.Net/Ai-for-Climate-Change-Innovative-Models-for-Predicting-Environmental-Impact/

Nagpal, M., Siddique, M. A., Sharma, K., Sharma, N., & Mittal, A. (2024). Optimizing wastewater treatment through artificial intelligence: recent advances and future prospects. *Water Science & Technology*, *90*(3), 731–757.

NASA. (2024). *How does Climate Change Affect Precipitation?* https://gpm.nasa. gov/resources/faq/how-does-climate-change-affect-precipitation

Nguyen, T. T., Grote, U., Neubacher, F., Rahut, D. B., Do, M. H., & Paudel, G. P. (2023). Security risks from climate change and environmental degradation: Implications for sustainable land use transformation in the Global South. *Current Opinion in Environmental Sustainability,* 63. https://doi.org/10.1016/j.cosust.2023.101322

Olaoluwa, E. E., Olorunsaye, O., Durowoju, O. S., Orimoloye, I. R., Daramola, M. T., & Ayobami, A. A. (2022). Understanding weather and climate extremes. In *Climate Impacts on Extreme Weather: Current to Future Changes on a Local to Global Scale*. https://doi.org/10.1016/B978-0-323-88456-3.00008-3

Olawade, D. B., Wada, O. Z., David-Olawade, A. C., Fapohunda, O., Ige, A. O., & Ling, J. (2024). Artificial intelligence potential for net zero sustainability: Current evidence and prospects. *Next Sustainability*, 4, 100041. https://doi.org/10.1016/j.nxsust.2024.100041

Osman, A., Afan, H. A., Allawi, M. F., Jaafar, O., Noureldin, A., Hamzah, F. M., Ahmed, A. N., & El-shafie, A. (2020). Adaptive Fast Orthogonal Search (FOS) algorithm for forecasting streamflow. Journal of Hydrology, *586*. https://doi.org/10.1016/j.jhydrol.2020.124896

Pattnaik, P., Kumar, R., & Pal, S. (2020). *Internet of Things and Analytics for Agriculture*. Springer Singapore.

Pierre, N., Ishimwe Viviane, I. V., Lambert, U., Viviane, I., Shadrack, I., Erneste, B., Schadrack, N., Alexis, N., Francois, K., & Theogene, H. (2023). AI based real-time weather condition prediction with optimized agricultural resources. *European Journal of Technology*, 7, 36–49. https://doi.org/10.47672/ejt.1496

Raja Shekar, P., & Mathew, A. (2024). *AI-Based Rainfall-Runoff Modelling for Sustainable Water Management in Potteruvagu Watershed, India* (pp. 95–107). https://doi.org/10.1007/978-3-031-52561-2_6

Rampal, N., Hobeichi, S., Gibson, P. B., Baño-Medina, J., Abramowitz, G., Beucler, T., González-Abad, J., Chapman, W., Harder, P., & Gutiérrez, J. M. (2024). Enhancing regional climate downscaling through advances in machine learning. *Artificial Intelligence for the Earth Systems*. https://doi.org/10.1175/aies-d-23-0066.1

Rustam, F., Ishaq, A., Kokab, S. T., de la Torre Diez, I., Mazón, J. L. V., Rodríguez, C. L., & Ashraf, I. (2022). An artificial neural network model for water quality and water consumption prediction. Water, *14*. https://doi.org/10.3390/w14213359

Sahour, H., Gholami, V., Vazifedan, M., & Saeedi, S. (2021). Machine learning applications for water-induced soil erosion modeling and mapping. *Soil and Tillage Research*, 211. https://doi.org/10.1016/j.still.2021.105032

Saxena, R., Srivastava, V., Bharti, D., Singh, R., Kumar, A., & Sharma, A. (2024). Artificial Intelligence for Water Resource Planning and Management. In A. Kumar, A. L. Srivastav, A. K. Dubey, V. Dutt, & N. Vyas (Eds.), *Innovations in Machine Learning and IoT for Water Management* (pp. 51–70). IGI Global Scientific publishing. New York, USA. https://doi.org/10.4018/979-8-3693-1194-3.ch003.

Schwefel, L. (2024). *AI-Driven Groundwater Monitoring: Exploring Cutting-Edge Technologies*. https://Www.Fulcrumapp.Com/Blog/Ai-Driven-Groundwater-Monitoring-Exploring-Cutting-Edge-Technologies/

Seneviratne, S. I., Zhang, X., Adnan, M., Badi, W., Dereczynski, C., Di Luca, A., Ghosh, S., Iskandar, I., Kossin, J., Lewis, S., Otto, F., Pinto, I., Satoh, M., Vicente-Serrano, S. M., Wehner, M., & Zhou, B. (2021). Weather and climate extreme events in a changing climate. In V. P. Masson-Delmotte, A. Zhai, S. L. Pirani, C. Connors, S. Péan, N. Berger, Y. Caud, L. Chen, M. I. Goldfarb, M. Gomis, K. Huang, E. Leitzell, J. B. R. Lonnoy, T. K. Matthews, T. Maycock, O. Waterfield, R. Y. Yelekçi, & B. Zhou (Eds.), *Climate Change 2021: The Physical Science Basis. Contribution of Working Group I to the Sixth Assessment Report of the Intergovernmental Panel on Climate Change* (pp. 1513–1766). Cambridge University Press, Cambridge.

Sheffield, J., Wood, E. F., Pan, M., Beck, H., Coccia, G., Serrat-Capdevila, A., &Verbist, K. (2018). Satellite remote sensing for water resources management: potential for supporting sustainable development in data-poor regions. *Water Resources Research*, 54, 9724–9758. https://doi.org/10.1029/2017WR022437

Suryavanshi, A., Mehta, S., Chattopadhyay, S., & Aeri, M. (2024). Navigating water scarcity with IoT: A smart management system approach. In *2024 4th International Conference on Intelligent Technologies, CONIT 2024*. https://doi.org/10.1109/CONIT61985.2024.10627282

Tank, A. (2024). *AI for Climate Change: Revolutionizing Solutions for a Sustainable Future. https://Transcendinfra.Com/Ai-for-Climate-Change/*

Tasoff, H. (2024). *Global Groundwater Depletion is Accelerating, but is Not Inevitable.* https://Www.Preventionweb.Net/News/Global-Groundwater-Depletion-Accelerating-Not-Inevitable

Tyagi, R., Kathuria, K., Sharma, P., & Sharma, P. K. (2024). Optimizing water management and conservation through IoT. In *2024 2nd International Conference on Disruptive Technologies, ICDT 2024* (pp. 654–659). https://doi.org/10.1109/ICDT61202.2024.10489231

UNESCO. (2024). *The United Nations World Water Development Report 2024: Water for Prosperity and Peace.* https://www.unesco.org/en/articles/united-nations- world-water-development-report-2024-water-prosperity-and-peace

United Nations. (2013). *What is Water Security? Infographic.* UN Water. https://www.unwater.org/publications/what-water-security-infographic

United Nations. (2024a). *Global Issues: Water.* https://www.un.org/en/global-issues/water

United Nations. (2024b). *Water – At the Center of the Climate Crisis.* Climate Action. https://www.un.org/en/climatechange/science/climate-issues/water

Vekaria, D., & Sinha, S. (2024). *ai*WATERS: An artificial intelligence framework for the water sector. AI in Civil Engineering, *3*. https://doi.org/10.1007/s43503-024-00025-7

Wang, F., Harindintwali, J. D., Wei, K., Shan, Y., Mi, Z., Costello, M. J., Grunwald, S., Feng, Z., Wang, F., Guo, Y., Wu, X., Kumar, P., Kästner, M., Feng, X., Kang, S., Liu, Z., Fu, Y., Zhao, W., Ouyang, C., … Tiedje, J. M. (2023a). Climate change: Strategies for mitigation and adaptation. *The Innovation Geoscience*, 1(1), 100015. https://doi.org/10.59717/j.xinn-geo.2023.100015

Wang, L., Wang, K., Xu, J., & Liu, J. (2023b). Ecological restoration process of watershed land space with intelligent IoT technology. *Water Supply*, 23, 3881–3898. https://doi.org/10.2166/ws.2023.202

Wang, Y., & Riaz, S. (2022). Retraction: accelerated iterative learning control for linear discrete time invariant switched systems. Mathematical Problems in Engineering, *2023*, 5738826.

Water Security & NbS. (2024). *Water Security &NbS.* The Nature Conservancy. https://resilientwatershedstoolbox.org/intro-wips/water-security-nbs

Woodrich, S. T., & Pape, T. (2024). Ecological restoration and artificial intelligence: whose values inform a project? Restoration Ecology, *32*. https://doi.org/10.1111/rec.14128

World Meteorological Organization. (2021, August 31). *Weather-Related Disasters Increase over Past 50 Years, Causing More Damage but Fewer Deaths.* World Meteorological Organization. https://wmo.int/media/news/weather-related-disasters-increase-over-past-50-years-causing-more-damage-fewer-deaths

World Wildlife Fund. (2024). *Why Are Glaciers and Sea Ice Melting?* World Wildlife (WWF). https://www.worldwildlife.org/pages/why-are-glaciers-and-sea- ice-melting

Xiang, X., Li, Q., Khan, S., & Khalaf, O. I. (2021). Urban water resource management for sustainable environment planning using artificial intelligence techniques. *Environmental Impact Assessment Review*, 86. https://doi.org/10.1016/j.eiar.2020.106515

Xiang, Y., Peng, T., Qi, H., Yin, Z., & Shen, T. (2024). Improving flood forecasting skill by combining ensemble precipitation forecasts and multiple hydrological models in a Mountainous Basin. Water, *16*. https://doi.org/10.3390/w16131887

Zhang, S., Li, M., & Fang, D. (2024). Study on integrated watershed management decision-making based on artificial intelligence. *Advances in Computer and Communication*, 4, 383–388. https://doi.org/10.26855/acc.2023.12.007

Zhang, F., Thapa, S., Immerzeel, W., Zhang, H., & Lutz, A. (2019). Water availability on the Third Pole: A review. *Water Security*, 7, 100033. https://doi.org/10.1016/j.wasec.2019.100033

Zhao, Y., Meng, X., Qi, T., Qing, F., Xiong, M., Li, Y., Guo, P., & Chen, G. (2020). AI-based identification of low-frequency debris flow catchments in the Bailong River basin, China. Geomorphology, *359*. https://doi.org/10.1016/j.geomorph.2020.107125

Zheng, X., Wen, J., Shi, L., Cheng, R., & Zhang, Z. (2020). A top-down approach to estimate global RO desalination water production considering uncertainty. Desalination. https://doi.org/10.1016/j.desal.2020.114523

Zheng, Z., Wang, Z., Zhao, J., Asce, A. M., & Zheng, H. (2019). Constrained model predictive control algorithm for cascaded irrigation canals. *Journal of Irrigation and Drainage Engineering*, *145*(6), 04019009.

17 Building Climate-Resilient Water Systems
Challenges, Solutions, and AI-Driven Strategies

Doaa A. El-Emam

17.1 INTRODUCTION

Climate change and water security are interlinked and pressing global challenges that require immediate attention. Climate change refers to significant, long-term alterations in Earth's climate patterns, including temperature fluctuations, changes in precipitation, and variations in other climatic variables (Campbell-Lendrum et al., 2023). The primary driver of climate change is the escalating concentration of greenhouse gases in the atmosphere, primarily resulting from human activities such as the combustion of fossil fuels, deforestation, and industrial processes (Singh, 2024).

These climate changes are leading to notable shifts in weather patterns across the globe. Countries are experiencing an increase in the frequency and severity of extreme weather events, including prolonged droughts that deplete water supplies, intense floods that can overwhelm infrastructure, and powerful storms that cause widespread destruction (Elemam and Eldeeb, 2023). Such variability in weather not only affects the immediate availability of water but also has long-term implications for water quality.

In many regions, especially those already vulnerable due to geographic and socio-economic factors, the intensified impact of climate change is exacerbating existing water scarcity issues. Droughts can substantially diminish water reservoirs, while floods can lead to contamination of freshwater sources, making them unsafe for consumption. Moreover, rising global temperatures contribute to the accelerated melting of glaciers and polar ice caps, leading to rising sea levels. This phenomenon poses significant risks for coastal communities, where higher sea levels increase the likelihood of flooding and saltwater intrusion into freshwater aquifers (Elemam and Eldeeb, 2023, 2024; Lakshmi, 2024).

Saltwater intrusion compromises the quality of drinking water, making it difficult for communities to access safe and reliable water supplies. The combined effects of these changes threaten agricultural productivity as well, jeopardizing food security and increasing competition for limited water resources (Elemam and Eldeeb, 2023, 2024).

DOI: 10.1201/9781003547099-20

The relationship between climate change and water security issues is critical; for instance, changing climate patterns can alter precipitation rates, leading to both droughts and floods that directly impact freshwater availability. Conversely, water scarcity can exacerbate the effects of climate change, as stressed ecosystems struggle to adapt (Nimma et al., 2024)

The urgency of addressing these interconnected issues cannot be overstated. As temperatures rise and weather patterns become increasingly unpredictable, the availability of clean and accessible water is at risk, which can lead to conflicts, economic instability, and severe impacts on agriculture and public health (Campbell-Lendrum et al., 2023).

To effectively tackle the myriad challenges posed by climate change and water scarcity, policymakers and communities must embrace comprehensive and sustainable strategies that seamlessly integrate climate adaptation with water management. This approach could involve several key initiatives. First, implementing innovative water conservation techniques is crucial. These techniques might include rainwater harvesting, wastewater recycling, and the use of smart irrigation systems that minimize water usage while maximizing agricultural productivity (Ali and Sang, 2023). Additionally, investing in durable infrastructure that can withstand the extremes brought about by climate change such as flooding, droughts, and heat waves is essential. This could involve enhancing existing dams, reservoirs, and pipelines, as well as developing green infrastructure such as permeable pavements and rain gardens that not only manage storm water but also replenish local aquifers. Promoting policies that encourage responsible water use is another vital component. This could mean establishing stricter regulations for industrial water usage, providing incentives for households to adopt water saving technologies, and supporting community level initiatives for conservation. Protecting natural water ecosystems, such as wetlands and rivers, is equally important, as these systems play a crucial role in maintaining biodiversity and providing natural water filtration. Collaboration across various sectors government agencies, nonprofit organizations, the private sector, and communities is essential in crafting effective solutions. International cooperation is particularly important, as many water bodies cross borders, requiring joint management strategies to ensure sustainable use (Paul et al., 2023). Education and awareness campaigns are also fundamental in fostering a culture of responsible water use. These initiatives can help individuals and communities understand the critical importance of water conservation, the impacts of their consumption habits, and the benefits of sustainable practices. By providing resources and training on how to save water at home and in businesses, we can empower individuals to contribute to broader conservation efforts (Ouyang et al., 2023).

Artificial intelligence (AI) can play a transformative role in addressing these challenges. By analyzing vast amounts of data from various sources such as satellite imagery, weather patterns, and consumption rates AI can improve the management of water resources and accurately predict future water availability. For example, AI-driven models can help identify potential droughts or floods, allowing for proactive measures to be taken. Furthermore, AI can support the optimization of water distribution systems, ensuring that every drop is utilized efficiently (Paswan et al., 2025). By integrating innovative techniques, durable infrastructure, responsible policies, collaborative efforts, and advanced technologies like AI, we can develop comprehensive solutions to ensure sustainable water management in the face of climate change. This holistic approach is vital not only for preserving our natural resources but also for supporting the resilience and wellbeing of our communities and ecosystems. This chapter emphasizes the importance of combining various strategies to address water management issues and highlights the role of technology and collaboration in achieving sustainable solutions.

17.2 CLIMATE CHANGE AND WATER SECURITY

Climate change is a significant issue in the contemporary policy agenda, with most states partially engaged in reducing greenhouse gas emissions and scaling up renewable energy. However, extreme reductions are needed to prevent severe climate damages and prevent irreversible chain reactions in the Earth's climate system. Climate change damages are typically modeled as a monotonic and increasing function of global mean temperature, which is closely connected to policies to limit temperature increases (Kumar et al., 2024). Modeling the impacts of climate change beyond the temporal and physical scale of global mean warming and linking temperature varying impacts to future policies is a challenging task (Noto et al., 2023). The global water cycle is one area where climate change impacts are more evident and likely to be more felt in the short run. Climate change interacts with current vulnerabilities, exacerbates challenges faced by governments and citizens due to an unpredictable water cycle, extreme events, and pressures on supply and quality. The availability and management of water resources matter in their own right, but the intersection of these vulnerabilities with climate change impacts will involve idiosyncratic bundles of threat and opportunity for each place and time at issue.

Managing watershed governance through examining the context in which markets, states, and people influence water use at different spatial and temporal scales is crucial, even as the character of water related challenges in regions and contexts changes with perceived climate futures (Jain et al., 2024).

17.2.1 Impacts of Climate Change on Water Resources

The most pronounced consequence of climate change on water resources is the alteration of hydrologic patterns accompanied by a significant intensification of the water cycle. This evolution in hydrologic behavior is reflected in an increase in the intensity and duration of droughts, as well as a rise in the frequency and severity of precipitation events. The implications of these changes are far reaching, affecting not only the provision of safe drinking water but also sanitation, agricultural practices, and the health of ecosystems. Moreover, these shifts can impact larger societal dynamics, including political stability and economic growth. Drought conditions can drastically reduce the availability of water resources (Figure 17.1). These extreme dry spells often lead to acute water shortages, exacerbating a range of interconnected crises. For instance, decreased water availability can result in land degradation, increase food insecurity, and, in dire scenarios, lead to famine (Abdullahi et al., 2024). Observed alterations in the hydrological system such as declining snowpack levels, a delayed onset and earlier conclusion of snowmelt, changes in the timing of streamflow, and reductions in summer low flows further heighten the vulnerability of both water supply systems and the ecosystems dependent on them. On the other hand, heightened precipitation variability brings its own set of serious challenges. Intense rainfall events can produce significant runoff, overwhelming urban infrastructure and leading to mixed sewer overflows. This situation is particularly concerning in densely populated urban areas, where it can lead to the proliferation of waterborne diseases, posing serious public health risks. As climate change continues to escalate, these diseases are becoming increasingly problematic, affecting marginalized communities the most. Additionally, more frequent and intense precipitation poses significant threats to municipal water management systems. Many of these systems are ill equipped to handle the sudden influx of water, which can overwhelm treatment facilities and distribution networks. As a result, the risk of contamination increases, leading to potential health crises. Most critically, communities that have previously faced climate induced hydrological events are likely to experience a rise in post-traumatic public health issues (Safi et al., 2024). As these communities grapple with the aftermath of natural disasters, they may struggle to secure adequate resources for recovery, further compounding the difficulties in addressing both immediate and long-term health needs. This cycle of vulnerability highlights the urgent need for comprehensive strategies to adapt to and mitigate the impacts of climate change on water resources.

17.2.2 Challenges to Water Security

While there are numerous challenges associated with securing our water resources, the most pressing issues in the context of climate change focus on the quantity of available water and the increasing frequency of water related natural disasters (Figure 17.2). Over the past century, significant advancements have been made to

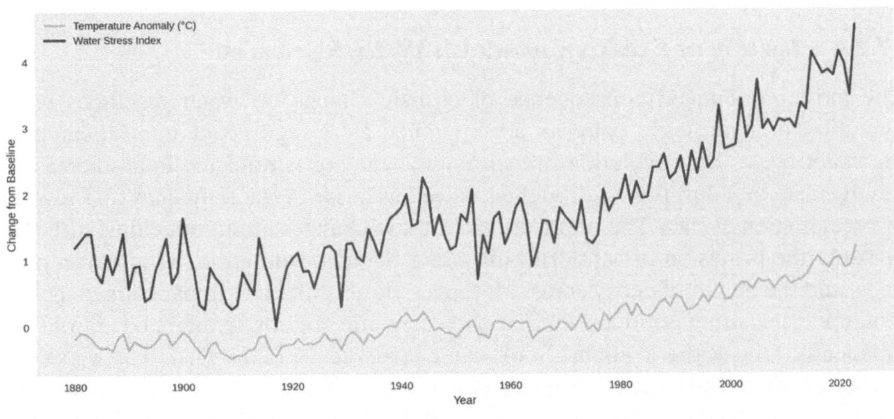

FIGURE 17.1 Global temperature anomalies and water stress (1880 to present). (Lenssen et al., 2024.)

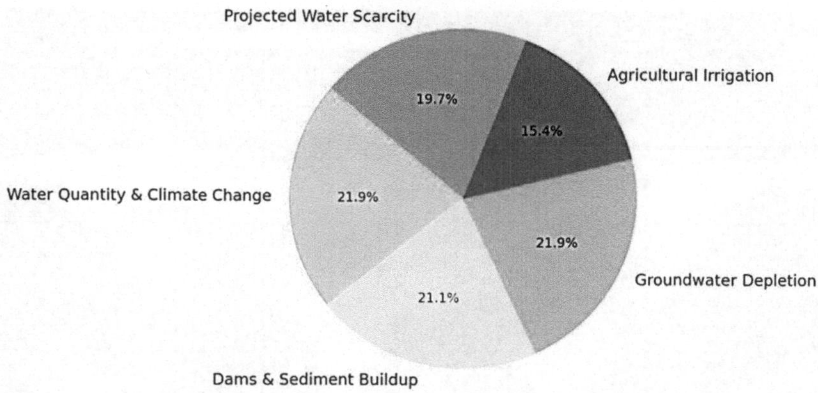

FIGURE 17.2 The key issues in water resource security related to climate change. (Rossi and Peres, 2023; Tsakiris and Loucks 2023.)

improve our water storage capabilities, particularly for industrial and agricultural uses. Hundreds, if not thousands, of large dams have been constructed around the globe with the primary goal of capturing river runoff, allowing us to store water for use during critical periods of shortage (Rossi and Peres, 2023). However, we are now confronting a serious dilemma: many of these dams are slowly becoming less effective as they accumulate sediment over time. This sediment buildup drastically reduces their water storage capacity, leading to a concerning estimate that dams worldwide could lose, on average, 4% of their water storage capability every decade (Djoukbala et al., 2024). This loss represents not just a decrease in water availability but also poses challenges for agricultural production, flood management, and local ecosystems that depend on stable water supplies. In addition to surface water storage, groundwater has historically served as a crucial "savings account" for numerous individuals and communities around the world. However, many regions are facing severe shortages of drinking water due to the near-total depletion of groundwater resources. This decline is primarily attributed to the over extraction of water from underground aquifers, which are being drawn down at an unsustainable rate much faster than the natural replenishment provided by the water cycle (Rossi and Peres, 2023). Disturbingly, approximately 70% of all groundwater extracted globally is used for agricultural irrigation, especially crucial for the cultivation of food crops during extended dry periods (Scanlon et al., 2023). If the current trend of excessive groundwater withdrawal continues, farmers and their communities will find themselves in an increasingly precarious situation, facing not only heightened risks but also a certainty of extreme water scarcity. Current projections suggest that without significant changes in water management practices, all nations will encounter chronic water scarcity within the next decade (Figure 17.2), with the notable exceptions being the Democratic Republic of the Congo and Papua New Guinea (Jaramillo et al., 2023). This stark reality calls for immediate and coordinated action to ensure the sustainable management of our vital water resources to secure food security and community resilience against climate change.

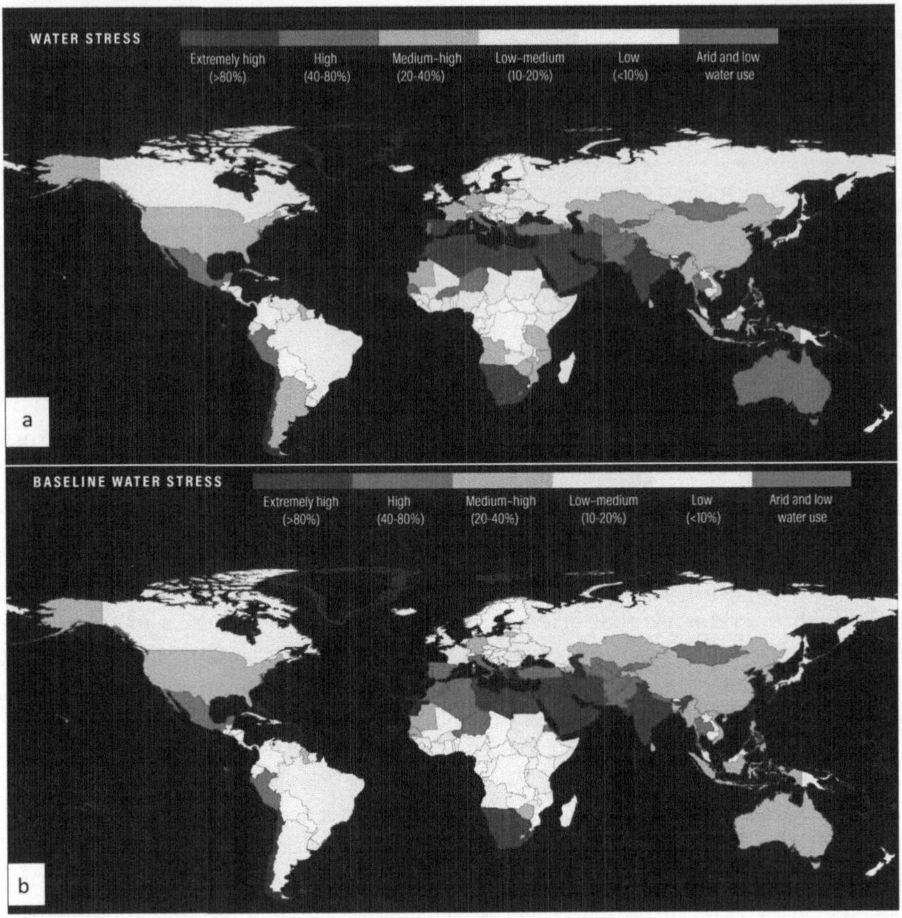

FIGURE 17.3 Global Water Stress Levels (a) Current Situation and (b) Projections for 2050. (World Resources Institute, 2024.)

17.2.3 Vulnerable Populations and Regions Affected by Water Insecurity

The increasing frequency and intensity of extreme weather events exacerbate water scarcity and contamination issues, disproportionately affecting marginalized communities and regions with limited resources.

One-quarter of the world's population resides in 25 nations that experience extremely high levels of water stress each year, depleting nearly all of their available water supply, according to the World Resources Institute's Aqueduct Water Risk Atlas. Approximately 4 billion people, or at least 50% of the global population, endure severe water stress for at least one month each year (Figure 17.3) (Gramlich and Walker, 2023; World Resources Institute, 2024). Water is essential for growing crops, raising livestock, generating electricity, maintaining human health, promoting equitable communities, and achieving global climate goals. This level of water stress jeopardizes people's lives, jobs, food security, and energy security. Water stress, defined as the ratio of water demand to renewable supply, measures the competition for local water resources. Areas more vulnerable to water shortages see a smaller difference between supply and demand. Climate change, population growth, and economic development are all likely to worsen water stress unless improved water management practices are implemented (Kuzma et al., 2023). Countries with the highest levels of water stress include Bahrain, Cyprus, Kuwait, Lebanon, Oman, and Qatar (Figure 17.4). This situation arises from limited water supply coupled with significant demand from domestic, agricultural, and industrial sectors. In the Middle East and North Africa, 83% of the population faces extremely high levels of water scarcity. Similarly, South Asia is concerning, with 74% of its population exposed to such conditions. Even if global temperatures rise by 1.3°C–2.4°C by 2100, an additional 1 billion people are predicted to live with severe water stress by 2050 (Figure 17.3). By then, it is anticipated that the number of watersheds experiencing high year-to-year variability will increase by 19%, while global water demand will rise by 20%–25%. By 2050, everyone in the Middle East and North Africa will be living under extremely high levels of water stress (Figure 17.5) (De Waal et al., 2023; Gabr, 2023; World Resources Institute, 2024). Water stress poses a significant challenge to the global economy and food security, with projections indicating that 31% of global GDP, equivalent to $70 trillion, may experience high water stress by 2050 (Barbier, 2024). More than half of this impacted GDP will come from Egypt, Mexico, India, and Turkey, an increase from $15 trillion in 2010 (Barbier, 2024; World Resources Institute, 2024). Water scarcity can lead to energy outages, industrial disruptions, and losses in agricultural output, as seen in India. By 2050, India, China, Central Asia, and much of Africa may experience GDP losses if improved water management policies are not enacted (Adom, 2024). Due to severe water stress, 60% of the world's irrigated agriculture is at risk, threatening global food security (Figure 17.6). To sustain a projected global population of 10 billion by 2050, food production must increase by 56% compared to levels in 2010 (Saleem et al., 2024; World Resources Institute, 2024). This challenge is made more difficult by increasing water scarcity and the impacts of climate related events such as droughts and floods. Countries need to adopt integrated water resource management, promote water efficiency in agriculture, strengthen water governance, and enhance water infrastructure through green infrastructure and nature based solutions. In exchange for investments in resilient infrastructure or biodiversity, international development banks and lenders should consider strategic debt relief initiatives. To prevent power outages caused by water shortages, policymakers in water stressed nations should prioritize water efficient energy sources like solar and wind. Urban areas should develop comprehensive action plans to enhance water resilience and implement strategies for the treatment and reuse of wastewater to create additional water resources. Farmers must adopt more efficient water management practices and establish scientifically informed water targets. To create a future where everyone has access to clean water, action must be taken at every level of government, society, and industry.

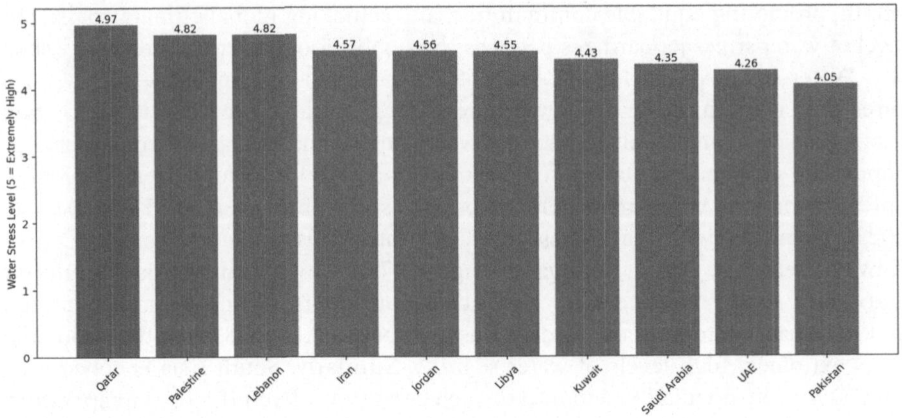

FIGURE 17.4 Ranking of the top 10 countries based on baseline water stress.

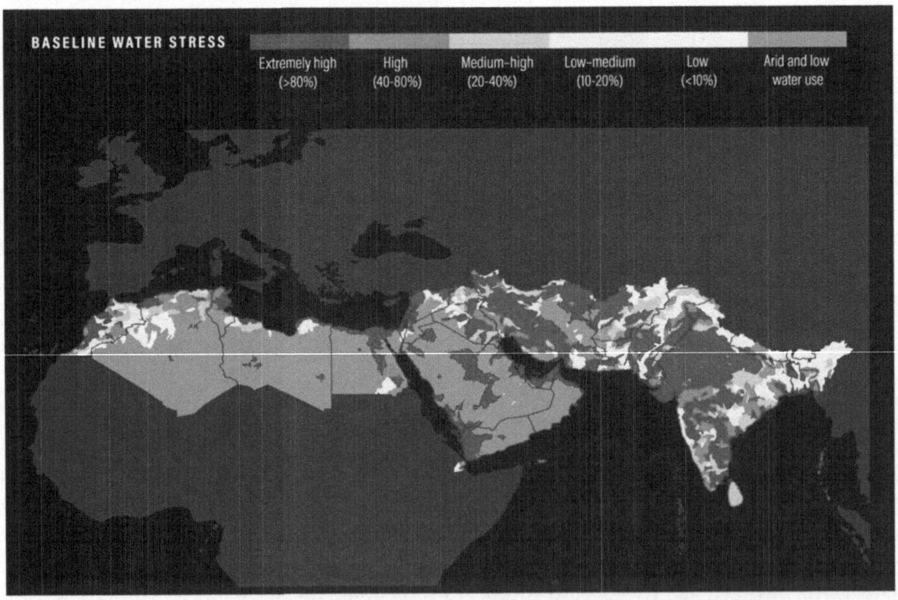

FIGURE 17.5 Water stress level of the middle east and north Africa region. (World Resources Institute, 2024.)

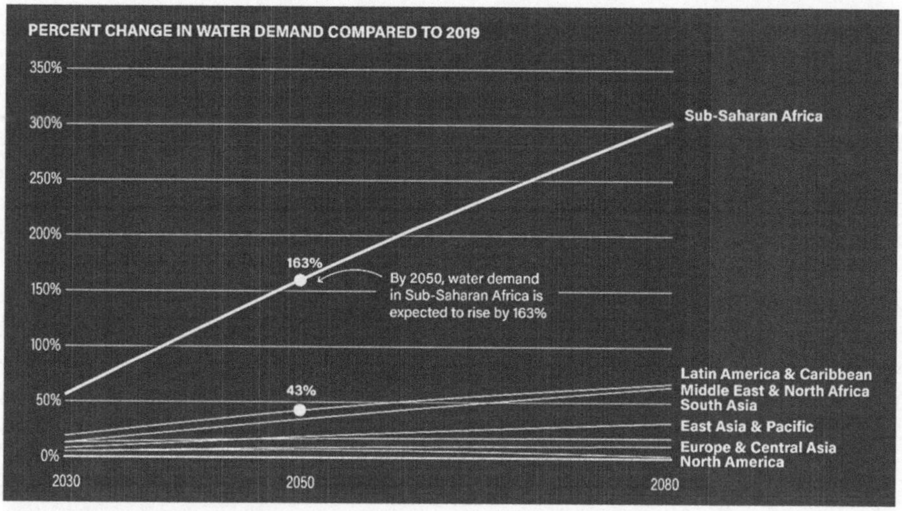

FIGURE 17.6 Projected Global Water Demand by 2050. (World Resources Institute, 2024.)

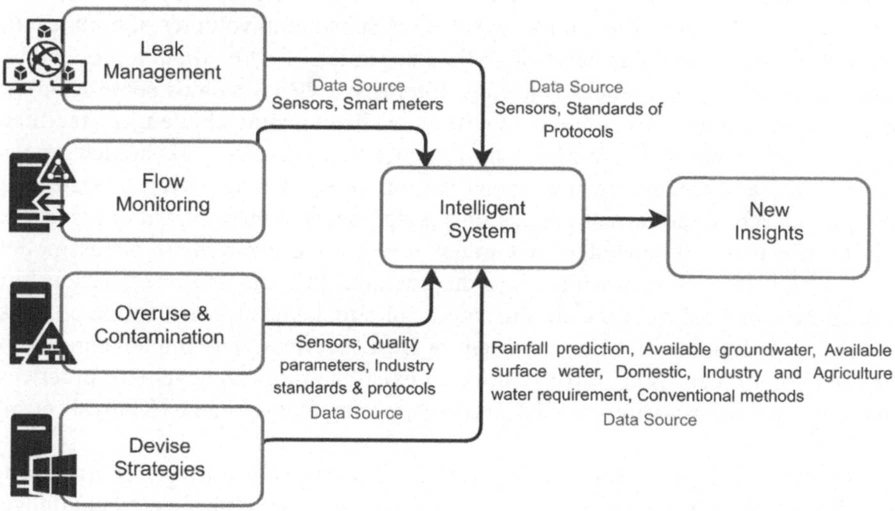

FIGURE 17.7 Harnessing intelligent systems for water resources management. (Krishnan et al., 2022.)

17.3 ROLE OF ARTIFICIAL INTELLIGENCE IN ADDRESSING CLIMATE CHANGE AND WATER SECURITY

Climate change poses a significant challenge for the coming decades, with many regions either currently facing or soon to experience increasing water scarcity. To address these uncertainties, innovative approaches to water management are essential. There is growing pressure to develop sustainable water management strategies, especially given the precarious state of existing water resources.

AI has emerged as a pivotal tool in the comprehensive analysis of extensive datasets, enabling the identification of critical patterns and trends pertinent to climate change and water security (Saleem et al., 2024). Through the application of machine learning algorithms, AI facilitates the forecasting of future scenarios and the development of innovative solutions aimed at mitigating the adverse impacts of these environmental challenges. Recent technological advancements in AI hold significant promise for enhancing climate change adaptation strategies and improving the management of water resources, ultimately contributing to the promotion and assurance of water security (Bamal et al., 2024).

AI has demonstrated substantial efficacy in equipping both citizens and decision makers with: (i) Monitoring Water Resources: The monitoring of water supplies, ecosystems, and communities can be achieved through the utilization of satellite and drone imagery, which allows for the prediction of dry and wet periods, soil moisture dynamics, drought occurrences, and other relevant metrics (Kamyab et al., 2023). The processing and analysis of the substantial volumes of high resolution Earth observation data necessitate the integration of AI methodologies and big data analytics; (ii) Awareness Building: Interactive visualizations serve as effective tools to enhance awareness regarding environmental challenges, facilitate the communication of complex scientific concepts to diverse stakeholder groups, and promote knowledge sharing among various constituencies (Metze, 2020); (iii) Mitigation: While the primary focus of this document centers on water resources, it is pertinent to acknowledge that water energy food ecosystems nexus modeling can provide valuable insights into the potential impacts and interrelationships among these critical resources in the context of climate change mitigation policies; and (iv) Coordination and Enhancement of Best Practices: The implementation of AI offers significant potential to optimize existing water management practices, including irrigation, water supply, and wastewater management (Kamyab et al., 2023).

Through the exploration of these avenues, AI can substantively contribute to the effective stewardship of water resources in light of the multifaceted challenges posed by climate change. Globally, we require optimized solutions that are tailored to the specific needs of different regions. The integration of AI in the water sector offers significant opportunities to tackle emerging water security issues and to bridge the existing gaps in water services. Whether addressing the challenges posed by water scarcity and pollution or committing to long-term water sustainability, AI interventions are both timely and highly relevant (Kamyab et al., 2023).

17.3.1 AI Techniques for Climate Change Adaptation

Climate change poses significant challenges that require innovative solutions across various sectors, with AI emerging as a crucial tool for adaptation strategies. Although current research on AI applications in climate adaptation is still limited, there is considerable potential for AI to analyze diverse climate models and enhance early warning systems. This capability can enable timely investments in conservation efforts. AI has already proven effective in previous applications related to large-scale conservation planning, such as vegetation mapping. Specialized remote sensing algorithms have been developed to identify areas affected by climate change. The integration of AI into climate adaptation strategies shows great promise, as it can improve environmental monitoring, enhance disaster response, optimize resource management, and support policy planning, ultimately fostering resilience. A landscape level tool has been successfully implemented to identify restoration opportunities, design climate adapted conservation networks, and assess environmental impacts using specific software. AI techniques have also been employed to identify restoration opportunities based on biophysical characteristics and connectivity in degraded watersheds. Additionally, various national research projects have used AI to monitor the effects of climate change on species, design climate adapted conservation priorities, and establish a corridor network aimed at mitigating the impacts of flooding and other climate-related challenges (Metze, 2020; Sun, 2023; Gryshova et al., 2024).

17.3.1.1 Environmental Monitoring

AI technologies enhance the ability to monitor environmental changes and assess climate impacts effectively. Key applications include:

- **Remote Sensing**: Machine learning algorithms analyze satellite imagery to track changes in land use, forest cover, and climate related variables. Convolutional Neural Networks (CNNs) are frequently employed for image classification and feature extraction, facilitating the assessment of deforestation and urbanization (Cao et al., 2020).
- **Climate Modeling**: AI techniques, particularly deep learning models, are utilized to improve climate predictions. Models such as Long Short-Term Memory (LSTM) networks can capture temporal dependencies in climate data, allowing for more accurate forecasting of extreme weather events (Zhao et al., 2024).
- **Biodiversity Assessment**: AI-driven tools, including acoustic monitoring and image recognition, enable the analysis of biodiversity changes and species distributions in response to climate shifts (Levy and Shahar, 2024).

17.3.1.2 Disaster Response and Risk Management

AI enhances disaster preparedness and response by facilitating predictive analytics and decision making processes:

- **Predictive Modeling**: AI models predict the likelihood of climate related disasters; such as floods or wildfires. For example, Random Forest and ensemble learning techniques are used to assess flood risk based on historical data and environmental parameters (Haggag et al., 2021).
- **Real-time Data Analysis**: AI systems are capable of processing real-time data from various sources, including social media and sensors, to provide timely information during disasters. This capability supports emergency management agencies in coordinating more effective responses (Jadhav et al., 2023).
- **Decision Support Systems**: AI algorithms assist in creating decision support systems that provide actionable insights for disaster management, optimizing resource allocation and risk mitigation strategies (Zolkafli et al., 2024).

17.3.1.3 Resource Management

Efficient management of natural resources is crucial for adaptation strategies, and AI contributes significantly to this process:

- **Water Resource Management**: AI techniques, including machine learning models and Internet of Things (IoT) integration, optimize water usage in agriculture and urban settings. For instance, predictive models can forecast water demand and supply, enhancing strategies for groundwater management (Haggag et al., 2021).
- **Agricultural Adaptation**: AI-driven precision agriculture techniques help farmers adapt to changing climate conditions. Decision trees and support vector machines analyze soil and weather data to inform planting schedules, irrigation practices, and crop selection (Gryshova et al., 2024).
- **Energy Management**: AI optimizes renewable energy production by predicting energy supply and demand patterns. Techniques such as reinforcement learning can enhance energy distribution systems, thereby supporting the transition to sustainable energy sources (Jadhav et al., 2023).

17.3.1.4 Policy Planning and Community Engagement

AI aids in formulating effective climate policies and fostering community resilience:

- **Scenario Simulation**: AI models can simulate the impacts of various policy scenarios on environmental outcomes, informing decision makers about potential strategies for adaptation (Nishant et al., 2020).
- **Stakeholder Engagement**: Natural language processing (NLP) tools analyze public sentiment and engagement regarding climate policies, facilitating community involvement in decision making processes (Nishant et al., 2020; Ouyang et al., 2023).
- **Education and Awareness**: AI-driven platforms enhance public awareness of climate issues by providing personalized content and learning resources, which can foster greater community resilience and participation in adaptation efforts (Ouyang et al., 2023).

Future research should prioritize the assessment of these technologies' effectiveness and address ethical concerns related to data usage and algorithmic bias.

17.3.2 Applications of AI in Water Resource Management

AI presents numerous opportunities within water resource management (Figure 17.7). Key applications include the efficient integration of diverse data sources, the development of predictive models to support decision makers, real-time detection of leaks and demand fluctuations, and the identification of treatment plant disruptions. Additionally, AI enhances water quality monitoring and can provide tailored water usage recommendations for households. Many utilities, along with hundreds of companies globally, are already implementing AI in various capacities (Kamyab et al., 2023). However, the water sector is generally trailing behind the energy sector, and scaling AI solutions can be relatively costly. As a result, the AI market in water management is smaller than that in energy. Currently, 23% of water technology start-ups incorporate AI, with 5% garnering significant media attention. The relationship between water technology and AI is progressively strengthening. As research and practical applications in this area continue to grow, we can anticipate an increasing number of successful case studies. This progress will create opportunities to apply insights gained from one project to another, further advancing the integration of AI in water resource management (Krishnan et al., 2022; Mohammadi Lanbaran et al., 2024).

Water management is crucial for conserving water resources, harvesting, planning, and distributing them to consumers. Traditional methods are insufficient, leading to water pollution in sectors like agriculture, wastewater, mining, industrial production, and untreated urban runoff. Machine learning algorithms can help address these challenges by expanding the learning process exponentially. In India, industries are the second highest consumers of water and one of the highest sources of pollutants. Artificial Neural Networks (ANN) and Federated Learning (FL) are effective AI methods used in wastewater treatment processes (Kimothi et al., 2022; Krishnan et al., 2022). AI algorithms can also predict water levels and destroy wastewater pollutants.

AI and machine learning are being integrated into smart water management systems to improve water supply and service delivery. These systems are used in wastewater treatment, water pollution control, smart agriculture, and water based agriculture (Lu et al., 2024). The IoT techniques have been proven to reduce challenges in water supply chain systems, such as water resource mismanagement in rural India. Smart cities focus on cost enhancement, better expectations for everyday comforts, water management systems, and faster exchanges in all fields.

Moreover, the integration of IoT and sensory systems plays a vital role in advancing sustainability in agriculture. Smart irrigation initiatives prominently align with the Sustainable Development Goals (SDGs) by improving farmers' understanding of their crops, minimizing environmental impacts, and conserving resources. IoT technology can facilitate reduced overall costs and enable real-time monitoring for both irrigation management and precision farming. In agriculture, smart water management for precision irrigation is essential for improving crop production, decreasing costs, and contributing to environmental sustainability. The Smart Water Management Platform (SWAMP) in Brazil addresses challenges such as information model, complexity, deployment, adaptability, and complexity (Kumar et al., 2024; Maurya et al., 2024). The IoT framework has potential for efficient water supply management in smart cities, with applications like the Supervisory Controller and Data Acquisition (SCADA) approach. The REFlex Water architecture provides low-cost and efficient solutions for controlling water distribution and monitoring in real time.

Groundwater management is crucial for urban systems, and IoT can improve farmers' understanding of their crops, reduce environmental consequences, and preserve resources (Kumar et al., 2023; Sushma et al., 2024). Cost efficient systems, such as light detecting devices and Dam Water Management systems, are being developed to reduce water wastage and improve water management. Taiwan has implemented Intelligent Water Management (IWM) to enhance water management in the country (Krishnan et al., 2022; Liu et al., 2024). Researchers have developed an IoT-based strategy to monitor water quality in cities, utilizing sensors such as water level, pH, flow, and control valve. The system uses Long Short-Term Memory (LSTM) and Convolutional Neural Network (CNN) to simulate water quality and levels. The researcher also developed a hydro informatics integration platform (IHIP) for online flood forecasting. They also introduced automatic water quality prediction models based on IWM. The system uses parameters like temperature, turbidity, and pH to analyze water quality. However, it is expensive to deploy in real-time scenarios. The Water Wise platform effectively manages and analyzes data from smart IoT systems, aiming to enhance productivity, minimize waste, and provide users with reliable information. Finally, a Complex Event Processing (CEP) technology model has been proposed to foresee risks in water supply management systems, complemented by a layered model focusing on the effective management of domestic water supply systems (Karbasi et al., 2024).

17.4 CHALLENGES IN AI FOR CLIMATE CHANGE ADAPTATION AND WATER SECURITY

As climate change exacerbates water scarcity and heightens the frequency of extreme weather events, the incorporation of AI technologies into strategies for adapting to climate change especially in the context of water security presents both considerable challenges and significant opportunities, while AI offers innovative solutions for improving climate resilience and water security, it encounters significant challenges, including data quality issues, ethical considerations, and implementation barriers (Amiri et al., 2024). The effectiveness of AI systems is contingent upon the availability and reliability of data, which is often lacking in many regions, particularly in developing countries. This data deficiency can impede the creation of effective AI models and their real world applicability. Ethical concerns are increasingly pertinent as AI systems influence water management and climate adaptation decisions. Issues such as bias, accountability, and transparency must be addressed to prevent unfair resource distribution and inadequate responses to climate threats, particularly affecting vulnerable populations. Establishing ethical frameworks is crucial for the responsible and equitable use of AI in water management. Furthermore, implementation barriers, including technical limitations and organizational resistance, hinder the widespread adoption of AI in climate adaptation strategies. Many water management agencies may lack the necessary technical expertise, resulting in underutilization of AI technologies. Collaboration among stakeholders' governments, NGOs, and private entities is vital for fostering innovation and sharing best practices in AI applications for water security (Jain et al., 2023).

17.4.1 Data Limitations

Data limitations present significant challenges for leveraging AI in climate change adaptation and water security. Addressing these limitations through investments in data infrastructure, improving data collection methodologies, fostering partnerships for data sharing, and enhancing the quality and coverage of datasets can help to unlock the full potential of AI in managing water resources sustainably (Jain et al., 2023; Ali and Thakkar, 2024).

One significant challenge is the **inadequate availability of comprehensive datasets**. In many regions, particularly in developing countries, there is a scarcity of extensive hydrological data, including historical records of rainfall, river flow, groundwater levels, and temperature. This lack of data restricts the ability to accurately model hydrological processes and predict future water availability, which is crucial for effective water management (Yaseen, 2023). Additionally, existing datasets may not provide adequate temporal and spatial coverage, leaving critical gaps in understanding local water dynamics and impairing the development of tailored adaptation strategies.

Another pressing concern is the **poor quality of existing data**. Data collected from varying methods and instruments may lead to inconsistencies and inaccuracies in datasets. The absence of standardized data collection protocols results in incompatible datasets that pose challenges for AI applications in this context (Agaj et al., 2024). Furthermore, historical data often contain errors due to human mistakes, instrument malfunction, or changes in measurement techniques over time. These inaccuracies can propagate through models, undermining predictions about future climate impacts on water resources.

The challenges of **data integration** also present significant obstacles. Effective management of water security relies on diverse types of data, including meteorological, hydrological, and socioeconomic information. Integrating these heterogeneous datasets is complex due to differences in format, scale, and methodologies (Vaidya et al., 2024). Moreover, capturing the multifaceted interactions within hydrological systems which are influenced by geology, land use, and climate factors requires a comprehensive array of datasets. Merging these various data sources to enhance model robustness remains a significant challenge.

In addition to these issues, there is often **limited access to real-time data**. In many developing regions, inadequate infrastructure for monitoring such as a lack of weather stations and IoT sensors hampers the collection of crucial real-time data on water quality and availability (Albahri et al., 2024). The absence of real-time monitoring makes it challenging to respond swiftly to changing conditions, particularly during water crises. Furthermore, the costs associated with advanced data collection technologies can be prohibitive for resource constrained areas, leading to underinvestment in essential monitoring efforts.

Finally, **data privacy and sharing issues** further complicate the landscape of data availability. Stakeholders may be hesitant to share data due to confidentiality concerns or competitive pressures, particularly among private sector entities. This reluctance can inhibit collaborative efforts necessary for gathering and analyzing comprehensive datasets for water management (Yaseen, 2023; Albahri et al., 2024). Additionally, stringent regulations regarding data sharing, especially concerning personal or sensitive information, can limit access to valuable datasets that are critical for robust AI modeling in climate adaptation and water security (Layode et al., 2024).

17.4.2 Technical and Computational Barriers

There are significant barriers to implementing AI solutions in water management. The high computational cost of AI algorithms is a major challenge, as training sophisticated machine learning models requires powerful GPUs and substantial memory. This can be especially hard for low income or rural areas where computing resources are limited (Drogkoula et al., 2023). The complexity of AI algorithms poses another significant barrier. Many advanced techniques, such as deep learning, require expertise in data science and machine learning, which may be lacking in the water management sector (Sudhakar, 2023). Rapid advancements in AI technology can also make it difficult for practitioners to stay current with best practices. Data processing and storage present further challenges. Climate monitoring generates vast amounts of data, and traditional processing frameworks often struggle to keep up, leading to analysis delays. This is critical in situations like flood monitoring and drought management, where timely responses are essential (Albahri et al., 2024). Limitations in cloud and local storage can also hinder the archiving and analysis of historical data, which is vital for training AI models and understanding long-term water trends. Integrating diverse data sources is another hurdle. Water resource management relies on various datasets, including satellite imagery and ground observations. However, differences in data formats and standards complicate the integration process, which is necessary for developing effective AI models. Eventually, maintaining and updating AI systems requires ongoing effort. AI models need regular updates and retraining to stay effective as environmental conditions change. This demands significant resources and expertise, often unavailable in smaller organizations. Additionally, models must be validated to avoid making decisions based on outdated or inaccurate information (Alzoubi and Mishra, 2024).

17.4.3 Algorithmic Bias and Interpretability

Algorithmic bias occurs when AI systems produce prejudiced results due to incorrect assumptions in the machine learning process. This bias can stem from skewed training data, unrepresentative sampling, and the limitations of the algorithms themselves. In water management, biased AI applications can lead to unequal resource distribution and marginalization of vulnerable communities. For instance, an AI model trained on urban data may overlook rural needs, exacerbating inequalities in water access (Li et al., 2024). Assessing the fairness of AI systems is challenging as training data may reflect historical disparities, thus perpetuating inequities in decision making. As water stress increases due to climate change, biased AI outputs can misallocate resources, prioritizing certain populations and potentially leading to social unrest. Another concern is the interpretability of AI models. Advanced AI techniques often function as "black boxes," making it difficult for stakeholders in water management to trust AI recommendations (Sudhakar, 2023). Without clear reasoning behind a model's outputs, stakeholders including policymakers and local communities may hesitate to implement suggested strategies. The challenge of interpretability can also hinder user engagement. If users do not understand AI models, they may distrust the technology and resist adopting it. To mitigate these issues, AI systems should prioritize fairness and transparency. This includes using techniques to minimize bias in training data and developing models with interpretability features, such as explainable AI (XAI) methods, to help stakeholders understand AI decision making (Mallick et al., 2024).

17.4.4 Capacity Building and Skill Gaps

As AI technologies become crucial for analyzing and managing water resources, the lack of necessary skills among the workforce poses a significant barrier to their successful adoption. Many water management professionals lack training in data science and AI, limiting their ability to apply these tools effectively. This knowledge gap can result in underutilization or misapplication of AI systems, leading to less effective decision making. To address this, targeted training programs are essential. These should focus on both technical skills, like data analysis, and practical integration of AI into existing water management practices. Collaboration among academic institutions, government, and industry can help develop tailored educational resources. Hands-on experience is also vital. Practical workshops, internships, and pilot projects allow individuals to apply their knowledge, enhancing their capability in AI technologies. Programs promoting mentorship between seasoned professionals and newcomers can further support skill development. Effective water governance requires collaboration across disciplines, such as hydrology and data analytics. Cross disciplinary training initiatives can bridge existing skill gaps, fostering innovative solutions. In developing countries, where educational resources may be limited, partnerships with international organizations can aid in capacity building and technology transfer (Sudhakar, 2023; Darko et al., 2024; Wei et al., 2024).

17.4.5 Ethical and Governance Issues

A primary ethical concern with AI is its potential to exacerbate inequalities. If AI systems are poorly designed, they may reinforce existing biases in data and decision making, leading to unequal access to water resources, particularly affecting marginalized populations and heightening their climate risks. Data privacy and consent are also critical issues. AI applications rely on large datasets that can contain sensitive information. It is essential to ensure that data collection, storage, and usage are transparent and respect individuals' privacy rights (Kamyab et al., 2023). Clear guidelines on data handling and consent are necessary to maintain trust between communities and organizations. Accountability in AI governance is vital. Determining who is responsible for AI-driven decisions impacting water resources developers, organizations, or policymakers is crucial. Without clarity, communities may face negative consequences without recourse. Governance frameworks should outline accountability mechanisms to address grievances. Transparency in AI decision making is essential. Many advanced AI models act as "black boxes," leading to distrust among users and affected communities. Promoting explainable AI (XAI) can help users understand AI outcomes, empowering them in decision making processes. Ethical considerations must also encompass sustainability and environmental justice. AI's role in water management should align with environmental goals and respect community rights affected by climate change (Mallick et al., 2024). Governance frameworks should ensure all stakeholders have a voice in decisions about their water resources. Interdisciplinary governance structures that include local communities, scientists, policymakers, and industry representatives are necessary for effective dialogue and joint decision making. Continuous evaluation and assessment of AI applications are necessary to ensure their effectiveness, equity, and sustainability. Developing metrics to measure these aspects aligns AI interventions with ethical principles and community needs. Ongoing monitoring and feedback mechanisms enable the identification and mitigation of ethical concerns as technology evolves (Boppiniti, 2023; Darko et al., 2024)

In summary, the intersection of AI, climate change adaptation, and water security presents a complex landscape that necessitates a balanced approach to harness AI's potential while addressing ethical, data related, and implementation challenges. By overcoming these obstacles, more resilient water management systems can be developed to effectively respond to current climate impacts and adapt to future uncertainties.

17.5 OPPORTUNITIES THROUGH CASE STUDIES

One of the most notable implementations of AI in climate adaptation can be found in Europe, where AI-powered flood prediction systems are transforming the way communities prepare for and respond to flooding events. Flooding poses a serious risk in many parts of Europe, particularly in low-lying areas and regions near rivers and coastlines. Traditional flood prediction methods often rely on historical data and static models, which may not adequately account for changing climate patterns or sudden weather events. However, with the advent of AI and machine learning algorithms, flood prediction has become more dynamic and accurate (Abegaz et al., 2024).

For instance, the European Space Agency (ESA) has been using AI to enhance flood forecasting by integrating satellite data with machine learning models. These systems analyze vast amounts of real-time data, including precipitation patterns, river flow rates, and land use changes, allowing for more precise predictions of potential flood events. According to a study published in the journal Nature Communications, the incorporation of AI in flood prediction increased the accuracy of forecasts by up to 30%, which significantly improves preparedness and response efforts. Moreover, this advanced forecasting capability enables local authorities to issue timely warnings to residents, potentially saving lives and reducing property damage (Schumann et al., 2023).

In Africa, the challenges posed by drought are becoming increasingly severe due to climate change. The continent is home to numerous regions that rely heavily on agriculture, making them particularly vulnerable to fluctuations in water availability. Here, AI is being harnessed to develop sophisticated drought management tools that help farmers and policymakers make informed decisions. For instance, the African Development Bank has launched an initiative that utilizes AI to analyze climate data and predict drought conditions. This initiative employs machine learning algorithms to process information from various sources, including satellite imagery and weather forecasts (Alasow et al., 2024).

By predicting drought conditions before they occur, farmers can adjust their planting schedules or select more drought resistant crops, which helps to mitigate the impact of water scarcity. Research has shown that regions implementing AI-driven drought management strategies have experienced a 15% increase in crop yields compared to those relying on traditional methods. This not only supports food security but also contributes to the economic stability of farming communities. The ability to proactively manage drought conditions through AI highlights the transformative potential of technology in addressing climate challenges in Africa (Ye et al., 2024).

In Asia, water quality monitoring technologies powered by AI are making significant strides in improving water security. Many countries in this region face serious water pollution issues due to industrial runoff, agricultural practices, and urban waste. Traditional water quality monitoring methods can be labor intensive and often fail to provide real-time data. However, with the integration of AI, innovative solutions are emerging (Ajakwe et al., 2023).

For example, in India, many projects have been conducted using AI to analyze data from various sensors placed in water bodies to monitor pollution levels. The system employs machine learning algorithms to detect patterns in water quality indicators, such as pH levels, turbidity, and contaminant concentrations. This real-time monitoring allows for rapid responses to water quality issues, enabling authorities to take corrective actions before pollution levels reach critical thresholds. A study published in the International Journal of Environmental Research and Public Health reported that the use of AI in water quality monitoring led to a 40% reduction in response time to pollution incidents, resulting in improved public health outcomes and enhanced water safety for communities (Kumar et al., 2024 (2024a), Kumar et al., 2024 (2024b).

These case studies collectively demonstrate how AI can effectively enhance water security and climate resilience, despite the existing challenges faced in implementing these technologies. Each example illustrates the potential for AI to provide innovative solutions tailored to specific regional needs, whether it be through flood prediction in Europe, drought management in Africa, or water quality monitoring in Asia. The evidence gathered from these initiatives reinforces the central thesis that AI is not just a technological advancement but a vital tool in the fight against climate change and its impacts on water systems (Ajakwe et al., 2023; Kumar et al., 2024; Ye et al., 2024).

Moreover, these case studies exemplify the theoretical foundations of AI applications in climate adaptation. The integration of AI into environmental management systems aligns with the principles of adaptive management, which emphasizes flexibility and continuous learning in response to changing environmental conditions. By utilizing AI, stakeholders can make data-driven decisions that enhance the resilience of water systems, ensuring that they can withstand the pressures of climate change.

In conclusion, the opportunities presented by AI through these case studies highlight the transformative potential of technology in addressing the challenges posed by climate change to water security. As we continue to explore and expand the use of AI in various contexts, it is crucial to remain mindful of the need for collaboration between technology developers, policymakers, and local communities. This collaboration will be essential for overcoming the barriers to implementation and for fostering a holistic approach to water management that prioritizes sustainability and resilience. Ultimately, by harnessing AI's capabilities, we can pave the way for a more secure and adaptive future in the face of an ever-changing climate.

17.6 FUTURE RESEARCH DIRECTIONS AND POTENTIAL ADVANCEMENTS

The convergence of AI, climate change adaptation, and water security offers numerous opportunities for future research and advancement. As the effects of climate change become increasingly severe, there is an urgent need for innovative solutions that improve water management strategies through the application of AI. Several key areas merit further investigation to optimize the potential of AI within this context. A significant avenue for research is the enhancement of data integration techniques. Current challenges in merging diverse data sources such as remote sensing data, ground based observations, and socioeconomic indicators can be addressed through sophisticated machine learning approaches, including deep learning and federated learning (Wani et al., 2024). Future research should focus on developing robust methodologies that enable the seamless integration of heterogeneous datasets, thereby facilitating the creation of comprehensive and dynamic models for water resource management. This improvement would not only enhance predictive accuracy but also support real-time monitoring and decision making in response to changing environmental conditions. Another crucial area for advancement is the formulation of explainable AI (XAI) frameworks specifically designed for water management applications. As AI systems evolve, ensuring interpretability will be essential for fostering trust among stakeholders and communities impacted by AI-driven decisions. Research efforts should concentrate on developing models that provide not only accurate predictions but also clear, actionable insights into their decision making processes. This could involve methodologies that deliver comprehensible explanations for AI outputs, thus empowering users to participate meaningfully in decision making concerning water resource allocation and climate adaptation strategies

(Popescu et al., 2024). The robustness and adaptability of AI models in the face of varying environmental and socioeconomic conditions represent a further significant research focus. AI systems must demonstrate resilience to uncertainty and be capable of managing a range of scenarios, particularly as climate change introduces variability in water availability and demand. Future studies may explore the development of adaptive AI models that can learn from new data and recalibrate their predictions accordingly. Integrating elements of reinforcement learning and scenario planning could enhance model flexibility and responsiveness, facilitating optimally adjusted water management strategies in real time (Drogkoula et al., 2023). Moreover, examining synergies between AI and traditional ecological knowledge can yield valuable insights for climate adaptation. Indigenous and local knowledge systems often provide in depth understandings of local water dynamics and ecosystems, refined over generations. Research that merges these perspectives with AI methodologies may lead to innovative strategies that are culturally sensitive and environmentally sound. This interdisciplinary approach has the potential to foster collaborative governance models that empower local communities while enhancing the effectiveness of AI-driven solutions (Bibri et al., 2024). Ongoing research is also necessary to address ethical and governance challenges in AI applications. As concerns about algorithmic bias, data privacy, and accountability persist, future studies should focus on establishing frameworks that ensure compliance with ethical standards and promote equitable outcomes. Investigating participatory governance models that incorporate diverse stakeholder perspectives in the AI design and implementation processes may provide critical insights on aligning technological solutions with societal values and community needs (Kulkov et al., 2024). Finally, the exploration of AI's role in promoting resilience against climate induced water crises represents a vital area for further inquiry. This includes investigating how AI can assist in the forecasting and management of extreme weather events, such as floods and droughts. The utilization of AI-driven predictive analytics can enable stakeholders to better prepare for and mitigate the impacts of climate related disruptions on water security. Research focused on developing AI-powered early warning systems could significantly enhance adaptive capacity and reduce vulnerability among at-risk populations (Jain et al., 2023). In conclusion, the potential of AI in climate change adaptation and water security is replete with opportunities for research and technological advancement. By addressing data integration, enhancing explain ability, improving model robustness, incorporating traditional ecological knowledge, tackling ethical governance issues, and exploring resilience building applications, the capacity of AI to support sustainable water management can be fully realized. Collaborative efforts among researchers, practitioners, policymakers, and communities will be crucial for effectively and equitably harnessing these advancements.

17.7 CONCLUSION

The interrelationship between climate change and water security presents a complex and urgent challenge that necessitates immediate and coordinated action. The increasing frequency and severity of extreme weather events, driven by climate change, exacerbate existing water scarcity issues, particularly in vulnerable regions.

This situation endangers not only the availability of safe drinking water but also affects agricultural productivity, food security, and public health. To effectively address these challenges, a comprehensive and multifaceted approach is required. This approach must integrate innovative water management strategies, durable infrastructure development, responsible policy implementation, and robust international cooperation. The potential of AI as a transformative tool in this context is evident, offering capabilities for enhanced monitoring, predictive analytics, and optimization of water resource management. However, the successful deployment of AI is contingent upon overcoming significant barriers, including data limitations, technical challenges, algorithmic bias, and ethical considerations.

Ultimately, fostering community engagement, investing in skill development, and ensuring equitable access to resources will be essential for identifying sustainable solutions to mitigate the impacts of climate change on water security. By adopting a holistic and inclusive strategy, it is possible to enhance resilience and safeguard water resources for future generations, thereby contributing to global sustainability objectives.

DECLARATION OF GENERATIVE AI IN SCIENTIFIC WRITING

During the preparation of this chapter, the authors used generative AI and AI-assisted technologies in the writing process (rephrasing, grammar checking, and chart production). After using this tool, the authors reviewed and edited the content as needed and took full responsibility for the content of the published chapter

REFERENCES

Abdullahi, A. M., Kalengyo, R. B., & Warsame, A. A. (2024). The unmet demand of food security in East Africa: Review of the triple challenges of climate change, economic crises, and conflicts. *Discover Sustainability*, 5(1), 1–15.

Abegaz, R., Wang, F., & Xu, J. (2024). History, causes, and trend of floods in the US: A review. *Natural Hazards*, 120, 1–41.

Adom, P. K. (2024). The socioeconomic impact of climate change in developing countries over the next decades: A literature survey. *Heliyon*, 10(15), e35134.

Agaj, T., Budka, A., Janicka, E., & Bytyqi, V. (2024). Using ARIMA and ETS models for forecasting water level changes for sustainable environmental management. *Scientific Reports*, 14(1), 22444.

Ajakwe, S. O., Ajakwe, I. U., Jun, T., Kim, D. S., & Lee, J. M. (2023). CIS-WQMS: Connected intelligence smart water quality monitoring scheme. *Internet of Things*, 23, 100800.

Alasow, A. A., Hamed, M. M., & Shahid, S. (2024). Spatiotemporal variability of drought and affected croplands in the horn of Africa. *Stochastic Environmental Research and Risk Assessment*, 38(1), 281–296.

Albahri, A. S., Khaleel, Y. L., Habeeb, M. A., Ismael, R. D., Hameed, Q. A., Deveci, M., … & Alzubaidi, L. (2024). A systematic review of trustworthy artificial intelligence applications in natural disasters. *Computers and Electrical Engineering*, 118, 109409.

Ali, A. H., & Thakkar, R. (2023). Climate changes through data science: Understanding and mitigating environmental crisis. *Mesopotamian Journal of Big Data*, 2023, 125–137.

Ali, S., & Sang, Y. F. (2023). Implementing rainwater harvesting systems as a novel approach for saving water and energy in flat urban areas. *Sustainable Cities and Society*, 89, 104304.

Alzoubi, Y. I., & Mishra, A. (2024). Green artificial intelligence initiatives: Potentials and challenges. *Journal of Cleaner Production*, 468, 143090

Amiri, Z., Heidari, A., & Navimipour, N. J. (2024). Comprehensive survey of artificial intelligence techniques and strategies for climate change mitigation. *Energy*, 308, 132827.

Bamal, A., Uddin, M. G., & Olbert, A. I. (2024). Harnessing machine learning for assessing climate change influences on groundwater resources: A comprehensive review. *Heliyon*, 17, e37073.

Barbier, E. B. (2024). Water and economic growth in developed and developing countries. *Water Economics & Policy*, 10(3), 1–24.

Bibri, S. E., Krogstie, J., Kaboli, A., & Alahi, A. (2024). Smarter eco-cities and their leading-edge artificial intelligence of things solutions for environmental sustainability: A comprehensive systematic review. *Environmental Science and Ecotechnology*, 19, 100330.

Boppiniti, S. T. (2023). Data ethics in AI: Addressing challenges in machine learning and data governance for responsible data science. *International Scientific Journal for Research*, 5(5), 1–29.

Campbell-Lendrum, D., Neville, T., Schweizer, C., & Neira, M. (2023). Climate change and health: Three grand challenges. *Nature medicine*, 29(7), 1631–1638.

Cao, R., Zhang, Q., Zhu, J., Li, Q., Li, Q., Liu, B., & Qiu, G. (2020). Enhancing remote sensing image retrieval using a triplet deep metric learning network. *International Journal of Remote Sensing*, 41(2), 740–751.

Darko, R. O., Odoi-Yorke, F., Abbey, A. A., Afutu, E., Owusu-Sekyere, J. D., Sam-Amoah, L. K., & Acheampong, L. (2024). A review of climate change impacts on irrigation water demand and supply-a detailed analysis of trends, evolution, and future research directions. *Water Resources Management*, 39, 1–29.

De Waal, D., Khemani, S., Barone, A., & Borgomeo, E. (2023). *The Economics of Water Scarcity in the Middle East and North Africa: Institutional Solutions*. World Bank Publications, Washington, DC.

Djoukbala, O., Djerbouai, S., Alqadhi, S., Hasbaia, M., Benselama, O., Abdo, H. G., & Mallick, J. (2024). A geospatial approach-based assessment of soil erosion impacts on the dams silting in the semi-arid region. *Geomatics, Natural Hazards and Risk*, 15(1), 2375543.

Drogkoula, M., Kokkinos, K., & Samaras, N. (2023). A comprehensive survey of machine learning methodologies with emphasis in water resources management. *Applied Sciences*, 13(22), 12147.

Elemam, D. A., & Eldeeb, A. R. (2023). Climate change in the coastal areas: Consequences, adaptations, and projections for the Northern Coastal Area, Egypt. *Scientific Journal for Damietta Faculty of Science*, 12(2), 19–29.

Elemam, D. A, & Hamam, E. (2024). Using the vulnerability index to assess the risk of climate change impacts on agricultural and food security. *Alfarama Journal of Basic & Applied Sciences*, 5(2), 293–308.

Gabr, M. E. (2023). Impact of climatic changes on future irrigation water requirement in the Middle East and North Africa's region: A case study of upper Egypt. *Applied Water Science*, 13(7), 158.

Gramlich, D., & Walker, T. (2023). Water risk modeling: A framework for finance. *Journal of Environmental Management*, 342, 117991.

Gryshova, I., Balian, A., Antonik, I., Miniailo, V., Nehodenko, V., & Nyzhnychenko, Y. (2024). Artificial intelligence in climate smart in agricultural: Toward a sustainable farming future. Access to science, business, innovation in the digital economy. *Access Press*, 5(1), 125–140.

Haggag, M., Siam, A. S., El-Dakhakhni, W., Coulibaly, P., & Hassini, E. (2021). A deep learning model for predicting climate-induced disasters. *Natural Hazards*, 107, 1009–1034.

Jadhav, A. R., Pathak, P. D., & Raut, R. Y. (2023). Water and wastewater quality prediction: Current trends and challenges in the implementation of artificial neural network. *Environmental Monitoring and Assessment*, 195(2), 321.

Jain, H., Dhupper, R., Shrivastava, A., Kumar, D., & Kumari, M. (2023). AI-enabled strategies for climate change adaptation: Protecting communities, infrastructure, and businesses from the impacts of climate change. *Computational Urban Science*, 3(1), 25.

Jain, S., Srivastava, A., Khadke, L., Chatterjee, U., & Elbeltagi, A. (2024). Global-scale water security and desertification management amidst climate change. *Environmental Science and Pollution Research*, 31, 1–25.

Jaramillo, L., Cebotari, A., Diallo, Y., Gupta, R., Koshima, Y., Kularatne, C., … & Yang, F. (2023). *Climate Challenges in Fragile and Conflict-Affected States*. International Monetary Fund, Worldwide.

Kamyab, H., Khademi, T., Chelliapan, S., SaberiKamarposhti, M., Rezania, S., Yusuf, M., … & Ahn, Y. (2023). The latest innovative avenues for the utilization of artificial Intelligence and big data analytics in water resource management. *Results in Engineering*, 20, 101566.

Karbasi, M., Ali, M., Bateni, S. M., Jun, C., Jamei, M., Farooque, A. A., & Yaseen, Z. M. (2024). Multi-step ahead forecasting of electrical conductivity in rivers by using a hybrid Convolutional Neural Network-Long Short-Term Memory (CNN-LSTM) model enhanced by Boruta-XGBoost feature selection algorithm. *Scientific reports*, 14(1), 15051.

Kimothi, S., Thapliyal, A., Akram, S. V., Singh, R., Gehlot, A., Mohamed, H. G., … & Noya, I. D. (2022). Big data analysis framework for water quality indicators with assimilation of IoT and ML. *Electronics*, 11(13), 1927.

Krishnan, S. R., Nallakaruppan, M. K., Chengoden, R., Koppu, S., Iyapparaja, M., Sadhasivam, J., & Sethuraman, S. (2022). Smart water resource management using Artificial Intelligence – a review. *Sustainability*, 14(20), 13384.

Kulkov, I., Kulkova, J., Rohrbeck, R., Menvielle, L., Kaartemo, V., & Makkonen, H. (2024). Artificial intelligence-driven sustainable development: Examining organizational, technical, and processing approaches to achieving global goals. *Sustainable Development*, 32(3), 2253–2267.

Kumar A, R., M Metwally, A. S., Ashraf, G. A., & Thamineni, B. L. (2023). Smart IoT-based water treatment with a Supervisory Control and Data Acquisition (SCADA) system process. *Water Reuse*, 13(3), 411–431.

Kumar, R., Goyal, M. K., Surampalli, R. Y., & Zhang, T. C. (2024a). River pollution in India: Exploring regulatory and remedial paths. *Clean Technologies and Environmental Policy*, 26, 1–23.

Kumar, S., Chatterjee, U., David Raj, A., & Sooryamol, K. R. (2024b). Global warming and climate crisis/extreme events. In: Chatterjee, U., Shaw, R., Kumar, S., Raj, A. D., Das, S. (eds) *Climate Crisis: Adaptive Approaches and Sustainability* (pp. 3–18). Cham, Switzerland: Springer Nature.

Kumar, V., Sharma, K. V., Kedam, N., Patel, A., Kate, T. R., & Rathnayake, U. (2024c). A comprehensive review on smart and sustainable agriculture using IoT technologies. *Smart Agricultural Technology*, 8, 100487.

Kuzma, S., Bierkens, M. F., Lakshman, S., Luo, T., Saccoccia, L., Sutanudjaja, E. H., & Van Beek, R. (2023). *Aqueduct 4.0: Updated Decision-Relevant Global Water Risk Indicators*. Washington, DC: World Resources Institute

Lakshmi, V. (2024). Enhancing human resilience against climate change: Assessment of hydroclimatic extremes and sea level rise impacts on the Eastern Shore of Virginia, United States. *Science of the Total Environment*, 947, 174289.

Layode, O., Naiho, H. N. N., Adeleke, G. S., Udeh, E. O., & Labake, T. T. (2024). Data privacy and security challenges in environmental research: Approaches to safeguarding sensitive information. *International Journal of Applied Research in Social Sciences*, 6(6), 1193–1214.

Lenssen, N., Schmidt, G. A., Hendrickson, M., Jacobs, P., Menne, M., and Ruedy, R. (2024). A GISTEMPv4 observational uncertainty ensemble. *Journal of Geophysical Research: Atmospheres*, 129(17), e2023JD040179. https://doi.org/10.1029/2023JD040179

Levy, O., & Shahar, S. (2024). Artificial intelligence for climate change biology: From data collection to predictions. *Integrative and Comparative Biology*, 64(3), 953–974.

Li, F., Yigitcanlar, T., Nepal, M., Thanh, K. N., & Dur, F. (2024). A novel urban heat vulnerability analysis: Integrating machine learning and remote sensing for enhanced insights. *Remote Sensing*, 16(16), 3032.

Liu, Z., Zhou, J., Yang, X., Zhao, Z., & Lv, Y. (2024). Research on water resource modeling based on machine learning technologies. *Water*, 16(3), 472.

Lu, Y., Li, T., & Deng, Y. (2024). Predication of Water Pollution Peak Concentrations by Hybrid BP Artificial Neural Network Coupled with Genetic Algorithm. *Applied Artificial Intelligence*, 38(1), 2341356.

Makhdum, M. S. A., Kousar, R., Ashfaq, M., & Behnassi, M. (2024). The water, food, and environmental security nexus. In: Behnassi, M., Al-Shaikh, A. A., Gurib-Fakim, A., Barjees Baig, M., Bahir, M. (eds) *The Water, Climate, and Food Nexus: Linkages, Challenges and Emerging Solutions* (pp. 17–32). Cham: Springer International Publishing.

Mallick, J., Alqadhi, S., Hang, H. T., & Alsubih, M. (2024). Interpreting optimised data-driven solution with explainable artificial intelligence (XAI) for water quality assessment for better decision-making in pollution management. *Environmental Science and Pollution Research*, 30, 1–22.

Maurya, B. M., Yadav, N., Amudha, T., Satheeshkumar, J., Sangeetha, A., Parthasarathy, V., … & Vellingiri, B. (2024). Artificial intelligence and machine learning algorithms in the detection of heavy metals in water and wastewater: Methodological and ethical challenges. *Chemosphere*, 353, 141474.

Metze, T. (2020). Visualization in environmental policy and planning: A systematic review and research agenda. *Journal of Environmental Policy & Planning*, 22(5), 745–760.

Mohammadi Lanbaran, N., Naujokaitis, D., Kairaitis, G., Jenciūtė, G., & Radziukynienė, N. (2024). Overview of startups developing artificial intelligence for the energy sector. *Applied Sciences*, 14(18), 8294.

Nishant, R., Kennedy, M., & Corbett, J. (2020). Artificial intelligence for sustainability: Challenges, opportunities, and a research agenda. *International Journal of Information Management*, 53, 102104.

Nimma, D., Devi, O. R., Laishram, B., Ramesh, J. V. N., Boddupalli, S., Tirth, V., & Arabi, A. (2024). Implications of climate change on freshwater ecosystems and their biodiversity. *Desalination and water treatment*, 321, 100889.

Noto, L. V., Cipolla, G., Pumo, D., & Francipane, A. (2023). Climate change in the Mediterranean Basin (Part II): A review of challenges and uncertainties in climate change modeling and impact analyses. *Water Resources Management*, 37(6), 2307–2323.

Ouyang, H., Tang, X., Zhang, R., Baklanov, A., Brasseur, G., Kumar, R., … & Luo, Y. (2023). Resilience building and collaborative governance for climate change adaptation in response to a new state of more frequent and intense extreme weather events. *International Journal of Disaster Risk Science*, 14(1), 162–169.

Paswan, A. K., Pathan, S. K., Agarwal, A., & Verma, V. (2025). Artificial intelligence in efficient management of water resources. In: Kumar D., Tewary T. and Shekhar S. (eds) *Data Analytics and Artificial Intelligence for Earth Resource Management* (pp. 185–206). Elsevier.

Paul, V. K., Rastogi, A., Dua, S., & Basu, C. (2023). *Healthcare Infrastructure, Resilience and Climate Change: Preparing for Extreme Weather Events*. Routledge, London.

Popescu, S. M., Mansoor, S., Wani, O. A., Kumar, S. S., Sharma, V., Sharma, A., … & Chung, Y. S. (2024). Artificial intelligence and IoT driven technologies for environmental pollution monitoring and management. *Frontiers in Environmental Science*, 12, 1336088.

Rossi, G., & Peres, D. J. (2023). Climatic and other global changes as current challenges in improving water systems management: Lessons from the case of Italy. *Water Resources Management*, 37(6), 2387–2402.

Safi, L., Mujeeb, M., Sahak, K., Mushwani, H., & Hashmi, S. K. (2024). Climate change impacts and threats on basic livelihood resources, food security and social stability in Afghanistan. *GeoJournal*, 89(2), 85.

Saleem, A., Anwar, S., Nawaz, T., Fahad, S., Saud, S., Ur Rahman, T., … & Nawaz, T. (2024). Securing a sustainable future: The climate change threat to agriculture, food security, and sustainable development goals. *Journal of Umm Al-Qura University for Applied Sciences*, 11, 595–611.

Scanlon, B. R., Fakhreddine, S., Rateb, A., de Graaf, I., Famiglietti, J., Gleeson, T., … & Zheng, C. (2023). Global water resources and the role of groundwater in a resilient water future. *Nature Reviews Earth & Environment*, 4(2), 87–101.

Schumann, G., Giustarini, L., Tarpanelli, A., Jarihani, B., & Martinis, S. (2023). Flood modeling and prediction using earth observation data. *Surveys in Geophysics*, 44(5), 1553–1578.

Singh, V. (2024). Global warming and climate change. In: Singh, V. (eds) *Textbook of Environment and Ecology* (pp. 283–295). Singapore: Springer Nature Singapore.

Sudhakar, M. (2023). Artificial intelligence applications in water treatment and water resource assessment: Challenges, innovations, and future directions. In: Mishra B. K., (eds) *Intelligent Engineering Applications and Applied Sciences for Sustainability* (pp. 248–269). IGI Global.

Sun, Z. (2023). Actionable AI for climate and environment. In: Sun, Z. (eds) *Actionable Science of Global Environment Change: From Big Data to Practical Research* (pp. 327–354). Cham: Springer International Publishing

Sushma, N., Suresh, H. N., Mohana, L. J., & Kumar, K. S. (2024). Experimental investigation on wireless integrated smart system for energy and water resource management in Indian smart cities. *Results in Engineering*, 23, 102687.

Tsakiris, G. P., & Loucks, D. P. (2023). Adaptive water resources management under climate change: An introduction. *Water Resources Management*, 37(6), 2221–2233.

Vaidya, M., Keskar, R., & Kotharkar, R. (2024). Classifying heterogeneous urban form into local climate zones using supervised learning and greedy clustering incorporating Landsat dataset. *Urban Climate*, 53, 101770

Wani, A. K., Rahayu, F., Ben Amor, I., Quadir, M., Murianingrum, M., Parnidi, P., … & Latifah, E. (2024). Environmental resilience through artificial intelligence: Innovations in monitoring and management. *Environmental Science and Pollution Research*, 31(12), 18379–18395.

Wei, H., Xu, W., Kang, B., Eisner, R., Muleke, A., Rodriguez, D., … & Harrison, M. T. (2024). Irrigation with artificial intelligence: Problems, premises, promises. *Human-Centric Intelligent Systems*, 4, 187–205.

World Resources Institute. (2024). *Aqueduct Water Risk Atlas*. https://www.wri.org/applications/aqueduct/water-risk-atlas

Yaseen, Z. M. (2023). A new benchmark on machine learning methodologies for hydrological processes modelling: A comprehensive review for limitations and future research directions. *Knowledge-Based Engineering and Sciences*, 4(3), 65–103.

Ye, Z., Yin, S., Cao, Y., & Wang, Y. (2024). AI-driven optimization of agricultural water management for enhanced sustainability. *Scientific Reports*, 14(1), 25721.

Zhao, T., Wang, S., Ouyang, C., Chen, M., Liu, C., Zhang, J., … & Wang, L. (2024). Artificial intelligence for geoscience: Progress, challenges and perspectives. *The Innovation*, 5, 100691.

Zolkafli, A., Mansor, N. S., Omar, M., Ahmad, M., Ibrahim, H., & Yasin, A. (2024). AI for smart disaster resilience among communities. In: Abdul Karim, S. A. (eds) *Intelligent Systems Modeling and Simulation III: Artificial Intelligent, Machine Learning, Intelligent Functions and Cyber Security* (pp. 369–395). Cham: Springer Nature Switzerland.

18 Future Trends and Emerging Technologies in the Integration of AI and Water Management

Soufiane Haddout and Tonni Agustiono Kurniawan

18.1 INTRODUCTION

Water, the most essential resource for human survival, has become one of the most precious and contested commodities globally. Due to climate change, population growth, urbanization, and increased pollution, water scarcity and management challenges are intensifying. With freshwater sources dwindling, smart management systems are crucial to address these issues (Ahmadov, 2020). The integration of artificial intelligence (AI) into water management represents a groundbreaking shift in the field. AI technologies, including machine learning (ML), neural networks, and deep learning, enable the optimization of various water-related processes. These technologies hold the promise of real-time data processing, predictive analytics, and autonomous systems, all of which are poised to revolutionize the way water is conserved, distributed, and treated (Krishnan et al., 2022). This chapter aims to provide a comprehensive overview of the future trends and emerging technologies that are set to shape the integration of AI in water management. We will explore how AI can improve water quality monitoring, wastewater treatment, smart irrigation, flood prediction, and overall water use optimization. We will also analyze the potential breakthroughs that will emerge, paving the way for smarter, more efficient, and sustainable water management.

18.2 PREDICTING FUTURE TRENDS IN THE INTEGRATION OF AI AND WATER MANAGEMENT

18.2.1 AI-Driven Decision-Making and Optimization

AI's ability to process and analyze vast amounts of data quickly makes it an invaluable tool in optimizing water resource management. Water systems are inherently complex, requiring real-time decision-making to handle varying demands, environmental conditions, and infrastructure limitations. AI, particularly ML algorithms, can analyze large datasets, including sensor data, historical records, and weather

DOI: 10.1201/9781003547099-21

forecasts to optimize water usage and distribution. Future trends indicate a growing reliance on AI-driven systems for real-time decision-making. For example, AI models could control water distribution in urban areas by optimizing flow rates based on current consumption, weather patterns, and future predictions. Similarly, these models could adjust wastewater treatment processes in real time, reducing energy consumption and ensuring the desired water quality is met without unnecessary use of chemicals (Parvaze et al, 2023).

18.2.1.1 Case Study: AI in Smart Water Grids

A practical example of AI in decision-making is its integration into smart water grids. Smart grids are designed to optimize water distribution by utilizing AI algorithms that adjust flow rates, pressure, and storage levels based on demand predictions and real-time sensor data. These systems use AI to predict where water shortages or overuse may occur, allowing for proactive measures to address potential inefficiencies (Bouramdane, 2023).

18.2.2 Predictive Analytics for Water Demand and Supply Forecasting

AI-driven predictive analytics is a critical area of development in water management. By utilizing ML techniques, predictive analytics can forecast water demand based on factors such as population growth, seasonal variations, and climate forecasts. In drought-prone regions, this capability is invaluable for predicting low-water periods and implementing water-saving measures well in advance. AI can also predict water supply availability by monitoring environmental variables such as rainfall, groundwater levels, and reservoir conditions (Kikon & Deka, 2022). By combining these predictions with historical data and climate models, water authorities can better plan for future water demands and potential supply shortages.

18.2.2.1 Case Study: AI in Water Demand Forecasting in California

In California, AI-driven predictive analytics have been used to forecast water demand during droughts (Ashoka et al, 2024). By analyzing historical usage patterns, weather data, and reservoir levels, AI models can predict future water demand with high accuracy. These predictions allow authorities to manage water resources more efficiently and enact conservation measures when necessary.

18.2.3 AI in Flood Prediction and Management

AI is also transforming flood prediction and management. Traditional flood forecasting systems, while valuable, often suffer from limitations in processing time, spatial resolution, and real-time accuracy. AI has the potential to overcome these limitations by integrating large datasets, such as satellite imagery, river gauge data, weather patterns, and topography, to create more accurate and timely flood predictions. ML models, especially deep learning networks, are already being used to predict flood risks by identifying complex patterns in hydrological data. These systems can forecast flood events with greater precision, leading to more effective early warning systems and risk management strategies.

18.2.3.1 Case Study: AI-Based Flood Forecasting in Southeast Asia

In Southeast Asia, AI models have been implemented to predict the likelihood and intensity of flooding during the monsoon season (Chitwatkulsiri & Miyamoto, 2023). Using satellite data and river gauges, AI systems predict flood-prone areas with high accuracy, allowing for timely evacuation plans and resource allocation. This has reduced the impact of floods and saved lives.

18.3 EMERGING TECHNOLOGIES AND POTENTIAL BREAKTHROUGHS

18.3.1 Autonomous Water Quality Monitoring

Water quality monitoring is crucial for ensuring safe drinking water and healthy aquatic ecosystems. Traditional methods of monitoring often involve manual sampling, laboratory analysis, and long delays in reporting, which can delay necessary interventions in cases of contamination. Emerging AI technologies allow for continuous, real-time monitoring of water quality using autonomous sensors. These sensors, coupled with AI algorithms, can detect changes in water quality indicators—such as pH, turbidity, and chemical contaminants—immediately. With ML algorithms, these systems can also predict potential contamination events, providing early warnings for interventions.

18.3.1.1 Case Study: Autonomous Water Monitoring with Drones

Drones equipped with AI sensors are being tested for water quality monitoring in remote areas. These drones autonomously collect water samples from various points and analyze them for contaminants in real time (Manjakkal et al, 2021). The AI system processes this data and sends alerts if contamination levels exceed safe limits, enabling quicker responses to water pollution events.

18.3.2 AI-Enabled Wastewater Treatment

AI has the potential to significantly improve the efficiency of wastewater treatment plants, reducing both costs and environmental impact. Traditional treatment systems rely on mechanical and chemical processes that consume significant energy and resources. AI can optimize these systems by adjusting operational parameters such as aeration, chemical dosing, and filtration in real time, based on incoming data. Reinforcement learning algorithms, a form of ML, are particularly effective in wastewater treatment. These algorithms can learn optimal operational strategies over time, continuously improving efficiency and reducing resource usage.

18.3.2.1 Case Study: AI in Wastewater Treatment in Europe

In several European cities, AI has been integrated into wastewater treatment facilities to optimize energy use and chemical consumption. By analyzing real-time data from sensors, AI systems can adjust the treatment process dynamically, reducing energy consumption by up to 30% while maintaining the required effluent quality standards (Cairone et al, 2024).

18.3.3 Smart Irrigation Systems

Agriculture is the largest consumer of freshwater, with irrigation accounting for a substantial portion of global water use. Smart irrigation systems, powered by AI, offer an efficient solution by using real-time data to determine the precise amount of water crops need. These systems can monitor weather forecasts, soil moisture levels, and crop water requirements, ensuring that irrigation is applied only when necessary. AI models can predict soil moisture depletion rates and adjust irrigation schedules accordingly, reducing water wastage and ensuring that crops receive optimal hydration.

18.3.3.1 Case Study: Smart Irrigation in Israel

Israel, a leader in water-efficient agricultural technologies, has implemented AI-powered smart irrigation systems in various regions. These systems use data from weather forecasts, soil sensors, and satellite images to optimize water use for crops (Laskar et al, 2024). The systems have led to a 20%–30% reduction in water usage while increasing crop yields.

18.3.4 Blockchain and AI for Transparent Water Management

Blockchain technology, when combined with AI, holds the potential to revolutionize water management systems by providing transparent, secure, and decentralized records of water usage. Blockchain's immutable ledger can track water transactions, distribution, and consumption, while AI algorithms can optimize water allocation based on this data. This combination can improve accountability, ensure equitable access to water, and enable real-time tracking of water resources, particularly in regions with water scarcity or inequitable access to resources.

18.3.4.1 Case Study: Blockchain and AI in Water Trading

In Australia, blockchain and AI have been explored for their potential in water trading markets. AI is used to predict future water availability, while blockchain tracks the sale and transfer of water rights (Satilmisoglu et al, 2024). This system improves transparency, ensures fair distribution, and helps optimize water usage during times of scarcity.

18.4 CONCLUSION

As the global water crisis intensifies, the need for smart, efficient, and sustainable water management systems becomes more critical. AI technologies, from predictive analytics and autonomous monitoring to smart irrigation and wastewater treatment, offer transformative solutions to these challenges. However, the integration of AI into water management will require overcoming significant hurdles, including data privacy concerns, system integration, and scalability issues. The future of AI in water management is full of promise, with numerous emerging technologies and breakthroughs poised to reshape the sector. By leveraging the power of AI, we can create more resilient, sustainable, and efficient water systems, ensuring that future generations will have access to the water resources they need.

REFERENCES

Ahmadov, E. (2020). Water resources management to achieve sustainable development in Azerbaijan. *Sustainable Futures*, 2, 100030.

Ashoka, P., Devi, B. R., Sharma, N., Behera, M., Gautam, A., Jha, A., & Sinha, G. (2024). Artificial intelligence in water management for sustainable farming: A review. *Journal of Scientific Research and Reports*, 30(6), 511–525.

Bouramdane, A. A. (2023). Cyberattacks in smart grids: Challenges and solving the multi-criteria decision-making for cybersecurity options, including ones that incorporate artificial intelligence, using an analytical hierarchy process. *Journal of Cybersecurity and Privacy*, 3(4), 662–705.

Cairone, S., Hasan, S. W., Choo, K. H., Li, C. W., Zarra, T., Belgiorno, V., & Naddeo, V. (2024). Integrating artificial intelligence modeling and membrane technologies for advanced wastewater treatment: Research progress and future perspectives. *Science of the Total Environment*, 944, 173999.

Chitwatkulsiri, D., & Miyamoto, H. (2023). Real-time urban flood forecasting systems for Southeast Asia – A review of present modelling and its future prospects. *Water*, 15(1), 178.

Kikon, A., & Deka, P. C. (2022). Artificial intelligence application in drought assessment, monitoring and forecasting: A review. *Stochastic Environmental Research and Risk Assessment*, 36(5), 1197–1214.

Krishnan, S. R., Nallakaruppan, M. K., Chengoden, R., Koppu, S., Iyapparaja, M., Sadhasivam, J., & Sethuraman, S. (2022). Smart water resource management using Artificial Intelligence – A review. *Sustainability*, 14(20), 13384.

Laskar, A. A., Kumar, K., Roy, P., Mazumder, A. S., & Das, B. (2024). Exploring the role of smart systems in farm machinery for soil fertility and crop productivity. *International Journal for Research in Applied Science and Engineering Technology,* 12(12), 2063–2075.

Manjakkal, L., Mitra, S., Petillot, Y. R., Shutler, J., Scott, E. M., Willander, M., & Dahiya, R. (2021). Connected sensors, innovative sensor deployment, and intelligent data analysis for online water quality monitoring. *IEEE Internet of Things Journal*, 8(18), 13805–13824.

Parvaze, S., Kumar, R., Khan, J. N., Al-Ansari, N., Parvaze, S., Vishwakarma, D. K., … & Kuriqi, A. (2023). Optimization of water distribution systems using genetic algorithms: A review. *Archives of Computational Methods in Engineering*, 30(7), 4209–4244.

Satilmisoglu, T. K., Sermet, Y., Kurt, M., & Demir, I. (2024). Blockchain opportunities for water resources management: A comprehensive review. *Sustainability*, 16(6), 2403.

Conclusion

The synergy between water management and artificial intelligence (AI) offers an unprecedented opportunity to address the critical and growing challenges associated with global water resources. As highlighted throughout this book, AI has the potential to revolutionize the water management, making systems more efficient, sustainable, and adaptive to the evolving demands of both the environment and society. The diverse range of applications—from predictive modeling and smart infrastructure to precision agriculture and climate change mitigation—demonstrates the transformative power of AI in securing water resources and promoting sustainability across sectors. AI's role in water management is not just limited to technological innovation but extends to shaping more resilient ecosystems, enhancing water quality monitoring, optimizing agricultural water use, and providing rapid disaster response systems. The real-world success stories and case studies shared throughout this book serve as compelling examples of how AI is already driving positive change, paving the way for broader adoption in water systems worldwide.

However, while the benefits are clear, the journey toward widespread AI adoption in water management is not without its challenges. Issues related to data privacy, ethical concerns, public perception, and the integration of AI into existing water governance frameworks require careful consideration. Additionally, overcoming the technical, social, and economic barriers to implementation necessitates collaboration among stakeholders, fostering innovation, and ensuring equitable access to AI-driven solutions for all communities, particularly marginalized groups. The intersection of AI and water management is not just a technological or environmental issue—it is deeply intertwined with policy, education, capacity building, and global cooperation. As highlighted in the chapters on sustainable development goals, gender perspectives, and international partnerships, the successful implementation of AI-driven water solutions requires a multi-sectoral approach that includes governance, public-private partnerships, and international standards.

Looking ahead, the future of AI in water management is promising, with new technologies such as biomimicry, emerging AI tools, and interdisciplinary research pushing the boundaries of what is possible. By continuing to innovate and collaborate, we can leverage AI to create smarter, more sustainable water systems, ensuring that future generations have access to clean, safe, and abundant water resources. In closing, the journey toward an AI-powered water future requires not only technological advancements but also a collective commitment to ethical, inclusive, and sustainable practices that will shape the future of our planet's most vital resource.

Index

M

N

P

Q

R

S

T

U

V

W